-FLUID POWER-
in Plant and Field

Second Edition

A Training Manual for Technicians and Servicemen Who Install, Operate, Maintain and Troubleshoot Hydraulic and Compressed Air Fluid Power Systems

Charles S. Hedges
Chief Engineer Womack Machine Supply Co.

and

Robert C. Womack
Member, President's Council, National Fluid Power Association
President and Board Chairman, Womack Machine Supply Co.
President and Board Chairman, AAA Products International
Member, Fluid Power Distributor's Association

Womack Educational Publications

DEPARTMENT OF WOMACK MACHINE SUPPLY COMPANY
2010 Shea Road, Dallas Texas 75235, U.S.A.
Phone: (214) 357-3871

Publisher's Information

FLUID POWER IN PLANT AND FIELD

Second Edition ©1985 by Womack Machine Supply Company — All Rights Reserved
This is the First Printing of the Second Edition — June 1985
Library of Congress Catalog Card No. 68-22573
ISBN No. 0-9605644-7-0
Printed in the U.S.A.

This book is intended as an educational tool and a reference manual for people who install, start up, operate, maintain, and troubleshoot air and hydraulic fluid power systems.

Please refer to Pages 6 and 7 for a brief resume of the contents of each chapter. A general index is on Pages 230 and 231.

For home or classroom study of the technology of fluid power, the reader is referred to Page 232 for a description of the Womack textbooks, Industrial Fluid Power — Volumes 1, 2, and 3.

LIABILITY DISCLAIMER

Please read this before using the book.

We have taken great care in the preparation of this edition to see that all information is accurate. It has been checked and re-checked. However, any segment of information may apply on one application, and for one reason or another may not apply on another application which seems to be similar. This may be due to differences in environment, personnel, operating conditions, materials, and components. And, no matter how carefully the material is checked there is always the possibility of printing errors, or errors in circuits, tables, or text. For this reason we are offering this book only for educational usage, to illustrate principles and techniques, and we do not assume any liability for the safe and/or satisfactory operation of any machine or installation designed from information in this book. To avoid accidents, please observe all safety precautions mentioned in the text.

HOWDY!
I'M Reddy Fluid. I'VE BEEN AROUND A LONG TIME PUSHING CYLINDERS AND TURNING MOTORS.
I HAVE MANY HELPFUL TIPS TO PASS ON TO MAINTENANCE AND SERVICE PEOPLE, STUDENTS, FLUID POWER DESIGNERS AND TROUBLE SHOOTERS
REDDY FLUID

INTRODUCTION

Fluid Power . . . the technology of hydraulics and pneumatics . . . has a long and colorful history. Over the years, fluid power has demonstrated its ability to handle everything from the brute-force jobs of mining and construction to the precise, lightening-speed actions of modern automation. New applications for the unique properties of pressurized liquids and gases for transmitting power are constantly being found.

As fluid power technology continues to develop, manufacturers are adding to the instructional material for fluid power products and their applications. However, while we look toward the future, we must remain in touch with the past.

There is a wealth of basic knowledge about fluid power, learned from past experience, that must be kept in mind to insure that all systems, new and old, perform to their full potential. Fortunately, the authors of this book have recorded (in a well-organized and easy-to-read format) many of these essential "lessons learned".

Whether used as an instructional text or on-the-job reference, this book is a valuable tool for those new to fluid power . . . and for those wishing to expand their present skills.

As one who has to learn lessons in fluid power the hard way, I cannot emphasize enough how fortunate we are to have "Fluid Power in Plant and Field" and the other Womack Educational Publications available to us. These books go a long way toward closing the gap between the rapid development of fluid power technology and the development of educational and training materials.

I wholeheartedly commend the authors of this book, Mr. Hedges and Mr. Womack, for this contribution to fluid power education.

James I. Morgan
President, National Fluid Power Association

REDDY FLUID SAYS:

"Information in this book is very practical and has been compiled from my experience in working with fluid power in the industrial plant and out in the field. It should be valuable to designers in avoiding potential problems during the design stage of new equipment".

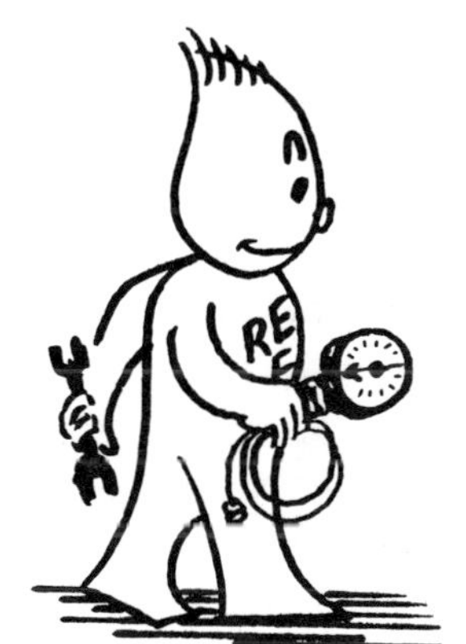

"I have added a great deal of information in this Second Edition on installation of components and systems — for their best location, and for their proper mounting and orientation. I can tell the installer many things which will keep him out of trouble. Much of this information is exclusive to this book".

"Notice the additional chapters in this edition on installation and troubleshooting of both hydrostatic transmissions and standard open loop systems. I have included routine and special maintenance, tips and checklists not only to assist in troubleshooting but to improve the performance and efficiency of fluid power systems already in operation".

"Hundreds of vocational and technical schools are using the Womack Industrial Fluid Power textbooks, Volumes 1, 2, and 3. In this present book I have included a lot of down-to-earth practical information from my own field experience which is not normally found in any textbook. It should be a valuable source of additional information to a student. See listings of textbooks inside the rear cover".

Outline of Book Contents

Chapter 1

Cylinders-Air & Hydraulic

Information in Chapter 1 concerns the proper installation of cylinders, and to the possible causes of improper operation, bending of the rod, premature mechanical failure, internal or external leakage, or otherwise unsatisfactory operation. For a detailed study of cylinder operation, size calculation, mountings, circuitry, and applications, refer to Volumes 1 and 2 "Industrial Fluid Power". Book descriptions and prices are inside the back cover of this book.

Cylinders are incidentally mentioned in other chapters of this book in association with other components being discussed.

INSTALLATION AND MOUNTING

Alignment. After mounting the cylinder, but before final tightening of the fastenings, check alignment of the piston rod with the movement of the work. Check in both retracted and extended positions. A good way to check alignment, if practical, is to detach the piston rod from the load and see if the two can be coupled without an undue amount of forcing. If the piston rod has a clevis or tang, see if the hinge pin can be inserted without forcing the rod to one side.

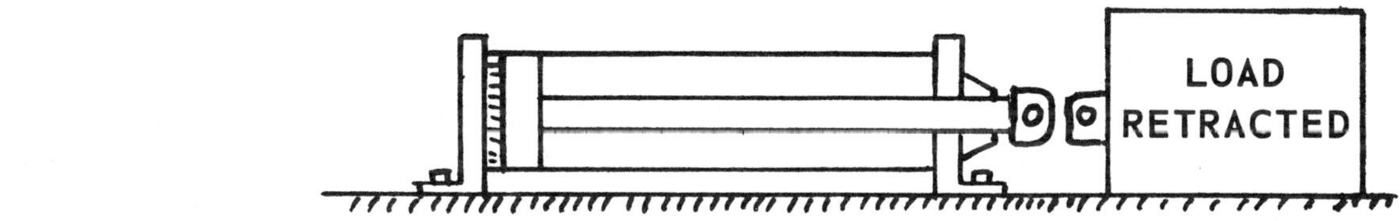

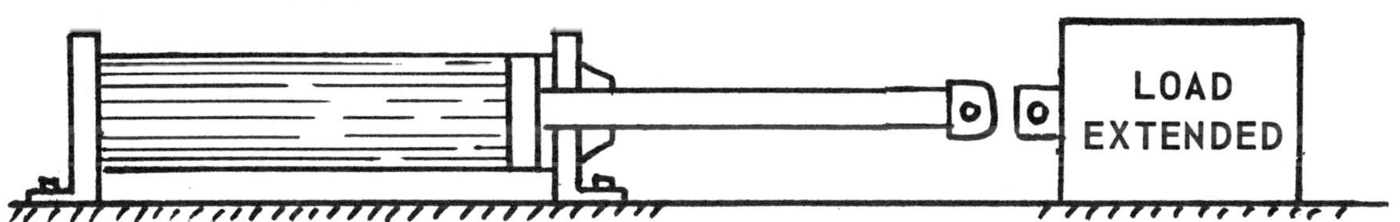

Uncouple Cylinder to Check Alignment in Both Retracted and Extended Positions.

On cylinders which are hinge mounted, either by a hinge on the rear end or by a trunnion, the alignment need be checked only in the plane of pivot action. On flange, lug, or foot mounted models, alignment should be checked in two planes at right angles to each other.

Accurate alignment is of great importance for a long and trouble-free cylinder life. Any side load against the rod will prematurely wear out the rod bearing, it may produce scratches in the cylinder barrel, and may wear out the piston and rod seals, causing frequent maintenance problems. And on long cylinders, mis-alignment at the extended position may lead to rod buckling. These are the effects of mis-alignment on the cylinder itself. There may also be damage to the machine to which the cylinder is coupled, depending on the nature of the machine.

In some cases, on existing machines, it may be possible to reduce the degree of mis-alignment by coupling the cylinder to the load with a universal joint.

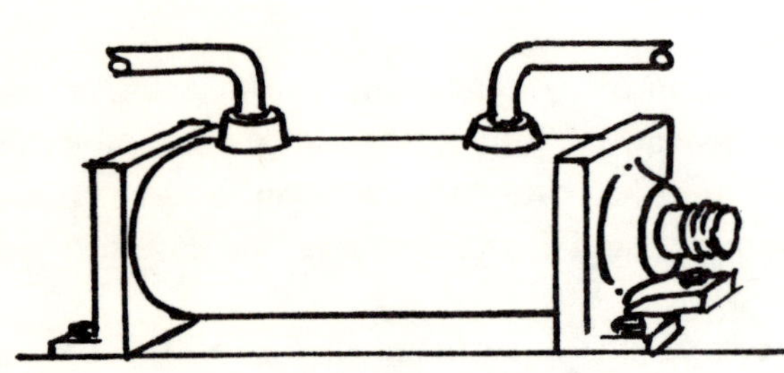
Mount Cylinder With Ports Up.

Cylinder Ports. In hydraulic circuitry, the preferred mounting position for horizontally mounted cylinders is with the ports up. The system can more quickly purge itself of air. All hydraulic systems will purge themselves in time, as the air gradually dissolves in the oil and is carried back to the reservoir. If air is not completely purged in a few hours, look for air leaking into the system. See chapter on reservoirs.

Some cylinders are built with air bleed holes for purging air before the system is started. If the cylinder is so equipped, these holes will usually be on each end of the barrel near the end caps and will be closed with a screw. Do not mistake cushion adjusting screws on the end cap for air bleed holes. Of course, the cylinder must be mounted with bleed holes up.

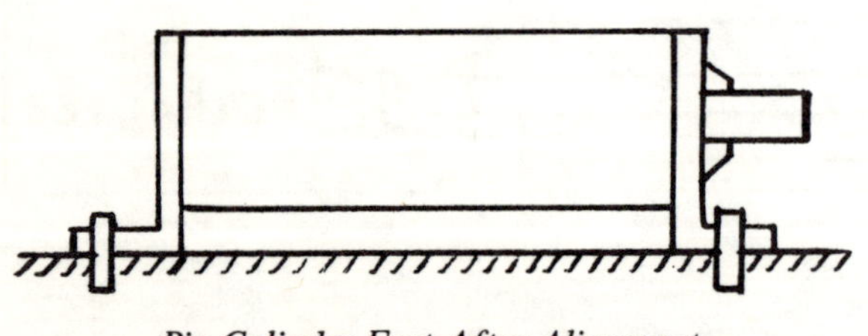
Pin Cylinder Feet After Alignment.

Foot Mounted Cylinders. Cylinders with lugs or feet welded to the end caps must be mounted on a stable surface. If the surface is out of plane, the cylinder may be distorted when bolted down. Distortion usually results in seal leakage and may cause premature wear on certain parts of the assembly.

Most foot or lug mounted cylinders have the outer surfaces of both front and rear foot machined, for positioning against a positive stop to absorb thrust against the load.

On cylinders larger than a 4-inch bore, especially those operating at 3000 PSI or higher, a center lug mounting style offers greater resistance to distortion caused by the bending moment created in a foot mount because the thrust against the load is not in line with the cylinder axis.

Shims can be placed under the front or rear foot, or both, to bring the cylinder into accurate alignment with the load. Then, as a further guarantee against any shifting of the cylinder, either the front or the rear foot can be dowelled to the mounting surface. Cylinders tend to elongate under pressure, so only one foot should be pinned or dowelled. If the higher pressure is required while the cylinder is extending, the rear foot should be pinned. If the higher pressure is required on retraction (as in a broaching operation) the front foot should be pinned.

Long, Slim Cylinders. There are two special problems in mounting long cylinders. (1), center sag and (2), barrel expansion. Center sag, on horizontally mounted cylinders, can be eliminated by placing a support block near the center of the barrel. The block should be contoured to fit the radius of the cylinder barrel, but should not be attached to it. This will permit the barrel to move slightly, without distortion under expansion caused by oil pressure or temperature.

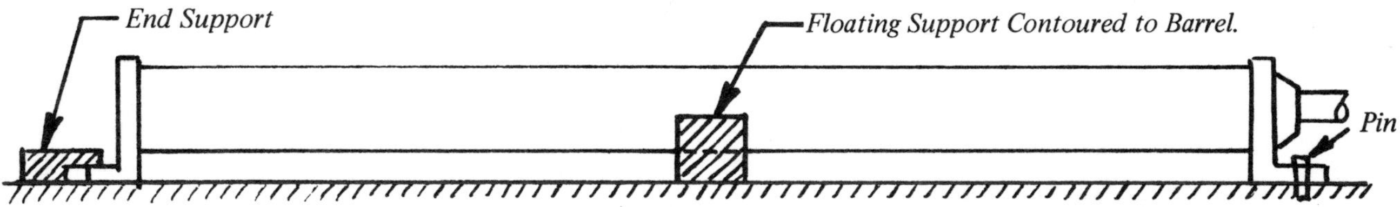

Mount Long Cylinder so Barrel Can Expand, and Use Floating Supports as Needed.

The second problem is that of expansion and contraction of the barrel because of oil compression, metal stretch, and temperature. This tends to break off lug or foot mounts, bow the cylinder barrel, bind the piston at certain positions, or cause leakage at the end gaskets. One end should be pinned or keyed. This should be the blind end if load is greater in the extension direction, or the rod end if load is greater in the retraction direction. Allow the other end to float longitudinally but provide guides to prevent it from shifting sidewise or up and down.

Hose connections to the cylinder ports are recommended. Rigid plumbing may eventually fatigue and rupture under the strains produced by expansion and contraction of the barrel.

Flange Mounted Cylinders. Standard industrial cylinders can be purchased with a mounting flange attached either to the rod or the blind end cap. For maximum mechanical strength in the machine on which the cylinder is mounted, the choice of a front or a rear flange mount should be considered in relation to the construction of the machine, and to the direction of motion in which the cylinder is under greater load.

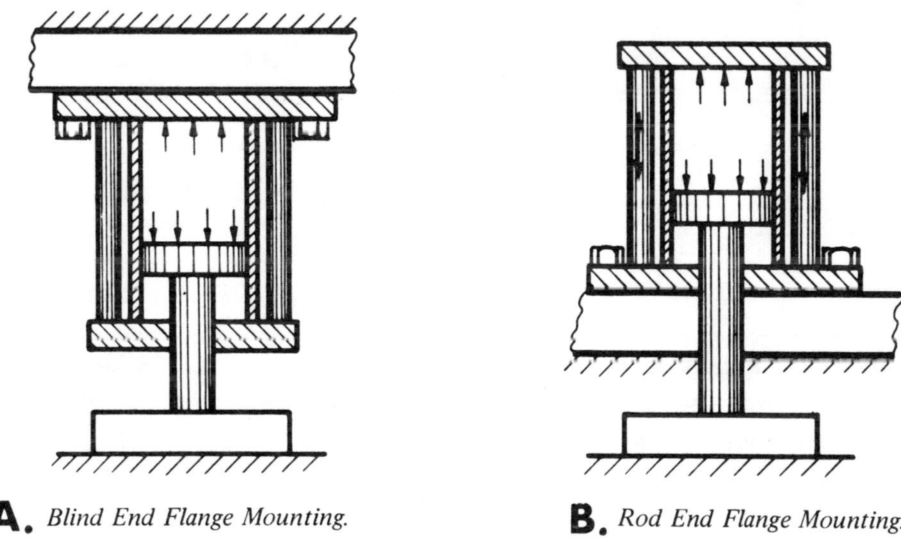

A. *Blind End Flange Mounting.*　　　**B.** *Rod End Flange Mounting.*

Cylinder Mounting Choice is Important, to Reduce Mechanical Stresses.

For example, if the heavier work is being performed while the cylinder is extending (see Part A of figure on preceding page), a rear flange mount is preferred, and this should be anchored to a back-up member of the machine. This removes all longitudinal stress from the barrel, tie rods, and mounting bolts, unless the piston is allowed to "bottom out" against its own rod end cap.

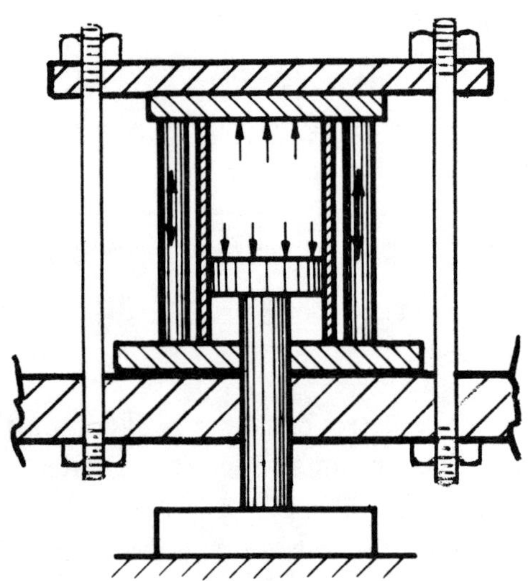

Strengthen Cylinder With Bolster Plate if Necessary.

If a rod flange mount were to be used (see Part B of figure on preceding page), stresses would be present in the barrel, tie rods, and mounting bolts during the entire extension stroke.

On the other hand, if the heavier work is to be done while the cylinder is retracting, a front flange mount (Part B) would be preferred because it would remove stresses in the barrel and tie rods except if the cylinder were allowed to "bottom out" against its own blind end cap.

If for some reason the preferred mounting is impractical, or if there is a problem on an existing machine of blowing seals on the cylinder barrel, over-stressing of tie rods, or a similar problem, a bolster plate can be placed to support the free end of the cylinder. As many tie rods as necessary to support the weight and stress can be used on the bolster plate.

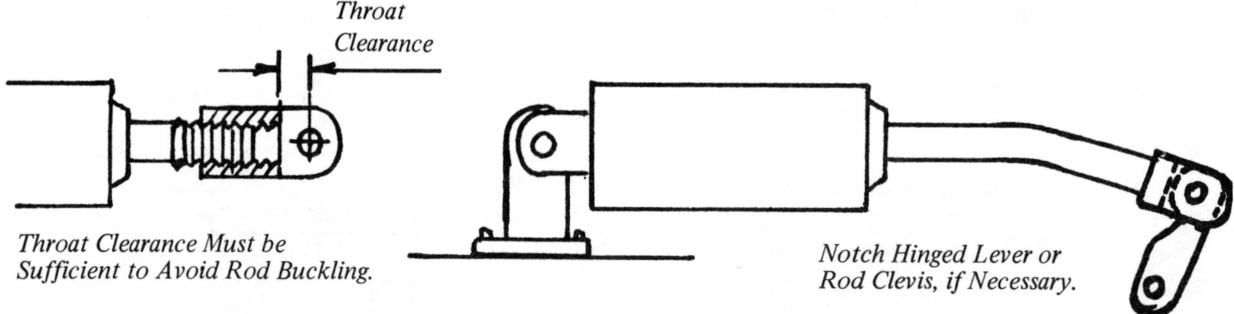

Throat Clearance Must be Sufficient to Avoid Rod Buckling.

Notch Hinged Lever or Rod Clevis, if Necessary.

Clevis Throat Clearance. Particular care must be taken when mounting hinged cylinders which swing through a wide angle, to be sure the clevis slot is deep enough to clear the mating bracket or lever at the extreme ends of the swing. This is a point which can pass unnoticed unless the installation mechanic is observant. If the bottom of the clevis slot hits the lever each time the cylinder makes a stroke, this will prematurely wear out the rod bearing and rod seal, and may put a permanent bend in the piston rod. It will also bind the piston and may damage the machine on which the cylinder is mounted. At installation, check clevis throat clearance with the clevis disconnected from the lever to see if it can be pinned to the lever without strain.

Wrench Flats. If original or replacement cylinders are ordered with "wrench flats" on the piston rod, this will simplify the job of screwing attachments on the end of the rod. Wrench flats are flat spots for placement of a wrench to keep the rod from turning. When specified on the purchase order, the manufacturer will provide these flats at little or no extra charge.

The rod is highly polished to minimize wear on the rod seal and to prevent seal leakage. It must be protected against nicks and scratches. Never use a pipe wrench unless absolutely necessary, and be sure to grip the rod in an area which does not go into the rod seal.

Caution! Do not use fluid pressure to hold the piston against the end cap as a means of preventing the rod from turning while screw-

Cylinders May be Ordered With Wrench Flats on Rod.

ing on a clevis. In some cylinders the mechanism which locks the piston to the rod may be damaged. In other cases, the piston rod may be accidentally unscrewed from the piston.

Hinge Mounted Cylinders. If the cylinder must swing through an arc while making its stroke, for example when moving a hinged lever, a hinge mounted cylinder model should be used. On most applications, if the back end of a cylinder is hinged, a clevis or tang must also be used on the end of the piston rod. Hinge mounted cylinders can also be used to advantage if the mounting surface is unstable or out of plane, making it difficult to accurately align the cylinder axis with the axis of load movement.

Trunnion Mount. Trunnion pivots are sometimes used with heavy cylinders which have a long stroke. They help to balance the weight of the cylinder and reduce danger of rod buckling when the cylinder is fully extended.

Caution! Trunnions are designed to support only side load. Mount trunnion support bearings as close to the barrel as possible to minimize bending stresses in the trunnion.

Hinged cylinders must be connected to the fluid supply through some sort of moving connection, and hose is ordinarily used. Special cylinders may have hollow trunnions through which oil (or air) is brought into and out of the cylinder through rotating joints.

Hinge-mounted Cylinder.

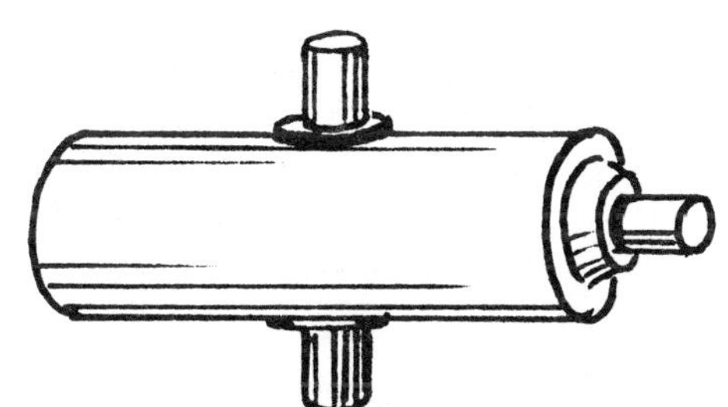

Trunnion-mounted Cylinder.

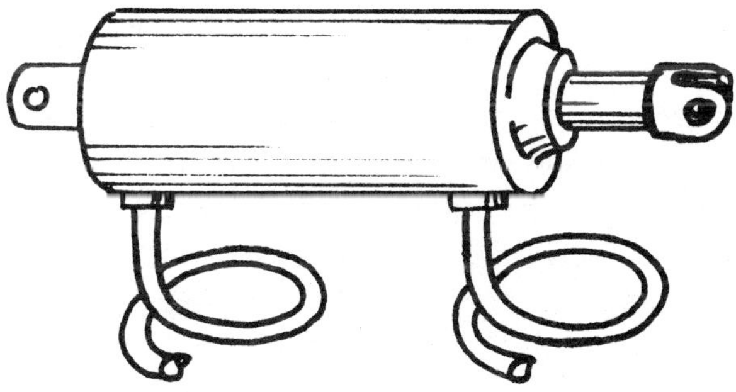

Use Hose Connections on Hinged Cylinders.

Protect Cylinder from Piston Impact. Standard catalog model cylinders are not designed to take a high impact against their end caps. If the piston is connected to a high-mass or fast-traveling load, an impact on each cycle can ultimately damage the cylinder. Damage can include separation of the rod from the piston, fracture of the end cap or the piston, stripping of threads on tie rods, and blow-out of barrel seals. On these kinds of loads a good way to prevent cylinder damage is to provide external positive stops just short of full stroke to absorb impact energy produced by the load.

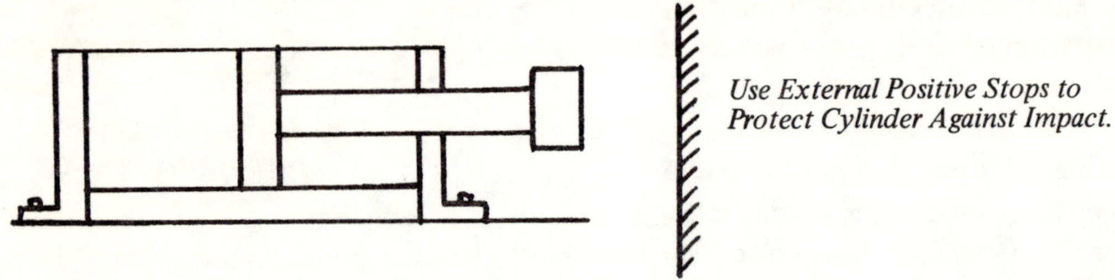

Use External Positive Stops to Protect Cylinder Against Impact.

If the work itself could be damaged by impact against positive stops, some sort of deceleration should be used. The simplest method for hydraulic cylinders is to order the cylinders with cushions on one or both ends. Refer to "Industrial Fluid Power - Volume 1" for description of cushions. Cushions provide deceleration over approximately the last inch of cylinder travel. On cylinders larger than 2½" bore the degree of cushioning is externally adjustable.

Cushions on air cylinders are of little practical value in decelerating momentum loads. Their only good application is in decelerating loads traveling at moderate speeds and consisting almost entirely of friction.

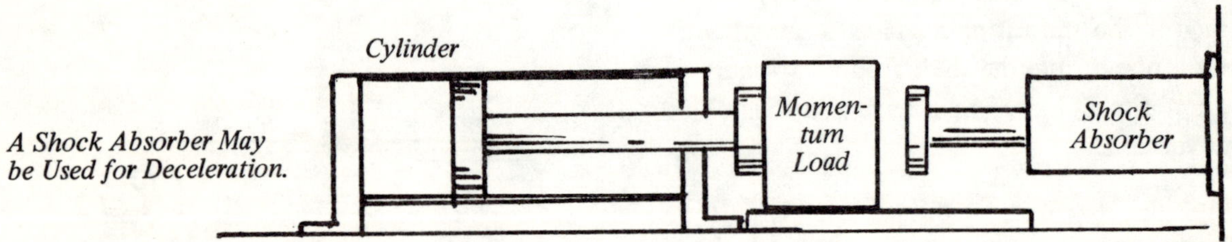

Even hydraulic cylinder cushions may sometimes be inadequate for safely decelerating extremely heavy loads, and a shock absorbing device may have to be used. This device looks like a cylinder with its rod extended, and has a series of specially designed internal orifices to give a steady rate of deceleration over the length of deceleration stroke which may be from 1 inch to almost any length.

The safest method of decelerating a cylinder which is carrying a massive load at high speed is to place a limit switch (on electrically operated circuits) at the point where deceleration should start. On non-electrical circuits a cam actuated pilot valve can be placed at the deceleration starting point. The location of the switch or cam valve can be experimentally adjusted to find a satisfactory point for the start of deceleration. Cams on the moving member which actuate a limit switch must be of a length and shape to protect the switch. See more information in the book "Electrical Control of Fluid Power". Circuits for deceleration switches and cam valves are described in "Industrial Fluid Power — Volume 2".

Speed Control of Cylinders. The speed of an AIR cylinder is usually controlled with a pair of flow control valves installed in the lines connecting control valve to cylinder, or by a pair of needle valves installed in the two exhaust ports of a 5-port, 4-way control valve. Because the air supply comes from an infinite source (the receiver tank) it may be necessary to limit cylinder speed in both directions.

The speed of a HYDRAULIC cylinder should be controlled with just one flow control valve placed in one of the lines connecting control valve to cylinder. Speed in the opposite direction will be limited by flow volume of the pump. If speed control is necessary in both directions, the pump has too high a flow and should be replaced with one of smaller size or its speed should be reduced.

Correct orientation of the flow control valves is important. On air circuits they are usually connected to meter air OUT of the advancing cylinder, and to permit free return flow into the cylinder. However, for some applications, better control may be obtained with meter-in speed control. For hydraulic cylinders, the oil is usually metered into the cylinder, but on selected applications, better control may be obtained by metering oil out of the cylinder. See more information later in this chapter and in "Industrial Fluid Power — Volumes 1 and 2".

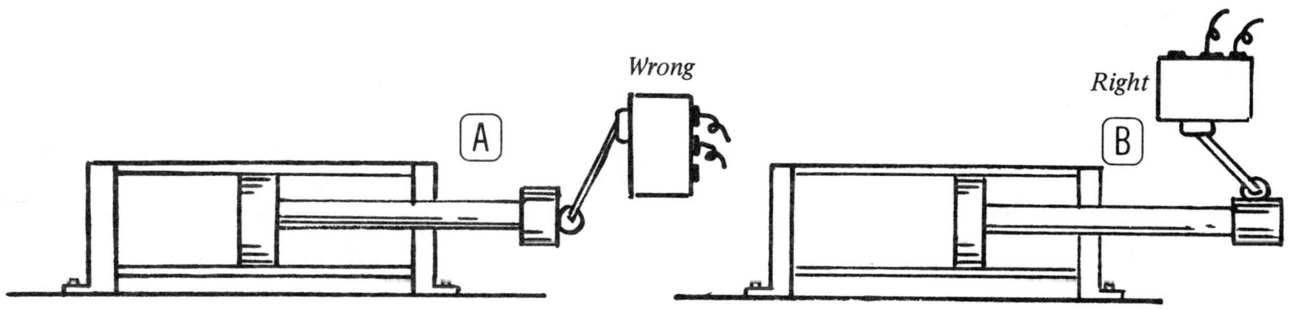

Mount Limit Switches and Cam Valves so They Will Not be Damaged by Cylinder Over-run.

Installation of Limit Switches. When installing electric limit switches, cam valves, air switches, or bleed buttons which are to be actuated either by the moving cylinder or by some other member of the machine, they should be mounted in such a way that they cannot be damaged if the cylinder should accidentally get out of control and travel past the desired position. For example, this might happen on an electrically controlled system if the cylinder were left in a partly extended position and the fluid pressure turned on without electricity being connected to the control circuit.

Diagram A illustrates an incorrect mounting. If the electric circuit is non-operative, the cylinder could overtravel and damage the switch. Diagram B shows a better mounting position to prevent overrun damage. Where switch overrun could occur, positive stops should be installed to prevent the cylinder from overrunning far enough to damage the switch.

Protect Cylinders From Heat. Where possible, shield cylinders, hydraulic power units, and hydraulic plumbing from sources of abnormal heat such as direct sunlight, furnaces, steam boilers, die casting machines, heat treating furnaces, etc. Heat from such sources

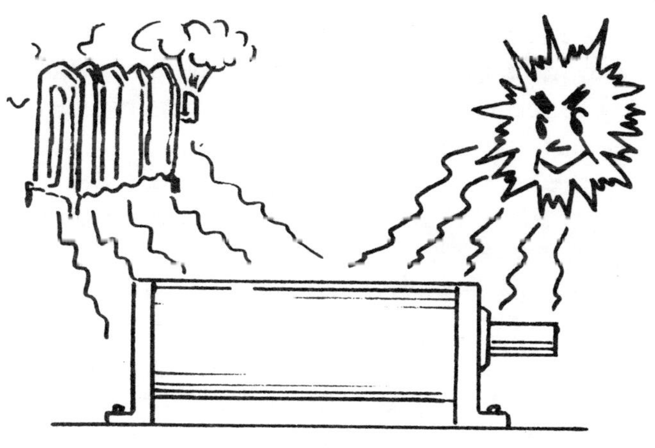

Shield Cylinder From Heat.

can be picked up in the hydraulic oil, causing overheating of the oil and damage to seals in all components.

To shield against radiated heat, sometimes a reflective baffle can be placed between the system and the source of heat. If a cylinder piston rod is coupled to a hot object, heat conduction can be minimized by using a coupling made of heat insulating material, by mounting the cylinder further away and using a rod extension, or by ordering the cylinder with an extra long rod extension. If the cylinder cannot be insulated from the heat source, it can be ordered with high temperature seals. Ordinary synthetic rubber will decompose more rapidly at temperatures over 200° F.

A solution to the heat problem which is often satisfactory is to install a water cooled heat exchanger to keep the hydraulic oil cool.

CAUSES OF JERKY CYLINDER MOTION

Bleed Air at Cylinder Ports.

Trapped Air. When the cylinder in a hydraulic system moves in an erratic or jerky manner the first thing we think of is trapped air. This certainly should be checked first on a new system just placed in operation. Carefully cracking fittings at the cylinder, in the pump line, or other places in the system will permit the immediate escape of much of the air and an immediate improvement in performance should be noticed. If desired, further steps can then be taken to bleed the system at other points.

Most hydraulic systems will purge themselves of air after operating for a few hours under pressure. The trapped air will dissolve in the oil, a little bit on each cycle, and will be carried back to the reservoir. When pressure on the oil is relieved, the air will bubble out of the oil in the same way gas will bubble out of a coke bottle when internal pressure is released by removing the cap. To assist in a more rapid purging, horizontally mounted cylinders should be mounted with oil ports up.

Since the entire hydraulic system, with the exception of the pump inlet line is under a certain amount of pressure above atmospheric, once air has been purged from the system, no more should enter unless reservoir oil level falls low enough to uncover the inlet strainer. If air continues to enter the system, check these possible sources: poor joints in the pump inlet line, a worn pump shaft seal, a cylinder rod seal (not necessarily a worn-out one), or a leak in an air-to-oil heat exchanger. Check also for mis-matched threads at the pump inlet, as when a tapered pipe fitting has been screwed into a straight thread port.

To check for an air leak at the pump shaft seal, squirt a little oil around the seal while the pump is running. If the oil is sucked inside the pump or if the pump noise decreases, this is positive evidence of an air leak through a worn-out shaft seal.

Cylinders mounted vertically with rod up may suck air in around the shaft seal while they are retracting if they are not properly counterbalanced. A counterbalance valve should be installed at the blind end port and adjusted to the least pressure which will support the load while the cylinder is retracting. If the cylinder drops faster than oil comes in from the pump, a partial vacuum will be produced in the rod end of the cylinder behind the rod seal, and on some cylinders this may suck in air through the rod seal.

__High Surface Friction__. An air cylinder is generally not suitable for slow feed of a load which slides on a large area of surface friction, for example a machine slide or lathe tailstock, even though the surface is well lubricated. It is well suited to moving the slide or tailstock quickly into or out

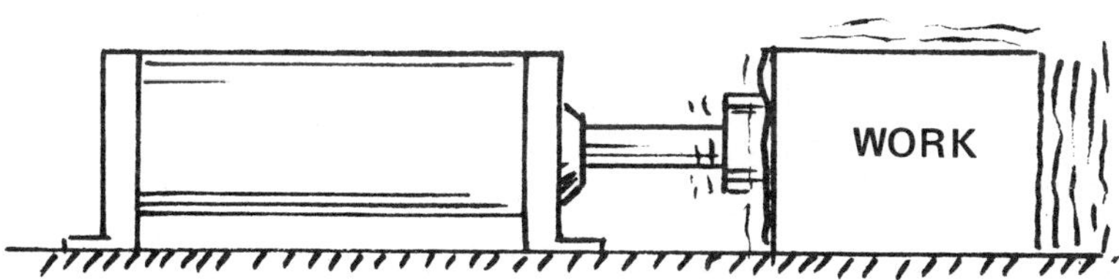

High Surface Friction May Result in Erratic Motion in Air Cylinder at Slow Speed.

of working position but cannot be expected to move it slowly at a smooth and constant rate. For applications like these, a hydraulic cylinder should be used. However, on some applications where moving friction is low and a constant feed rate is not critically important, an air-over-oil system as described in the book "Industrial Fluid Power — Volume 2" may give satisfactory results. These systems are powered with shop air but have hydraulic oil in the speed metering loop.

If the surface friction can be reduced by using a ball bearing slide, and if the speed rate does not have to be constant, an air cylinder may give acceptable results.

The cause of uneven action of an air cylinder on a high friction load is the difference between the high coefficient of breakaway (static) friction compared to the lower coefficient of running friction, and the normal compressibility of air. When air is admitted to the cylinder, pressure builds up behind the piston high enough to overcome the load plus breakaway friction. The piston starts to move, and because friction resistance decreases, the cylinder jumps ahead of the air pressure. As pressure behind the piston decreases, the cylinder stalls and waits for the air pressure to build up again to breakaway level. This action is repeated over and over through the entire stroke. An air cylinder is satisfactory on a fast moving application because the cylinder, when it once breaks away, cannot jump ahead of the incoming air pressure.

Short-stroke hydraulic cylinders, because the compressibility of hydraulic oil is much less than that of air, are not as subject to this erratic action, but oil compressibility on long-stroke cylinders can cause the same erratic action although to a lesser degree.

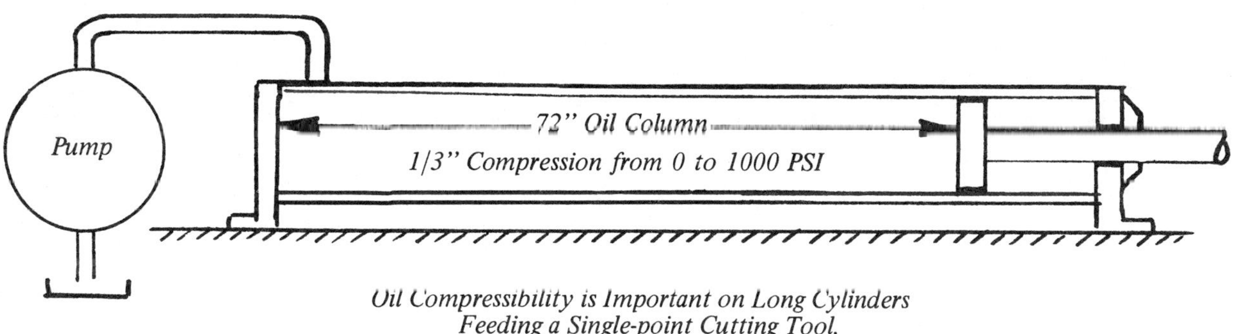

Oil Compressibility is Important on Long Cylinders Feeding a Single-point Cutting Tool.

__Hydraulic Cylinder Feed__. On most applications hydraulic oil compressibility can be disregarded, but on certain applications, particularly those in which a cylinder is directly pushing a metal cutting tool, especially a single-point cutting tool, or pushing a machine slide or lathe tailstock or similar application with high surface friction, the same type of chatter can be produced as previously described for air cylinders. Compressibility is particularly troublesome on servo applications. Here, it is important to have a very rigid system. Servo systems, including choice of components and plumbing layout should be designed by experts who can take into account the effect of oil compressibility as well as other important factors.

To calculate the oil compression which can be tolerated is an involved calculation and also a matter of judgement, but the actual amount of compression can be calculated by using the example on the preceding page. In this cylinder which has a 72" stroke, according to a common rule-of-thumb, the oil column will shorten by approximately 1/2% when the fluid pressure is increased from 0 to 1000 PSI. In this case the 72" oil column (with the piston near full extension) would shorten by about 1/3 inch, and the cylinder would be likely to chatter on a slow-feed, high-friction application.

Please note that compression shortening of the oil column does not relate to cylinder diameter, nor to actual cubic inch volume inside the cylinder, but is proportional only to the length of oil column under compression. The oil column in a 1" bore cylinder would shorten the same amount when subjected to 1000 PSI internal pressure as will the oil column in a 10" bore cylinder. On servo systems, the length of oil column between servo valve cylinder ports and the cylinder piston is the significant distance to keep short. On ordinary (non-servo) systems the significant distance is between the pump outlet port and the cylinder piston. The shorter these distances, the more rigid the system. Although trapped volume does not determine shortening of the oil column under pressure, it may be significant for other reasons when action of the entire system is considered.

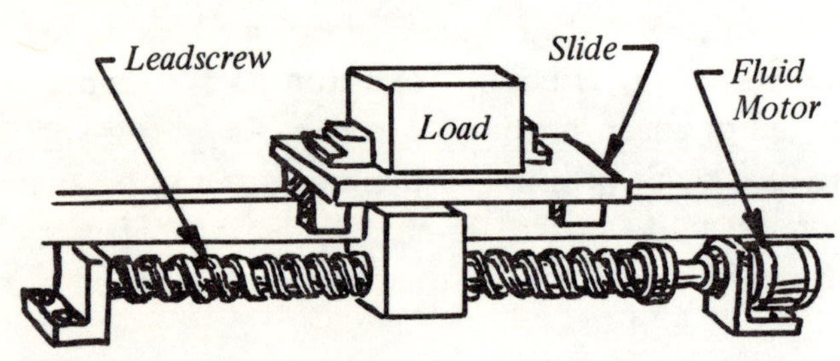

A Hydraulic Motor, With Leadscrew, May be Used to Drive a Machine Slide.

Remedies for Hydraulic Cylinder Chatter. While in most ordinary systems oil compressibility and cylinder line length are not important factors, they may be quite important for precise applications such as tracing, contour milling, planer bed reciprocation, etc., which require a long feed stroke. Hydraulic motors rather than cylinders are often used for such applications. These are coupled to the lead screw of the machine or sometimes by other means such as rack and pinion drive. The purpose of using a hydraulic motor is to reduce compression in the oil by shortening the oil column.

Whatever means is used to couple the hydraulic motor (or cylinder) to the machine, oil feed lines should be kept as short as possible. This may require changing the physical layout, moving the hydraulic pumping unit closer to the rest of the system, and reducing line length wherever possible. On servo systems, to get maximum rigidity, the control valve can be mounted directly on the frame of the cylinder or motor so that connecting line length can be reduced to only a few inches.

Where lines connecting valve to cylinder cannot be made extremely short, rigidity can only be improved by re-designing the system to produce the necessary force and speed with a cylinder of larger bore operating at a lower pressure. Operation at lower

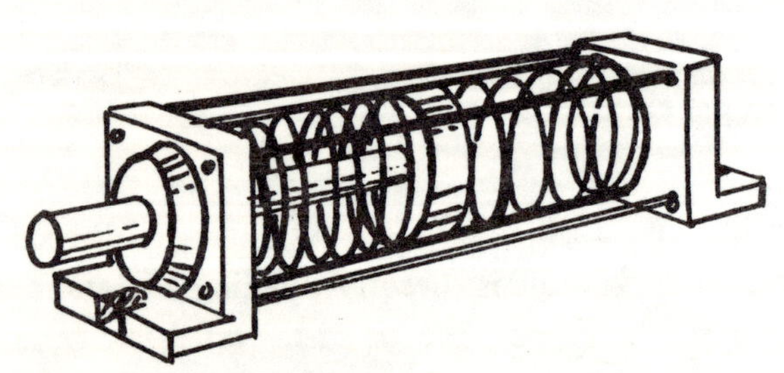

The Oil Column on Each Side of the Piston Acts Like a Stiff Spring.

pressure will reduce oil compression which, in turn, will improve rigidity. To obtain the same cylinder speed the system pump will have to be replaced with one having a higher flow volume.

Keep the cylinder stroke (and overall length) as short as possible. The piston is in effect being pushed by a heavy spring, so the shorter the oil column the stiffer the spring. When the oil column is considered as a stiff spring, it is apparent why chatter can develop even in a hydraulic system. Cylinders should be ordered with air bleed screws so air can be purged easily from the system by cracking fittings and opening bleed screws.

Speed Control as Related to Chatter. On systems operating on higher than 1000 PSI, pressure compensated flow control valves, rather than needle valves, will improve performance on applications where load resistance may increase or decrease during the stroke. On systems of lower pressure, pressure compensated flow control valves may introduce a power loss higher than acceptable.

By-pass Speed Control Generates Less Heat.

By-pass or bleed-off type of speed control diverts a part of the pump flow to tank to reduce cylinder speed. Less heat is generated in the system *but only during periods of less than full load operation.* But because of poorer speed regulation (more variation in speed with changes in load), chatter may be worse than with series meter-in or meter-out speed control.

Series meter-in control will give better speed regulation (less change in speed due to changes in load) than by-pass control, and probably with less chatter, but may generate more heat in the oil *during periods of less than full load operation.* This method is preferred on hydraulic cylinder applications, especially where there may be an overhung load which might cause damaging pressure intensification in the cylinder if meter-out control were to be used. Also, this method is preferred on most hydraulic motor applications.

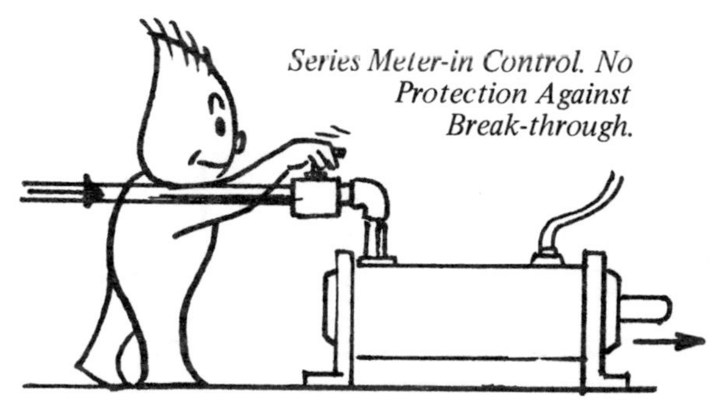

Series Meter-in Control. No Protection Against Break-through.

Series meter-out control is usually preferred for air cylinders, and for those hydraulic cylinder applications where there is no great

Series Meter-out Control. Best for Most Applications.

amount of overhung load. It may give better control of ''lunge'' when load is suddenly removed, as by a hydraulically driven tool breaking through the work. It is not normally used for hydraulic motors.

We recommend in cases of cylinder chatter that a different method of speed control be tried experimentally. Because of variations in parameters it usually is difficult to predict during the design stage the speed control method which will give the best overall performance with the least chatter.

Comparing the various speed control methods, the series meter-out method will usually result in the least cylinder chatter if it is compatible with other requirements of the circuit.

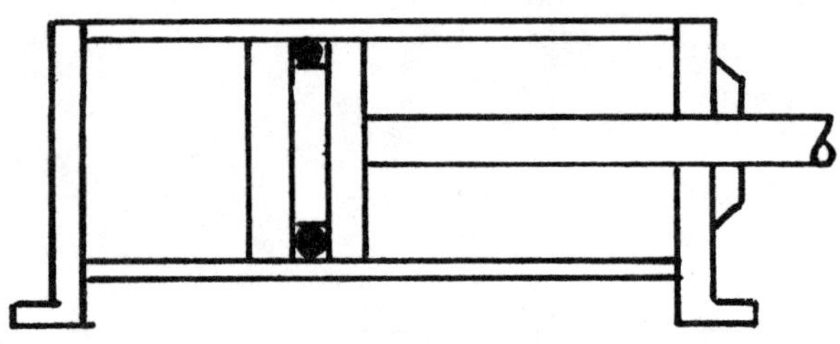

High-friction Packings May Cause Chatter.

Piston Seals. Chatter in an air cylinder is sometimes caused by high breakaway friction in the piston seals. O-rings make tight piston seals but their breakaway friction is high, and this may cause chatter especially in small bore cylinders operating at low pressure, or in vacuum cylinders. On these applications the breakaway friction is relatively high in relation to the force output from the cylinder.

If O-rings are used as piston seals, make sure the groove width is sufficient to permit the rings to roll slightly as the piston starts to move. Groove width should be about 1½ times the O-ring thickness.

Leather or soft synthetic rubber cups make better piston seals than O-rings or reinforced rubber. Chevron rings make good seals but are more expensive than cup seals and more difficult to replace.

Hydraulic cylinders which have an adjustable rod gland for tightening rod seals may chatter if the gland is over-tightened. This sometimes happens on large diameter sections of telescopic cylinders.

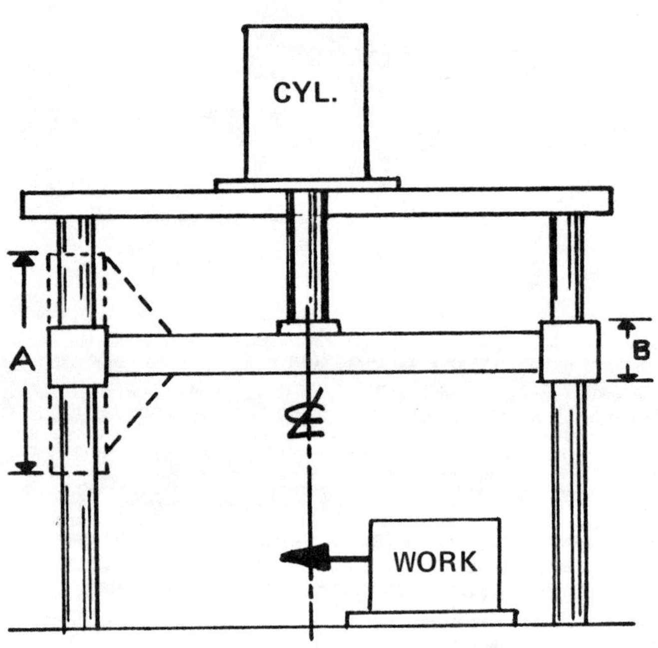

Use Long Guides and Center Work Under Press.

Mechanical Bind. Machines which have two or more moving members operating in slides or guides may be subject to mechanical bind in the guides which will cause the platen to chatter or to actually lock up. Four-post presses and press brakes are examples of this kind of chatter. If bind does occur, check alignment and lubrication of guides.

Bind in guided members of presses may be due to bad mechanical design. This illustration is a good example of poor design. The guide length, B, is too short in relation to the span between the guides. A rule-of-thumb is to make the guide length, A, at least equal to the span. If this is not possible, shorter guides may be satisfactory if the guides and followers are accurately machined to close tolerances.

Always keep the work as well centered in the press as possible. Off-center work is a frequent cause of mechanical bind in the press, and this will either lock up the press, cause cylinder chatter, or will damage the press frame.

CAUSES OF CYLINDER BURSTING

It is easy to understand that a cylinder can burst because of being subjected to excessive pressure. This can happen either because of excessively high steady pressure, from high transients, caused by shock conditions although the steady pressure may be within the cylinder rating, or from an unauthorized person tampering with the pump relief valve. But occasionally we have a case of mysterious breakage in a system where there appears to be no good reason. These failures are usually charged off to a "manufacturing defect" but may instead be due to one of the following causes:

Pressure Intensification. The pressure in the rod end of a hydraulic cylinder can be intensified far beyond the system relief valve setting under certain conditions. Intensification may burst the cylinder barrel, blow out seals at the ends of the barrel, blow out the rod seals, or may burst plumbing lines or components in other parts of the system.

The cause of pressure intensification is the difference in exposure area between the full piston area on the blind end and the net area around the piston rod. The larger the diameter of the rod, the smaller the exposure area

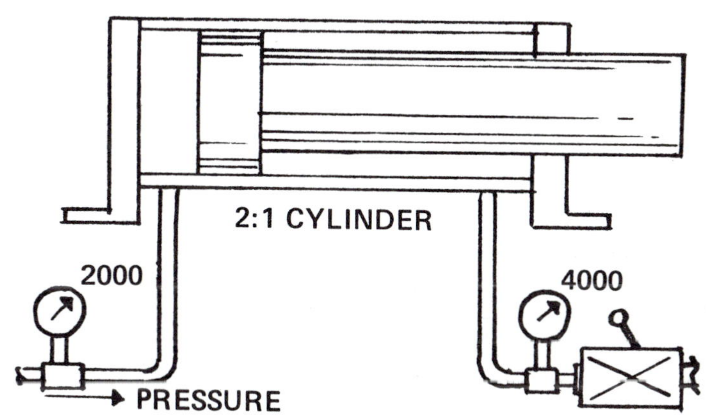

Pressure Intensification May Cause Breakage.

around the rod and the greater the difference between full piston area and net area. Therefore, intensification is a greater danger in hydraulic cylinders which have oversize or 2:1 ratio piston rods. Incidentally, intensification is of no concern in air cylinders. Catalog model cylinders are offered in several rod diameters. The smallest listed rod is considered the standard rod. The largest rod offered in catalog cylinders (2:1 ratio) has a rod area approximately 1/2 the full piston area. Therefore, the net area will also be 1/2 the full piston area.

Referring to the illustration, rod end pressure can rise to twice the pump (and relief valve) pressure if flow from the rod end is blocked or severely restricted as by a flow control valve or a pilot-operated check valve. Intensification always occurs in the rod end of a cylinder and will affect the cylinder or components connected to the rod end loop.

In the illustration, which shows a 2:1 ratio cylinder, and assuming an inlet pressure of 2000 PSI (relief valve setting), to the blind end, pressure in the rod end can intensify to 4000 PSI if outlet flow is blocked by closing the valve. The phenomena of intensification is explained in Volumes 1 and 2 "Industrial Fluid Power". This calls for caution in raising the relief valve setting of any hydraulic system in which cylinders with oversize or 2:1 ratio rods are used, particularly if there are meter-out flow control valves, pilot-operated check valves or other restrictions in the rod end plumbing.

Heat. Hydraulic oil expands when its temperature is raised. If a mobile cylinder, for example, is full of oil and then disconnected from its lines by quick connect couplers, the trapped oil will expand and burst the cylinder if it is left in the sunlight for several hours. This sometimes occurs with farm machinery or mobile equipment. Some manufacturers add a blow-out device to protect against this

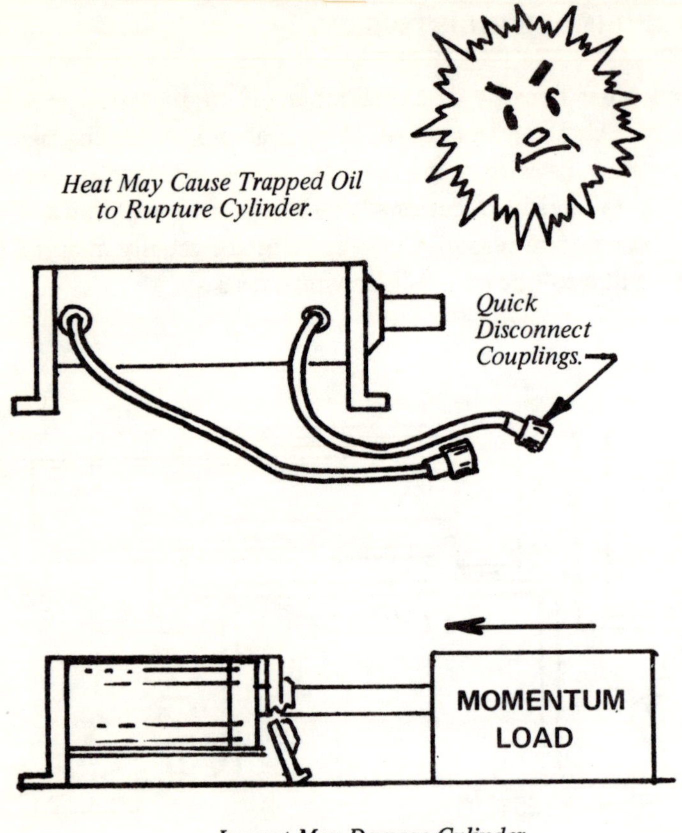

Heat May Cause Trapped Oil to Rupture Cylinder.

Impact May Damage Cylinder.

condition. This device is called a thermal relief valve. It is a miniature, non-adjustable relief valve. When pressure increases to its cracking pressure level, a few drops of oil are spilled on the ground to relieve the pressure.

If a cylinder is left coupled to the machine, damage seldom occurs because control valve leakage will prevent internal pressure from reaching the bursting point.

Impact. A heavy load traveling at high velocity carries a lot of momentum energy which is released when the load is stopped. The more abruptly the load is stopped, the higher is the momentum peak.

A sudden impact as the load is abruptly stopped by a cylinder piston hitting its own end cap may damage the cylinder by fracturing the piston or end cap or breaking the rod loose from the piston.

Positive external stops may be installed to stop the load, and this will protect the cylinder, but the impact may damage the machine on which the positive stops are mounted.

The impact from slowly moving cylinders is seldom destructive, but on rapidly moving massive load applications, cushioning devices or deceleration circuits should be installed. See Volumes 1 and 2 "Industrial Fluid Power" for information on cushioning and deceleration devices.

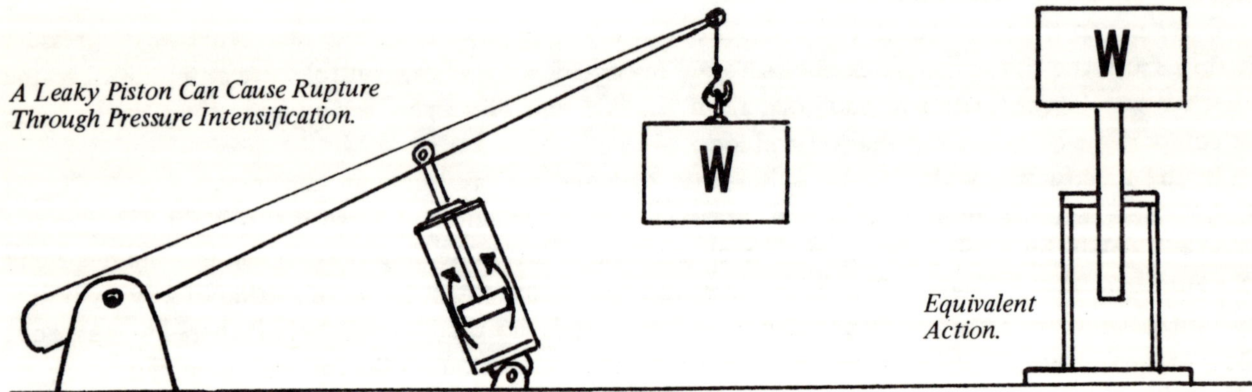

A Leaky Piston Can Cause Rupture Through Pressure Intensification.

Piston Leakage. Cylinders have been known to burst after the machine on which they were mounted was shut down for a while, and if the cylinder was left partially extended, supporting a heavy reactive load as in the illustration.

Breakage will occur at the weakest point in the cylinder or in the associated valving or plumbing.

It is caused by pressure intensification working through leaky piston seals. Older equipment on which cylinders are more likely to have well-worn and leaky seals are more subject to this phenomenon. The operator of a machine should never leave heavy loads suspended, supported by a hydraulic cylinder.

In the illustration a crane has been stopped with the beam in a partially raised position. The hydraulic system has been sized, for example, so it takes 2000 PSI to raise the beam with the load shown. This means that when stopped the weight of the load may generate up to 2000 PSI static pressure in the blind end of the cylinder. This would be all right if the piston were leaktight. But if the piston leaks there will be a transfer of a small amount of oil to the rod side which will produce intensification. The equivalent diagram of this situation is shown on the right side. In effect, the full load will be supported on the *rod* area of the cylinder. Standard cylinders have a rod area of about 1/4th the full piston area. This means the pressure in the rod end can intensify to 4 times the pressure generated by load weight in the blind end of the cylinder, or an intensified pressure of about 8000 PSI could be produced. This is enough to rupture most mobile cylinders.

Fortunately, in most systems, intensification damage does not develop because leakage across the control valve spool exceeds piston seal leakage, and the beam simply sags slowly to the ground.

On idle hydraulic machinery, when not in use, the beam should be supported or should be lowered to the ground. A particularly dangerous situation is a beam on a moving vehicle in a partially raised position while the vehicle is moving. Road shocks can easily generate enough intensified pressure to destroy the cylinder and allow the beam to collapse.

Another potentially dangerous situation is the use of a pilot-operated check valve installed in the rod end plumbing to prevent sag of a cylinder caused by a leaky control valve spool. Pilot-operated valves are frequently installed to correct an undesirable sag. This is all right if the possibility of pressure intensification has been considered, and the components have a pressure rating high enough so no damage can occur.

OTHER CAUSES FOR CYLINDER FAILURE

Material relating to cylinder failure can be found in various places in this book. Also consult index for information on related subjects. A great deal of cylinder information is in Volumes 1 and 2 "Industrial Fluid Power". See book listings on inside of back cover.

Refer to preceding pages in this chapter for suggestions and precautions on mounting, and for protection of cylinders against physical abuse. Refer to Chapter 7 for proper conditioning of compressed air to cylinders, and lubrication of small bore, short stroke cylinders. Refer to Chapter 7, Page 153 for vacuum cylinders.

PROTECTION OF CYLINDERS FROM PHYSICAL DAMAGE

This section discusses precautions to be taken to protect a cylinder from physical abuse and against dirt. Other chapters cover water damage, proper lubrication, cleanliness of the hydraulic oil, etc.

Falling Objects. If necessary, provide baffles or shields to protect piston rod and cylinder barrel against nicking or denting by falling objects. Dents and scratches on the rod act like a saw, wearing rod seals rapidly, and causing oil leakage around

Protect Cylinder From
Falling Objects.

the rod seal. A dent in the barrel will cause rapid wear of the piston seals and may even cause the piston to stall at that point if fluid can by-pass the piston.

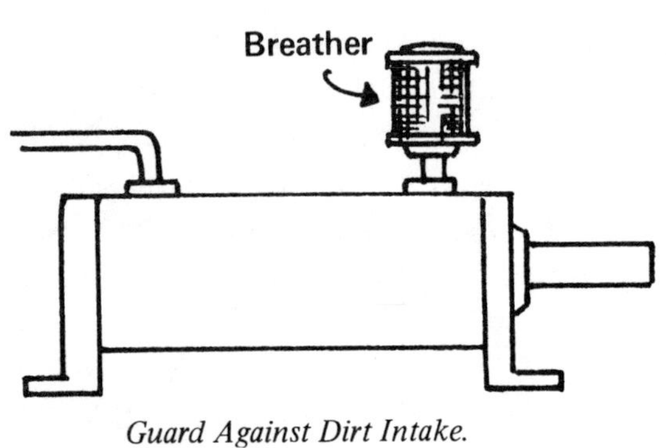

Guard Against Dirt Intake.

Dirt Intake. Almost any double-acting cylinder can be used for single-acting service (power in only one direction) if there is some means to return it to starting position. This could be gravity, load reaction, or an external spring.

When so used, an air cylinder should have a high quality air breather installed in the unused port to prevent ingestion of atmospheric dirt. This is very important because the dirt will lodge in the piston seals and will cut longitudinal scratches in the barrel which will, in turn, wear out the seals, cause piston leaks, and possibly loss of power.

For air cylinders a good air muffler with felt or cotton element, a regular spin-on hydraulic filter with 10μm rating, or a filter as used in a trio assembly can be used. For hydraulic cylinders, the unused port should be connected through a return line to the reservoir above oil level. In case the piston seal should leak, oil will not be spilled around the machine.

Impact Protection. As explained earlier in this chapter, cylinders should be protected from destructive impacts. External positive stops or cushioning devices are desirable in most cases to prolong the mechanical life of cylinders, but sometimes may be difficult to install. Just how much impact a cylinder can tolerate and still have a good life expectancy must be decided by the user, keeping these points in mind:

(1). Most cylinders can tolerate more impact at the end of the retraction stroke than at the end of the extension stroke. A piston rod is more likely to tear loose from the piston on the extension stroke.

(2). Cylinders mounted with a rear flange can usually tolerate more impact on the retraction stroke, while front flange mounted cylinders can tolerate more impact at the end of the extension stroke.

(3). Use speed control valves on air cylinders and do not operate the cylinder faster than necessary to satisfactorily perform the work. Adjust speed individually in each direction.

(4). When designing with hydraulic cylinders, consider the use of a built-in cushion on one or both ends to provide deceleration and reduce impact. Cushions must be specified when cylinder is purchased; they cannot be added in the field. Refer to Volume 1 "Industrial Fluid Power" for a description of cylinder cushions. Refer to Volumes 1 and 2 for other deceleration methods.

(5). The type of load being moved has a lot to do with whether the impact is destructive. If the load has little weight (mass) and consists mainly of friction, the cylinder can bottom out at a fairly high speed with very little destructive force. If the load is massive, and if moving at only a moderate speed, the momentum energy released when the cylinder impacts can be very destructive.

(6). When selecting a cylinder to be used on an application involving high speed and possible destructive impact, stay away from cheap, light-duty models. Select a heavy duty model which is ruggedly built and which will stand a lot of abuse.

Falling Dust. Cylinders located in areas of falling dust or metal chips should have their rods suitably protected. Select a cylinder model with a good rod wiper. Provide a mechanical deflector to shield against gravel, metal chips, and other large particles.

Rod wipers, while desirable, may not give complete protection, as grit may become embedded in a homogenous wiper and remain there to continue its damage. We feel that a bronze rod scraper may provide better protection than a synthetic rubber wiper.

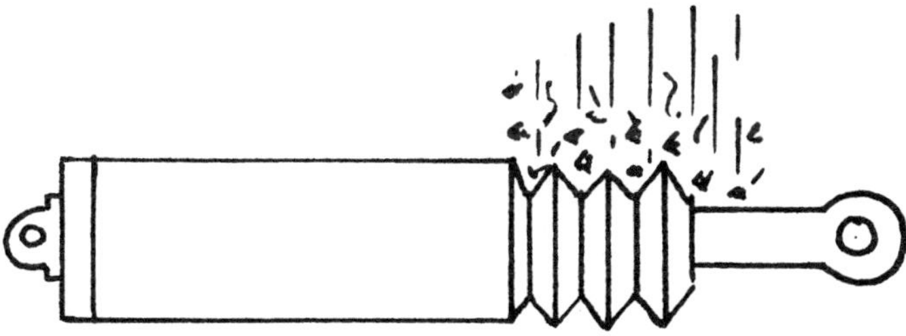

Protect Rod Against Falling Dirt.

In locations where there is fine atmospheric dust, such as in mills and foundries, a rubber accordian-type boot can be installed over the rod. One caution in the use of these boots should be mentioned. The air volume inside the boot changes as the cylinder extends and retracts, and for complete protection, an air vent line from inside the boot must be run to an area of clean air or must be piped through a 10μm breather. A boot without an external air vent will give only partial protection.

Collision by Vehicle. If the cylinder is working in an area where a vehicle could collide with it due to carelessness or accident, guard rails or posts should be installed. Moving machinery such as cranes, hoists, lift trucks, overhead conveyors or cranes, and dump trucks present possible hazards.

Small, light-duty cylinders can possibly be protected with heavy sheet metal shields.

Protect Against Personnel Carelessness.

CYLINDER DRIFT PROBLEMS AND THEIR SOLUTION

There are certain conditions in a hydraulic cylinder circuit which may cause a cylinder to slowly drift (creep) when its 4-way control valve is in center neutral position. The cause (in the fluid circuit) for cylinder drift is the unbalance between areas on opposite sides of the piston. Oil which leaks across the spool of a 4-way valve, under pressure, acts on unequal areas of the piston to develop a force unbalance which may cause unwanted piston movement unless there happens to be sufficient dead load or reactionary load against the piston rod to restrain piston drift. Oil leaking internally across the piston seals can also cause drift as discussed later.

Cylinders with oversize or 2:1 ratio rods have a greater tendency to drift than cylinders with standard (smallest) rod. Drifting force is equal to system pressure times rod area. Therefore, on circuits where drifting could be a problem, the cylinder should have the smallest rod which has sufficient column strength.

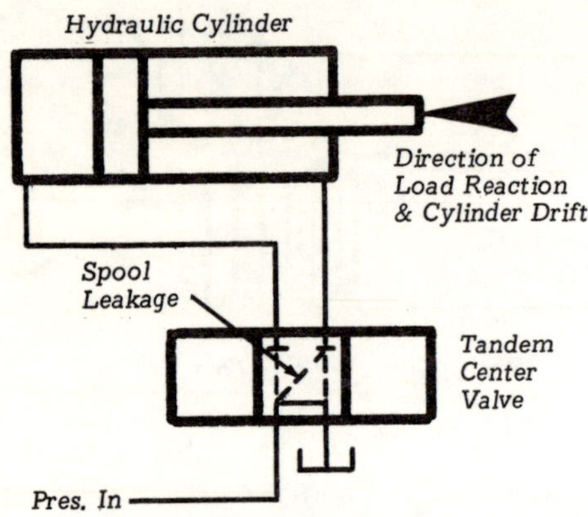

*Drift due to valve spool
Leakage and a reactionary load*

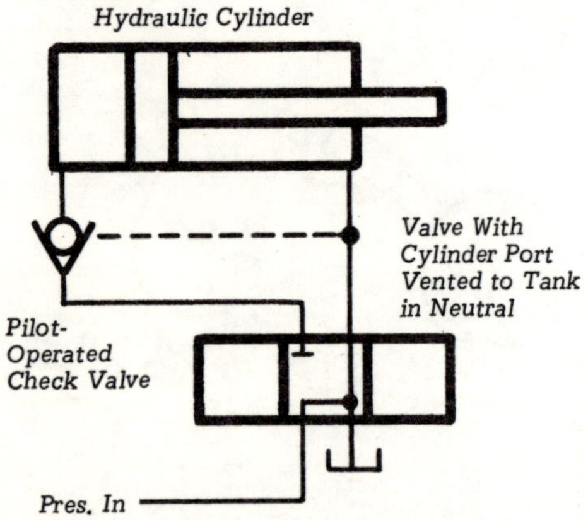

*A pilot-operated check
valve eliminates valve leakage*

Reactionary Load. If the cylinder must support a heavy weight or reactionary force while stopped, internal spool leakage of the 4-way valve will permit the cylinder to drift. In this diagram, internal leakage paths through the valve are shown in dotted lines.

This kind of drift can usually be stopped by placing a pilot-operated check valve in the appropriate cylinder line, as shown in the next diagram, to prevent escape of oil from the cylinder. Refer to "Industrial Fluid Power — Volume 2" for information on these valves. They must be piloted from the opposite cylinder line, and this line must be vented to tank through the spool of the 4-way valve in neutral position to prevent a pressure build-up on the pilot (from valve spool leakage) which would cause the check valve to open and the cylinder to drift.

Note that neither a pilot-operated check valve nor any other external device will prevent cylinder drift if the drift is due to leaking piston seals.

If the reactionary load (while the cylinder is stopped) is pulling instead of pushing on the piston rod, the pilot-operated check valve must be placed in the rod port, and the opposite cylinder port on the 4-way valve must be in a vented condition in neutral. Since pressure intensification can develop in the rod end of a cylinder when the rod line is blocked (as by a pilot-operated check valve), the application must be evaluated to be sure pressure intensification will not cause damage.

Tandem Center System. If two or more cylinder branch circuits are operated from one pump, the upstream cylinder (Cylinder 1 in this diagram) may drift when the downstream cylinder is operating under high pressure.

Leakage across Valve 1 spool, entering the cylinder blind end will cause pressure intensification in the rod end of the cylinder. This intensified pressure, being higher than pump pressure, may leak across the valve spool into the pump line, causing forward drift of

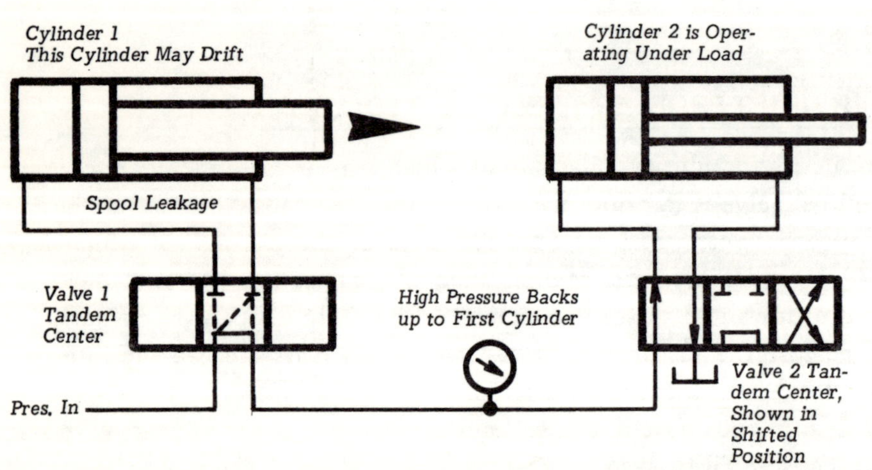

*On an open center system, the upstream cylinder may drift
due to valve leakage when the downstream cylinder is operated.*

the cylinder unless the reaction or gravity load is sufficient to prevent it. To calculate drifting force, multiply rod area times inlet pressure to Valve 1.

Closed Center System. A closed center system is one which includes two or more branch cylinder circuits operating in parallel from one pump. The pressure port on all directional control valves, whether they are individual valves or a bank valve, is closed to oil flow when the valve spool is in center neutral position. Pump pressure builds up to relief valve or compensator level and is maintained during the time when all valves are in neutral position.

To avoid power waste and heat build-up in the

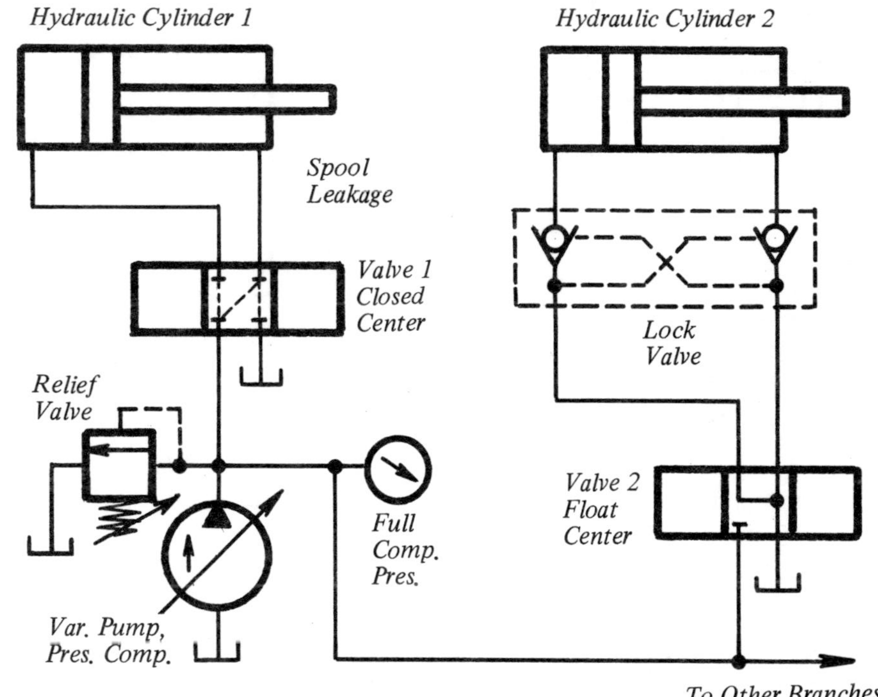

Closed Center System With Two Branch Circuits

oil, a variable displacement pump with pressure compensator is usually employed in a closed center system. When all valve spools are centered and system pressure is maintained at compensator level, the leakage across the 4-way valve spools is at a maximum and may cause one or more cylinders to drift, especially those with large diameter rods and those with little reactionary load against them.

The possible leakage paths across the spool of Valve 1 are shown in dotted lines. Leakage into the blind end of Cylinder 1 will cause pressure intensification in the rod end. The intensified pressure, being higher than pump pressure, may leak across the valve spool and into the pressure line, causing the cylinder to drift forward. However, if there is a dead load against the piston rod greater than the drifting force calculated by multiplying rod area times pump pressure, the cylinder will not drift.

A suggested solution for closed center systems is shown for Cylinder 2. A lock valve can be placed in the lines to the cylinder. A lock valve is a double section pilot-operated check valve. In addition to the lock valve, the spool in Valve 2 must be a float center type, venting both cylinder ports to tank when the spool is centered. The lock valve will prevent cylinder drift in either direction. The vented cylinder ports on the 4-way valve will prevent spool leakage from building up on the pilots of the check valve and holding them open. Any spool leakage will simply bleed off to tank.

Piston Seal Leakage. If the piston seals are not leaktight, the cylinder may drift under either the influence of valve spool leakage or reactionary force against the piston rod. This type of drift can sometimes be stopped or minimized under certain conditions of operation:

If the reactionary or weight load is tending to pull the rod out as in the left view (see illustration on next page), there is no way to arrange the external circuit by valving or otherwise to guarantee there will be no drift in either a closed center or tandem center valve system. A pilot-operated check valve

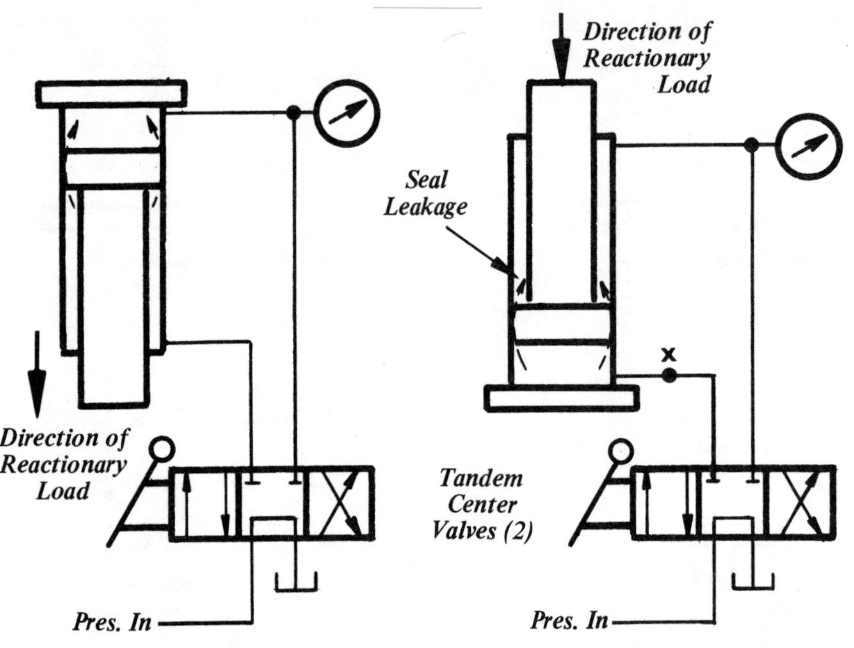

Load Pulling Against Rod   *Load Pushing Against Rod.*

installed in the rod end piping would not prevent internal oil transfer from one side of the piston to the other. The only remedy seems to be to select a cylinder with leaktight piston seals. O-rings or multiple V (chevrons) seal tighter than other types. Even this may not completely eliminate drift caused by valve spool leakage. If drift develops after a period of use, the piston seals should be replaced.

However, if the weight or reactionary load is pushing against the piston rod, as in the right view, drift can be minimized by using a 4-way valve with blocked cylinder ports in neutral. It is interesting to note that leaky piston seals cannot cause drift in this situation. The space opened up above the piston, if drift were possible, would not be sufficient to accept the leakage oil from under the piston because of the difference in volume in the two cavities. A pilot-operated check valve installed at Point x would eliminate downward drift by preventing valve spool leakage but the tandem center valve would have to be replaced by an open center valve.

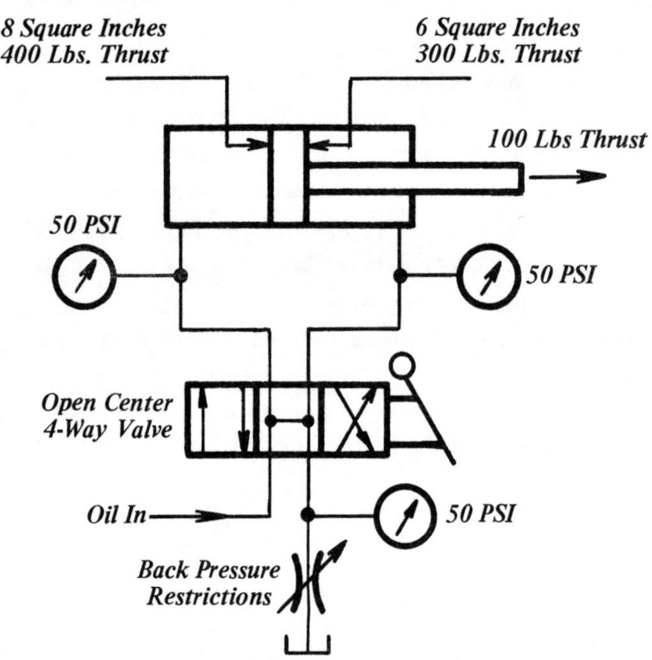

*Back pressure on an open center valve
may cause cylinder to drift.*

Open Center Valve. An open center valve spool is one in which all ports are open to each other and to tank in center neutral position. Although it can be used to control cylinders, it is primarily a control valve for hydraulic motors, and a variation of this spool is called a motor spool.

Great care must be used in applying this valve to the control of single-end-rod cylinders, that is, cylinders which have a piston rod out one end only. In this cylinder, the areas on opposite sides of the piston are unequal. Therefore, if the same pressure is applied, at the same time, to both cylinder ports, a drifting force will be produced.

This illustration shows how a drifting force can be inadvertently produced. The open center 4-way valve is in neutral and pump oil is flowing through to tank. There will always be at least a small pressure drop in the return-to-tank line, and for illustration we have assumed a back pressure of 50 PSI. This 50 PSI will appear on both sides of the cylinder piston. Assuming the full piston area is 8 square inches, a drifting force of 400 lbs. will be produced. This will be counterbalanced by a retraction force of 300 lbs. produced by 50 PSI working on a 6 square inch area. This leaves a net 100 lbs. which tends to cause the cylinder to move forward. If there is a reactionary force of more than 100 lbs., the piston cannot creep forward, in fact it may creep backward.

An open center spool should not normally be used to control a single-end-rod cylinder, or on one which must hold position if stopped at a mid point in its stroke. If necessary to use this spool, the following points should be considered to minimize the danger of drift:

(1). The tank return line from the open center valve must not be combined with the return oil from any other component nor with component drain lines. Even a momentary discharge from some other component through a common drain line could cause the cylinder to move slightly.

(2). Enlarge the diameter of piping in the tank return line of the open center valve. This will reduce the back pressure.

(3). Remove all other components from the drain line. This means heat exchangers, micronic filters, back pressure devices for obtaining pilot pressure for solenoid valves, etc. These devices should be installed elsewhere in the circuit.

(4). Re-route the tank return line, if practical, to shorten its length and to remove as many bends as possible.

(5). Back pressure can be reduced by using a 4-way valve of greater capacity.

(6). On some applications it may be possible to mechanically lock the load while it is stopped, by an electrically or hydraulically operated brake or other means.

(7). Install a counterbalance valve in the line to the rod end of the cylinder, and set the opening pressure to the minimum which will stop the drift.

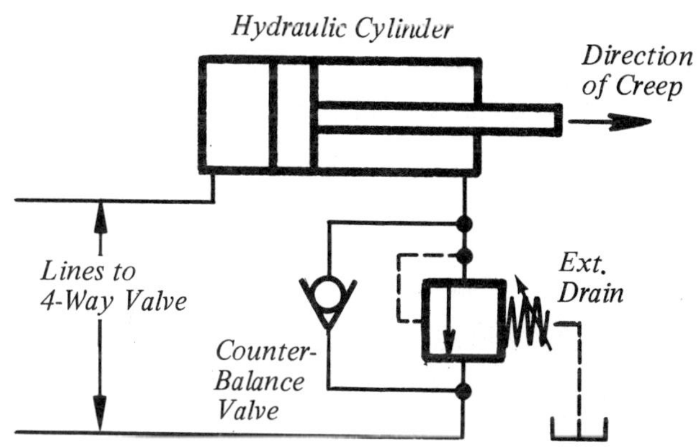

A counterbalance valve can be set high enough to stop cylinder drift.

Central Hydraulic System. Quite often, cylinder creep occurs in a central hydraulic system using 3-position 4-way valves. See diagram on the next page.

A central hydraulic system is one which operates much like an air system in the sense that one pump supplies hydraulic oil under pressure to several independent machines or branch circuits. The pump runs under load, delivering oil under pressure to a receiver tank (accumulator) until sufficient oil is stored in the receiver to raise its pressure to a pre-set high level. When the high level is reached, a pressure signal causes the pump to be unloaded but to continue running in an idle condition. When sufficient oil has been used from the receiver to cause its pressure to fall to a pre-set low level, the pump is again connected to the receiver and runs in a loaded condition until sufficient oil has been stored in the receiver to bring its pressure up to the high level again. Usually the low level is about 80% of the high (fully charged) pressure level. For details of accumulators which can be used as receivers, refer to Volume 1 "Industrial Fluid Power".

In a central hydraulic system, full pressure is maintained at the pressure ports of all 4-way valves. Since all 4-way valves have at least a small amount of leakage, creeping occurs because of spool leakage from the pressure port into one or both cylinder port cavities. Arrows show leakage paths in this illustration. Leakage will produce the same kind of creep as shown on Page 27 for Valve 1. However in this case creep may be much greater because the pressure is maintained on the inlet port at or near maximum at all times.

Central hydraulic systems may use either a 2-position 4-way valve without a center neutral, or if the cylinder must be stopped at an intermediate position, a 4-way valve with a closed pressure port in neutral must be used. This valve will usually have either a closed center or float center spool.

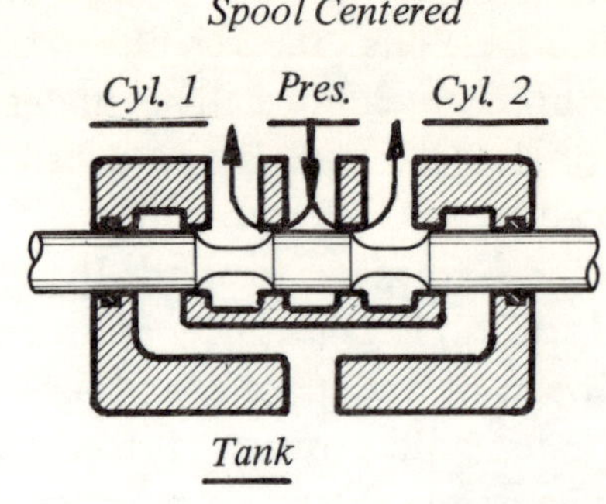

Spool Slippage in Closed Center Valves.

To avoid potential problems from cylinder creep, the following points should be considered before designing such a system.

(1). Avoid the use of 3-position valves with neutral position if possible. Use 2-position valves and let the cylinder stall against a positive stop at the end of its stroke. Of course, some means, other than through the valve spool, must be provided for pump unloading if the cylinder must remain in this stalled condition for more than a few moments at a time.

(2). If circuit action permits, a float center spool in the 4-way valve is preferred to a closed center spool. This will prevent leakage oil from producing creeping pressure against the cylinder piston.

(3). If a closed center spool (all ports closed in neutral) must be used, creeping can be prevented by using a double-end-rod cylinder (rods of the same diameter extending from both end of the cylinder). With equal areas on opposite sides of the piston, creeping forces in each direction will cancel, providing there is leakage into both cylinder port cavities. Or, use a cylinder with smallest diameter rod which has sufficient column strength. This will at least minimize creep.

(4). On some applications, a counterbalance valve installed in the rod line to a single-end-rod cylinder, as shown on the preceding page, may solve a creep problem.

(5). Drift may sometimes not be a problem in a central hydraulic system when components are new and may show up later as valve spools become worn and leakage develops or increases.

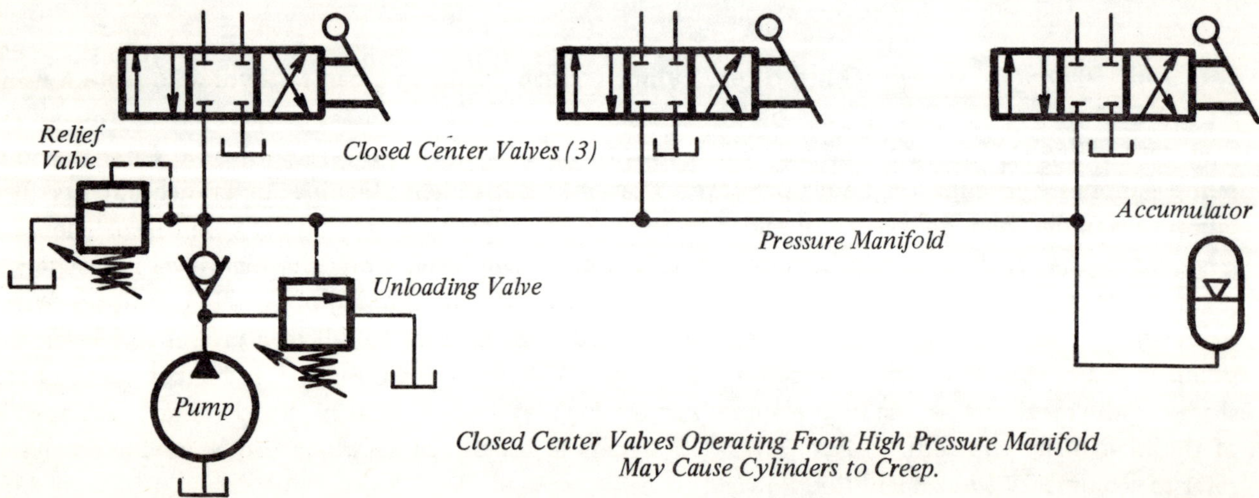

Closed Center Valves Operating From High Pressure Manifold
May Cause Cylinders to Creep.

HOW TO INCREASE CYLINDER SPEED

Compressed Air Cylinders. Since air cylinders are receiving air from an almost unlimited supply drawn from the receiver tank at the compressor, circuits can be designed for very high speed. It is simply a matter of having lines and valves large enough to move a high volume of air to and from the cylinder.

The following is a brief demonstration of factors which influence air cylinder speed. A more complete demonstration will be found in Volume 2 "Industrial Fluid Power".

The first illustration is that of a single-acting cylinder operating vertically and working against a load weight of 1000 lbs. This includes both gravity weight and mechanical friction loss in the cylinder. The cylinder has a 4" diameter piston with 12.57 square inches of exposure. It will require an air pressure of 80 PSI to just balance the reactionary load force.

The circuit below and the ones on the next page are schematic diagrams showing the cylinder operating with various levels of design pressure and how the design pressure affects cylinder speed.

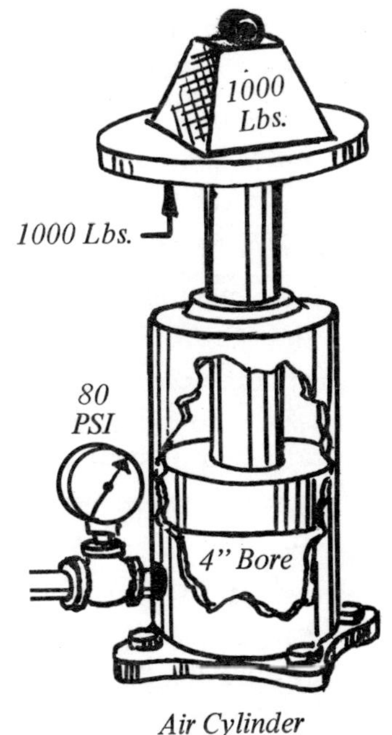

Air Cylinder

This cylinder is in a balanced condition with the load. Additional air pressure must be supplied before it will start to move.

Air System With 80 PSI Design Pressure. In the illustration below, our cylinder is connected into an air circuit in which the pressure regulator has been set to deliver a maximum of 80 PSI. This pressure, working on a 4" bore cylinder will produce exactly the force to balance a 1000 lb. load resistance (consisting of load weight plus friction). The cylinder will support the load but will neither extend nor retract. Under this condition, there will be no air flow through the piping and valves. And with no air

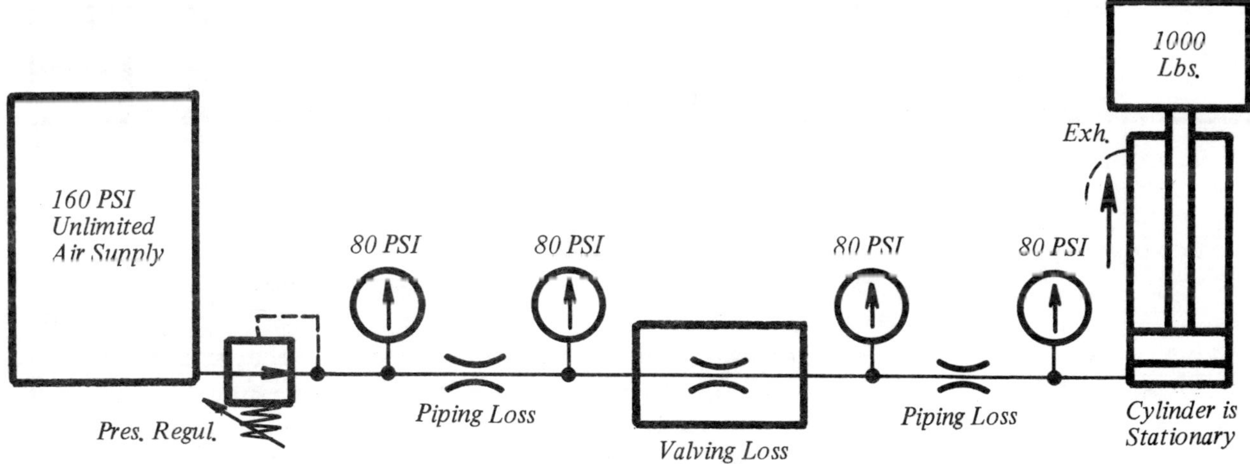

AIR SUPPLY 80 PSI. Cylinder is balanced with the load. No pressure loss in piping or valving.

flow there will be no pressure loss in the circuit and the entire 80 PSI will be present at the cylinder port. This is in accordance with Pascal's Law. We also want to point out that, although the cylinder is exerting its full 1000 lbs. force, since there is no movement of the load, there is no power being transferred through the circuit. In order for the cylinder to start extending there will have to be some additional pressure available to push the air through flow resistances in the circuit. Cylinder speed will depend on the amount of additional pressure available. A higher level of pressure will push more air to the cylinder and it will travel faster.

Air System With 120 PSI Design Pressure. In the illustration below the same cylinder is working in the same air circuit but the setting of the pressure regulator has been increased to 120 PSI. This provides not only the 80 PSI to balance the load but gives an additional 40 PSI to be used for pushing air through flow resistances in piping and valves to cause the cylinder to extend.

The cylinder starts to move and accelerates to a speed where pressure losses through the circuit are exactly equal to 40 PSI, the excess pressure. This is as fast as it can move. If it must move faster, more additional pressure must be provided to push a greater volume of air through circuit resistances. Just what the speed will be with 40 PSI above load balance pressure cannot easily be calculated because the flow resistances cannot be exactly determined.

Look at the gauge readings in the circuit. Design pressure at the outlet of the pressure regulator is 120 PSI. The pressure steadily drops due to flow resistance and drops to 80 PSI at the cylinder port. As long as the cylinder is free to move against the 1000 lb. load, its port pressure can be neither higher nor lower than 80 PSI. If the cylinder should reach the end of its stroke and stall, its port pressure would rise to 120 PSI as air movement through the circuit ceased. Remember that the load resistance shown as 1000 lbs. includes mechanical friction in the seals as well as load weight.

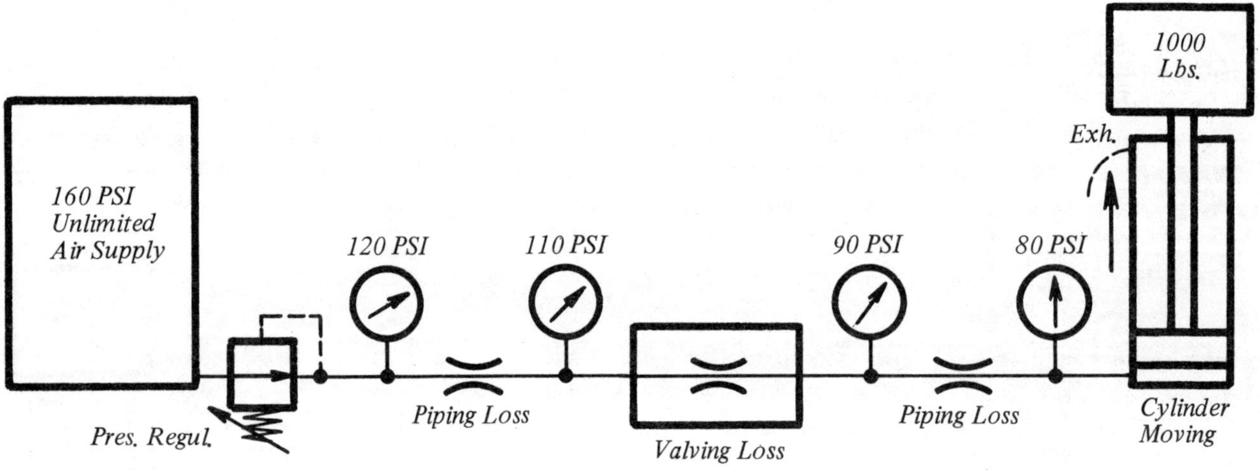

AIR SUPPLY 120 PSI. The cylinder is moving at a moderate speed.

Air System With 160 PSI Design Pressure. The same cylinder is shown in the same circuit in the illustration at the top of the next page, but the pressure regulator has been removed, allowing the circuit to be exposed to the full 160 PSI. Of this pressure, 80 PSI is required to balance the load, leaving an excess pressure of 80 PSI to push air through the circuit to extend the cylinder. Of course the cylinder can now move faster because there is more pressure to push a greater air volume to the

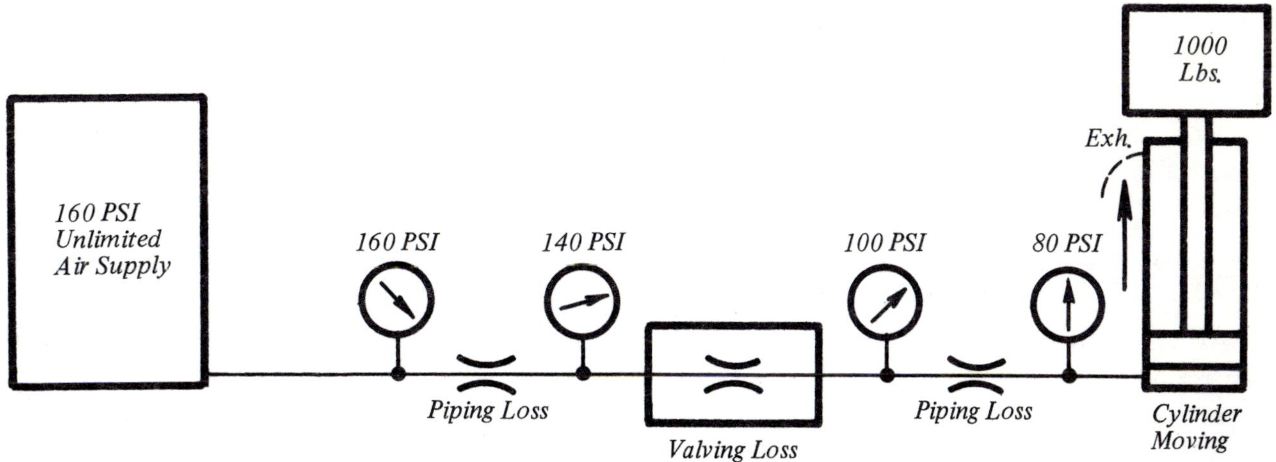

AIR SUPPLY 160 PSI. The cylinder moves at high speed.

cylinder. As in the preceding circuit, the pressure reading at the cylinder port will be no higher or lower than 80 PSI as long as the cylinder is working against a 1000 lb. load resistance. If it should stall, the full 160 PSI would appear at the cylinder port and would produce 2000 lbs. force on the load. On some applications this might crush the load.

Note: On those applications where the cylinder moves in free travel then encounters a high static load near the end of its stroke, and where very little additional movement is required against the load, line size and valve size can be small, smaller than would be required with a moving load, because during the free travel part of the stroke, very little pressure is required against friction load of the cylinder seals, and most of the pressure is used to push air through the circuit.

Increasing Air Cylinder Speed. As demonstrated in the preceding examples, cylinder speed can only be increased by modifying the circuit in some way to get a larger air flow from the receiver tank, through the air circuit, and to the cylinder. There are several modifications which can be made.

(1). Supply Pressure. Perhaps the easiest way to increase air flow is to crank the pressure regulator to a higher outlet pressure, if higher pressure is available, and if this can be done without endangering personnel or equipment. Warning! Care must be used in circuits which use certain kinds of solenoid valves. If pressure is increased above pressure rating of the valve, the solenoid may shift without being energized, and this may cause an industrial accident.

A knowledge of the pressure needed to operate the machine is quite important. If the pressure may fluctuate during the day, the lowest pressure observed during the day should be taken as the inlet pressure to the circuit. Then, a cylinder of suitable bore diameter should be selected which will work at substantially less than the inlet pressure because some of the pressure must be reserved for movement of air.

(2). Piping and Valve Size. Any reduction of flow resistance anywhere in the circuit will increase cylinder speed. Piping and valve size should be selected for a good balance between flow resistance and first cost. A rule-of-thumb is to take the cylinder port size as a reference. Air cylinders up to and including 3'' bore usually have a port size equivalent to 1/4'' NPT. Piping and valves should have a flow capacity at least equal to cylinder port size. Larger sizes will, of course, give faster cylinder speed.

If two cylinders are fed through one control valve, flow capacity of the valve and plumbing to and from the valve should be twice the capacity that would serve one cylinder.

Note: Most of the flow resistance in an air circuit occurs in the cylinder line which is exhausting air from the cylinder. Flow resistance is proportional to flow velocity, and flow velocity is much higher in the exhaust line because the air has expanded to a much larger volume. Use a generous line size on the port of the cylinder which handles exhaust air in the direction in which an increase of speed is desired.

(3). Cylinder Piston Area. Piston area must be large enough to produce the required load force at only 1/2 to 3/4 the inlet pressure if a reasonable speed is to be obtained.

(4). Air Muffler. If an exhaust air muffler is used, remove and discard it. Replace it with a much larger one or pipe the exhaust air to a remote area through a large pipe.

(5). Flow Control Valves. Remove flow control valves from the system. Their internal orifices are much smaller than their connection size and restrict the air in both the "free flow" and "controlled flow" directions even with the needle valve wide open. If speed control is necessary, use flow control valves several sizes larger than the original ones.

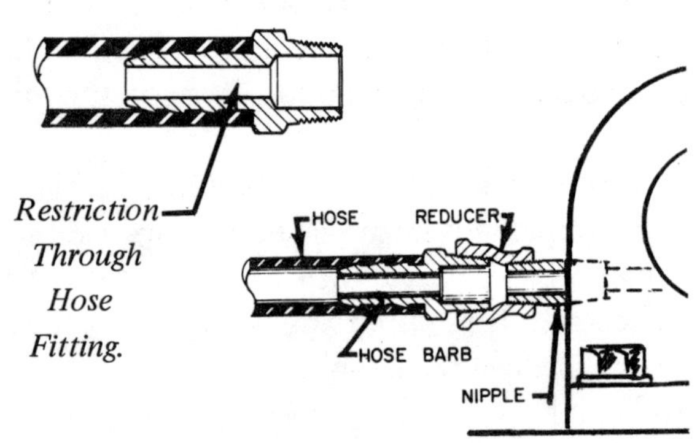

Restriction Through Hose Fitting.

Restriction in Hose Fitting Eliminated.

(6). Pipe Diameter. Enlarge the inside diameter of all piping in the system. Shorten lines if possible, and eliminate unnecessary bends. Eliminate elbows and tees, and unnecessary valving. The illustration shows a possible trouble spot where hose is used. Every hose end slightly restricts the flow. Loss through a single hose is not great but if there are a large number of hoses the cumulative loss can be important. To eliminate these losses, use hose of larger inside diameter, then use a reducer coupling and a close nipple for connection into cylinder and valve ports.

(7). Quick Exhaust Valves. Install a quick exhaust valve directly at the cylinder port. Install one at each port of a double-acting cylinder.

(8). Control Valve. If necessary, replace the 4-way control valve with one of larger flow capacity.

(9). Cylinder. The cylinder can be replaced with one of larger bore so less pressure is needed to balance the load but line and valving size may have to be larger to reduce loss for the higher flow required.

Air-Over-Oil Systems. Speed Increase. An air-over-oil system is a cross between an air system and a hydraulic system. Shop air of 80 to 150 PSI is used for pressure, working through a pressure tank as shown in the next illustration. A flow of hydraulic oil is produced by air pressure for operation of a hydraulic cylinder or other device. The purpose of such a system is to gain stability on applications where the cylinder is used for moving a high friction load at a moderate to slow speed. Flow control valves in the oil circuit are used for speed control of the cylinder. Power transfer in an air-over-oil circuit is the same as for a straight air circuit working at the same pressure. Inlet air, usually from the 4-way control valve is introduced to the top of the pressure tank. A hydraulic oil flow at the same pressure is obtained from the bottom of the tank. Please refer to Volume 2 — "Industrial Fluid Power" for a detailed description and applications for air-over-oil systems.

These circuits are very slow compared to an equivalent air circuit. This is due to the relatively low pressure available to move the oil through the piping and flow control valves. Increasing the air pressure will give some increase in speed, but it is usually necessary to reduce flow restrictions in the oil part of the circuit. This can be done in several ways:

(1). Replace all piping in the hydraulic section with larger sizes, bushing down to mate with port openings of smaller size.

(2). Shorten all lines and eliminate bends in the hydraulic part of the circuit. Eliminate elbows and tees, and any unnecessary valving.

(3). Remove flow control valves from the hydraulic circuit unless absolutely needed. These valves have internal orifices much smaller than their connection size, and are usually the major cause of sluggish operation. Larger size flow control valves can be used, bushing them down to mate with other connections of smaller size.

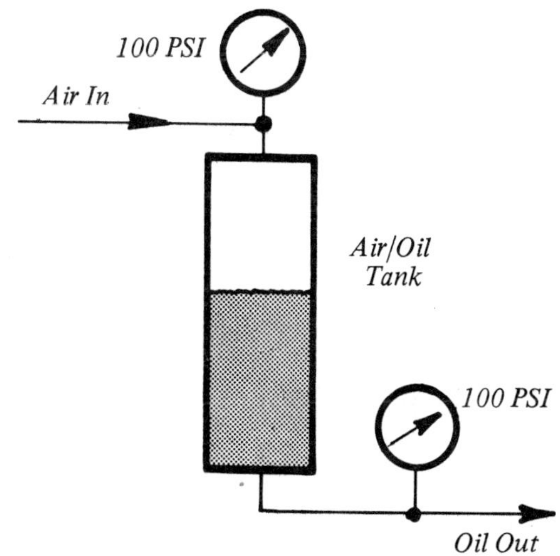

Operating Principle of Air-over-oil System.

Increasing the Speed of a Hydraulic Cylinder. Unlike an air system, the speed of a hydraulic cylinder is determined by the flow volume of the pump and cannot be increased by simply raising the setting of the pressure relief valve. However, the same principle applies — the pump must produce sufficient pressure to balance the load plus enough additional pressure to push the oil through circuit flow resistance to the cylinder. If there is not sufficient pressure to satisfy both these requirements, the pump will operate at relief valve setting; the cylinder may move at less than full speed with a part of the pump flow discharging to tank across the relief valve. The cylinder may start and may accelerate up to a speed where all the additional pressure above the load balance pressure is used up, and there is still not sufficient pressure to push all of the pump flow through circuit resistance. Raising the relief valve pressure, assuming there is sufficient drive horsepower available, will provide the additional pressure needed to push the full pump volume to the cylinder.

Operation in this "gray" area will soon cause the system to overheat. After sufficient pressure has been provided to balance the load and to overcome flow resistance, cylinder speed is determined by pump volume alone, and will not increase at higher relief valve settings. Speed can only be increased by increasing pump flow.

In a hydraulic system which appears to be running too slowly, a simple test can be made to see whether the full pump flow is getting to the cylinder. To be really valid, this test must be run with the cylinder working against full load. Oil slippage from the system may not show up while the cylinder is running idle.

First, the pump RPM must be known and this can be determined from the driving motor nameplate. Then, the pump flow at this RPM must be determined from the pump catalog rating. It will be stated in GPM (gallons per minute) and should be converted to cubic inches per minute by multiplying times 231. Also, the cylinder piston diameter (and area) must be known.

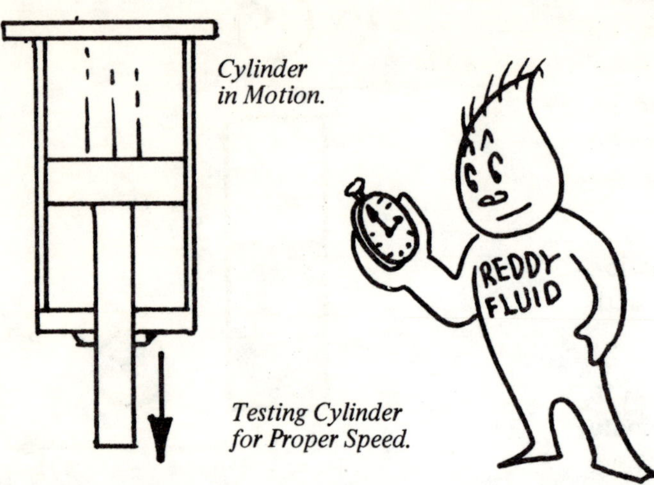

*Cylinder
in Motion.*

*Testing Cylinder
for Proper Speed.*

Start the cylinder extending and time its travel with a watch until it reaches the end of its stroke. Knowing the length of stroke, in inches, a simple calculation will tell if it is traveling at full speed. For example, if the cylinder traveled 12 inches in 10 seconds, it would have traveled 72 inches in 60 seconds (1 minute). Therefore its speed is 72 inches per minute.

Next, calculate from pump flow volume and piston area how far the piston should have traveled in a time of 1 minute. For example, suppose pump flow (from the catalog) is 15 GPM. This is 231 x 15 = 3465 cubic inches per minute. Travel distance in 1 minute on a cylinder with 6'' diameter piston (28.27 square inch area) would be 3465 ÷ 28.27 = 123 inches per minute. On a cylinder with 4'' diameter piston (12.57 square inch area), travel speed would be 3465 ÷ 12.57 = 276 inches per minute. A small difference can be expected between measured speed and calculated speed, but if it is significantly less, the fault may be one of the following:

(1). Worn Piston Seals. On a hydraulic cylinder, if piston seals are badly worn, oil can slip past the piston. Cylinder speed may be slow when the cylinder is moving against a heavy load with full pressure, but may be much faster when the load is removed. A procedure is given on Page 207 for testing cylinder and valve seals.

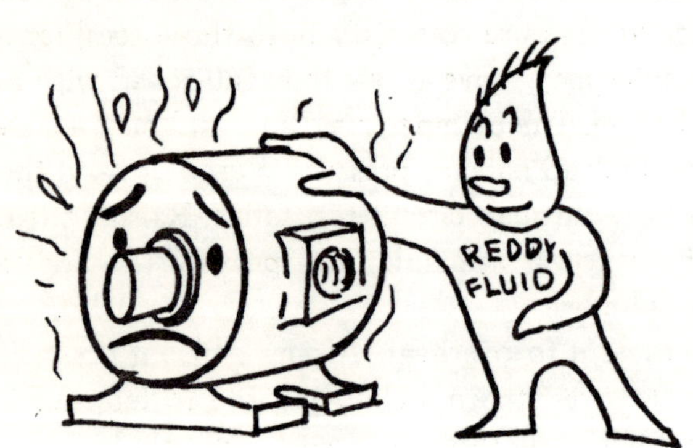

*Slippage due to worn pump or oil that
is too thin may cause pump to run hot.*

(2). Pump Slippage. When the internal parts of a hydraulic pump become badly worn, leakage may increase to the point that cylinder speed is affected. Leakage increases with pump outlet pressure, so a cylinder may travel with near normal speed when moving in free travel, but may move more slowly against a load.

Slippage in the pump creates heat which may be especially noticeable around the pump shaft and bearing. Pumps normally run with a case temperature a little higher than reservoir temperature, but a defective pump can sometimes be spotted by an extremely high case temperature.

Pump slippage is greater with oil of low viscosity. Check the viscosity of the system oil against the viscosity recommended in the pump instruction manual.

(3). Overheated Oil. Hydraulic oil becomes less viscous as its temperature increases; that is, it becomes thinner at higher temperatures. Positive displacement hydraulic pumps become less effective in pumping oil as it becomes thinner. Internal slippage produces even more heat and more loss of viscosity. This condition can become so aggravated that all or most of the pumped oil slips back to the inlet internally and little or no oil is delivered to the external circuit.

As a rule-of-thumb, if the oil becomes so hot, after a few hours operation, that a person cannot comfortably hold the palm of his hand against the side of the reservoir, the oil is operating at too high a temperature. A heat exchanger should be added to the system.

<u>(4)</u>. Relief Valve. Cylinder slow-down may be caused by the relief valve being adjusted slightly too low. This condition was explained on Page 35. There may not be quite enough pressure available to balance the load and to move the full pump flow through the circuit. Setting the relief valve just a *little* higher may solve this problem.

Note: We discourage tampering with the system relief valve setting by operators who may not be familiar with possible safety hazards to personnel or damage to equipment at higher pressures. All adjustments to system pressure should be made by persons authorized to do so.

<u>(5)</u>. Accumulator Pre-Charge. On hydraulic systems using accumulators, a cylinder slow-down for part or all of the stroke may be the result of pre-charge pressure having dropped too low for proper operation. Refer to Chapter 5 for information on charging and gauging accumulators.

On newly designed accumulator systems, a decrease in speed at a certain point in the stroke may indicate insufficient accumulator capacity. If this should be the case, additional accumulators can be connected in parallel with the original one.

Increasing the Speed of a Hydraulic Cylinder.
After checking all the points described in preceding paragraphs, and if the cylinder, although receiving the full flow from the pump, is still moving too slowly, greater speed can only be obtained by increasing the pump flow relative to the cylinder bore. This may be done in one of several ways:

<u>(1)</u>. Pump Size. Replace the pump with one of

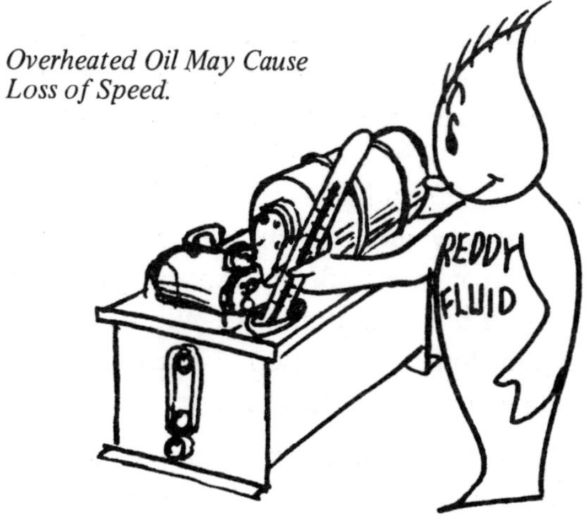

Overheated Oil May Cause Loss of Speed.

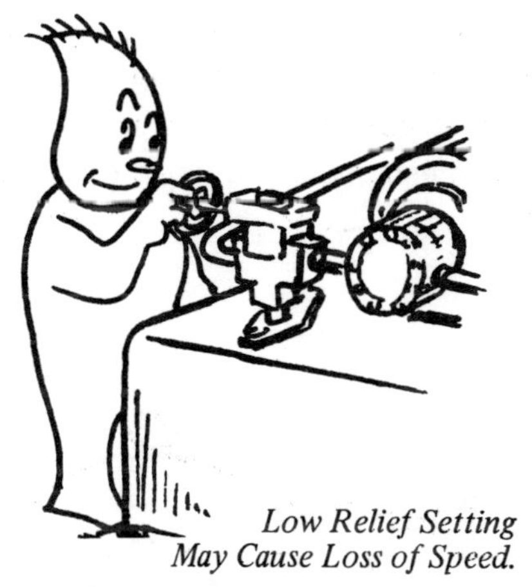

Low Relief Setting May Cause Loss of Speed.

Slow-down may be caused by low pre-charge or insufficient accumulator capacity.

greater displacement. Of course this brings up problems of physical interchangeability, but can introduce other problems, too. The rest of the system should be examined to be sure the ratings of other components are compatible with increased flow. The pump suction strainer, for one thing, should be increased in size to avoid cavitating the new pump. The size of micronic filters in the pump pressure or in the return line may have to be increased. Piping and valving may be able to handle a nominal increase in flow but with a little more flow resistance. In turn, this may mean the relief valve will have to be set slightly higher to overcome the increased flow resistance.

Make very sure that the electric motor or engine which drives the system has sufficient horsepower for a larger pump. Before replacing the pump check electric line current with a loop ammeter while the system is operating with full load. Compare with nameplate amperage rating to see if the motor has additional horsepower capacity not now being used. Remember that input horsepower requirements will go up in direct proportion to the increased flow of oil.

(2). Electric Motor Size. On belt drive systems, pump speed can be increased by changing the belt ratio, but remember that additional horsepower will be required in proportion to the increase in pump speed.

The motor itself can be replaced with one of higher speed. An increase of speed from 1200 to 1800 RPM will produce 50% more oil but will require 50% more drive horsepower. Several additional problems may be created by speeding up the pump. The increased flow may overload existing filters, valves, and piping. The pump itself may cavitate at higher RPM. It may not produce enough suction or its inlet port may be too small to draw in a larger flow of oil. Check its specifications. There may also be mechanical problems of balance, vibration, and bearing loads. Pump noise will increase with shaft speed, so be prepared to accept a higher noise level.

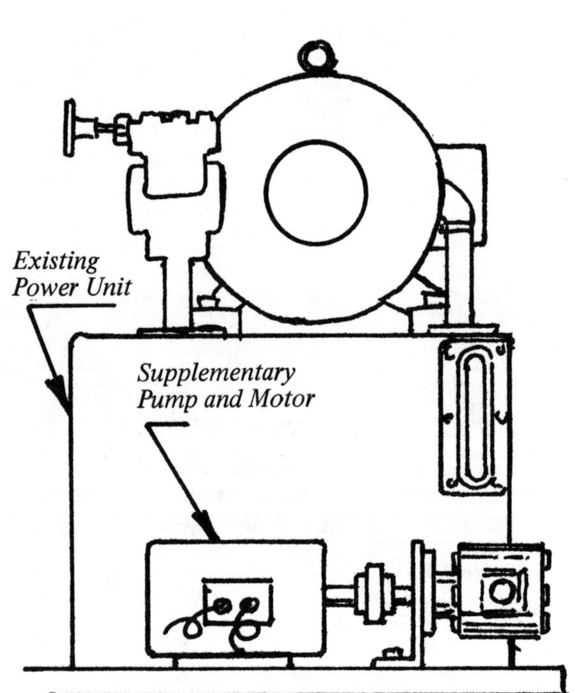

Add Supplementary Pump to Existing System.

(3). Supplementary Pump. A very good way to increase the hydraulic speed is with the addition of another pump driven with its own motor. The output of the new pump can be joined with the output from the original pump.

The new motor and pump can be mounted on a base plate and installed alongside the existing power unit. A "picture frame" can be welded from angle iron and covered with a hot rolled metal plate for the mounting base.

It is imperative that the new pump draw from and discharge back into the same reservoir as the existing pump, for obvious reasons.

The new pump should have its own intake strainer. No relief valve is needed for the new pump if its output is teed into the line with the original pump. If the new pump increases the total flow more than 25%, it may be necessary to replace the original relief valve with a larger one.

All components downstream of the pumps should be evaluated for their capacity to handle an increased flow of oil. This is particularly important in the case of micronic filters. They have paper or fiber elements which could rupture under excessive flow and allow trapped dirt to flow downstream.

Another advantage to the auxiliary pump system is that if one pump should break down, the system could continue in operation at a reduced speed with the other pump. Check valves should be used to join the flows from the two pumps. If one pump is shut down, the check valves will prevent loss of oil from the other pump. Be sure, when adding check valves to the pump circuit, that access to the relief valve is not cut off from any pump.

MISCELLANEOUS TIPS ON CYLINDER OPERATION

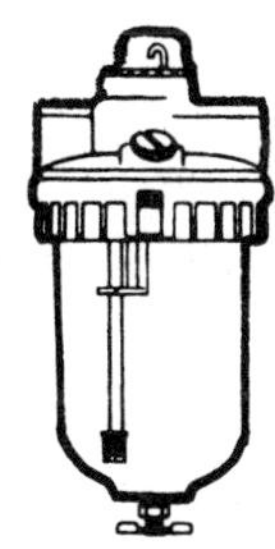

Air Line Lubricator

Lubrication. Air Cylinders usually receive lubrication from an oil mist injected into the compressed air stream by an air line lubricator. This lubricator is part of an air processing "trio" described in Chapter 7. Lubrication reduces wear, helps protect the highly polished barrel from corrosion by entrained water, reduces breakaway and running friction, and makes a tight fluid seal between piston and barrel. Air cylinders require very little lubrication, and the lubricator, as far as the cylinder is concerned, should be set for very low feed. Of course hydraulic cylinders are lubricated by the hydraulic oil.

Pre-lubrication of Cylinders Which Are Seldom Operated.

Pre-Lubrication. Most manufacturers do pre-lubricate air cylinders at assembly. If a cylinder is used in an application where an air line lubricator cannot be used, it can be disassembled occasionally and the inside of the barrel coated with a molybdenum disulfide grease such as "Moly-Kote". On remote applications where a cylinder operates a pipe line plug valve, for example, and where operation is infrequent, it can be internally coated about once a year or every 1000 to 5000 cycles.

Short Stroke Cylinders. Small bore and/or short stroke cylinders with long connecting lines may develop certain problems if the internal volume of the cylinder is about the same or less than the volume in the connecting lines to the valve. Fresh fluid may never reach the cylinder; it may simply circulate back and forth in the connecting lines. This may be a problem with either air or hydraulic cylinders.

Short Stroke *Hydraulic* Cylinders. Oil which may circulate back and forth in the connecting lines without ever being allowed to return to tank may become overheated if cycling is rapid and continuous. Hot oil causes rubber seals and hose to deteriorate rapidly. There may also be a hazard to personnel who may accidentally touch the cylinder or the overheated lines.

The circuit at the top of the next page will eliminate this problem. Two lines connect to the rod end and two to the blind end of the cylinder, and they are isolated with check valves. Oil flowing toward the cylinder follows one of the Paths A. Returning oil follows one of the Paths B. Since the check valves prevent the oil from flowing in reverse, a new charge of oil enters the cylinder on every cycle.

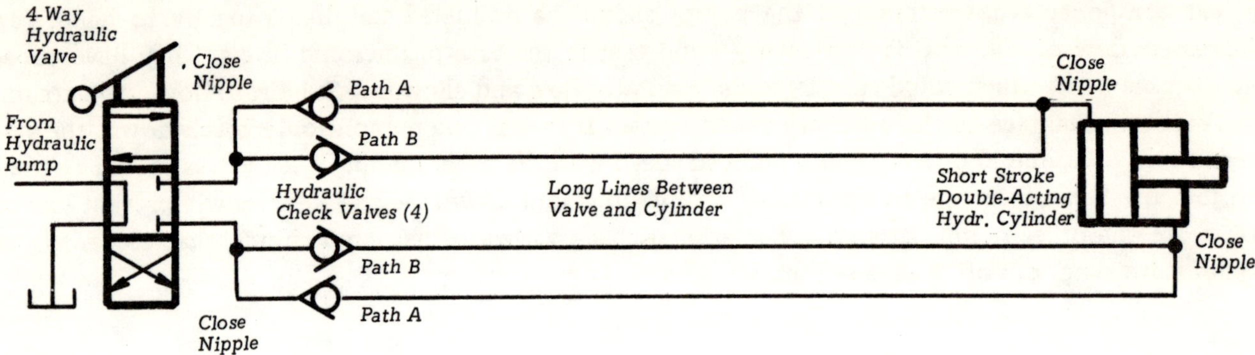

Replenishment of oil in lines connecting valve to hydraulic cylinder.

Short Stroke *Air* Cylinders. Short stroke and/or small bore air cylinders may also have a similar problem of air circulating back and forth in the connecting lines with no new air entering the cylinder. Air lines are not likely to overheat, but the oil mist delivered by the lubricator may never reach the cylinder; it may simply be discharged to atmosphere on every return stroke of the cylinder. Lack of lubrication will cause excessive wear, and create excessive breakaway and running friction.

The double lines and check valves shown for the hydraulic cylinder will also work very well for an air system, but quick exhaust valves may be a better solution. They not only will allow lubricant to enter the cylinder, they increase cylinder speed. The illustration below, showing a double-acting air cylinder, uses a quick exhaust valve in one or both cylinder ports. They should be installed with close nipples at the cylinder ports.

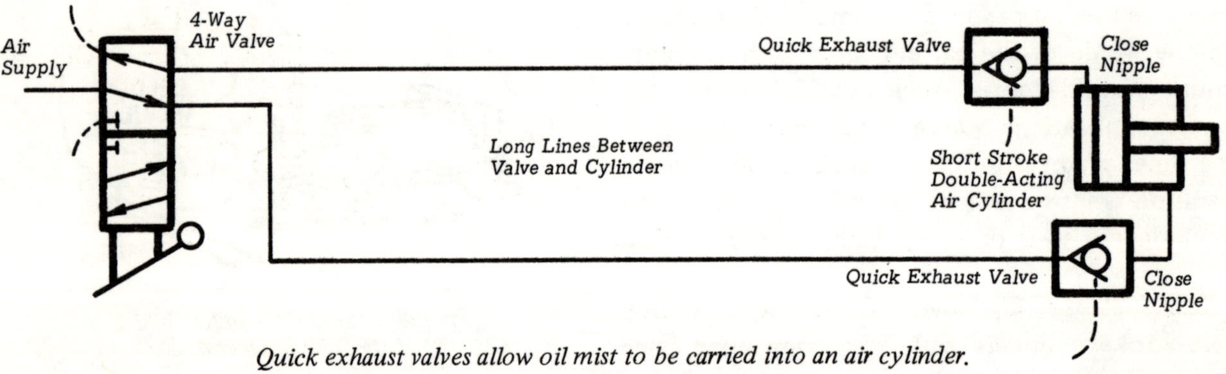

Quick exhaust valves allow oil mist to be carried into an air cylinder.

Exhaust air from the cylinder does not return to the 4-way control valve; it vents directly to atmosphere, allowing a new charge of lubricated air to enter the cylinder on every stroke. A quick exhaust valve will solve the lubrication problem on single-acting air cylinders also.

As in any air system, the exhaust air contains the excess oil mist and in some areas or environments could present a human health problem. If this should be a problem, the mist can be collected and precipitated. In any case, keep lubricator oil feed very low. We feel that a slight amount of under-lubrication may be better than over-lubrication.

Valve Flow Capacity. Some brands of 4-way valves have a greater flow capacity from one cylinder port than from the other due to internal construction. If your valve has this difference in flow capacity, connect blind end of cylinder to valve cylinder port with the higher capacity.

Chapter 2

Valves - Air & Hydraulic

Valve information in this book is aimed toward correct installation and piping, solutions to certain field problems, suggestions for improving performance, and practical solutions to problems encountered in the operation of air and hydraulic valves.

For a study of valve types, graphic symbol construction, how they work, and additional valve circuits, refer to Volumes 1 and 2 "Industrial Fluid Power".

INSTALLATION AND MOUNTING

Mounting Surface. Spool-type valves should be mounted on a smooth surface which is not subject to distortion or warping from heat or mechanical stresses. In the machine tool industry, solenoid valves usually have spools selectively fitted into the body. Clearances are very small and any distortion created by mounting screws may cause the spool to bind. On precision valves the spool may be lapped into the body only after the body has been bolted to a machined surface and mounting screws have been tightened to the specified torque.

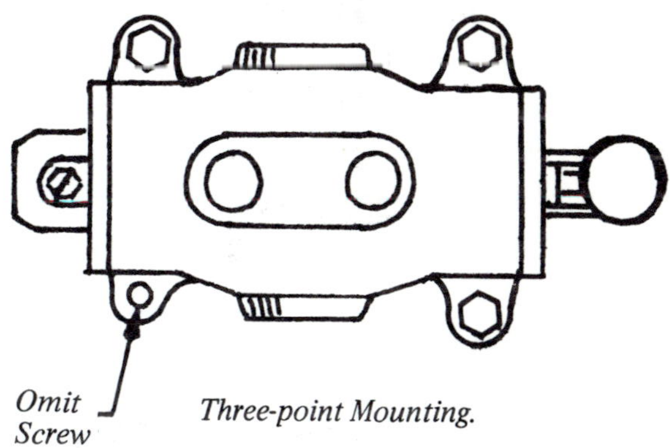

Three-point Mounting.

In ordinary industrial and mobile work the spool fit is usually not this close, but if the valve is mounted on an uneven surface spool binding may still be a problem. Spool bind can also be caused on valves which have tapered pipe thread ports in the body, from over-tightening the pipe threads. Valves which mount on sub-bases, or have straight thread ports are not so subject to body distortion and spool bind. If spool binding appears to be a problem, loosen several of the mounting screws to see if binding is caused by uneven mounting. Slightly loosen tapered pipe fittings to see if it is the result of over-tightening pipe threads.

Preferably all spool-type valves should be mounted on a machined surface. But if this is impractical, a 3-point mounting can be used providing the three corners used for mounting are those which will not distort the body when their screws are tightened. Some mobile-type manual valves are manufactured with 3-point mounting for the specific purpose of eliminating spool bind.

If the mounting feet are cast into, or attached to, the valve end caps, a level mounting may possibly be achieved by slightly loosening the end cap screws (which attach the end caps to valve body) while tightening the foot mounting screws. If mounting feet are cast into the valve body, the valve can be custom fitted to the mounting surface by grinding off a small amount of metal from the highest foot.

Mounting Position. A good practice to follow is to mount spool-type valves, whenever possible, with their spool in a horizontal position. This is particularly important on 2-position types such as double solenoid, double piloted, button bleeder, etc., which have no springs to retain the spool in a normal position. Dynamic unbalance created by fluid flow is the main cause of spool self-shift, and is described later in this chapter. Spool weight and machine vibration also contribute to self-shift. Most industrial valves now have either ball and spring or friction detents on the main spool to prevent accidents due to spool self-shift.

If necessary, small size spring return or spring centered valves can be mounted in other than a horizontal position, but accidents can happen even to these valves if a spool spring should break. Spool weight is insignificant on small valves but can be considerable on valves of 1¼'' size and larger, and these should always be mounted in a horizontal position.

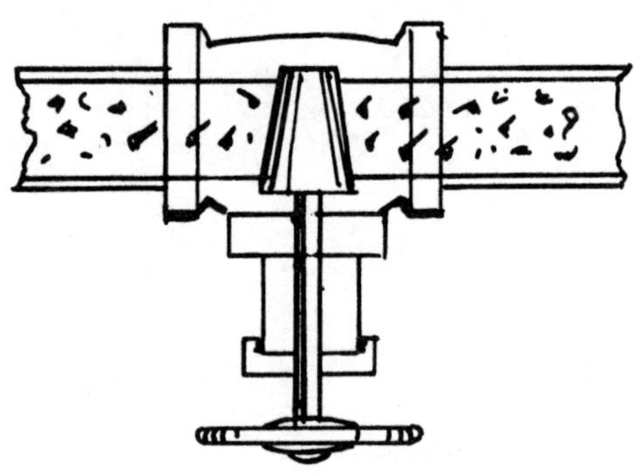
Gate Valve on Slurry. Mount Upside Down.

Gate Valve Mounting. Most gate, globe, plug, and needle valves can be mounted in any position. An exception is a wedging-type gate valve handling a slurry or any liquid which contains solid particles. If mounted upright, solid particles may gravitate into the gate slot while the gate is open. This could prevent leaktight closing of the gate. Mounting the valve upside down will usually remedy this problem.

The same principle may apply to other types of valves in which solid particles could collect around the valve seat and prevent full closing.

Subplate Mounted Valves. Both air and hydraulic valves are available in manifold mounting models. The valve is bolted to a manifold or subplate. All or most of the fluid connections come in on the subplate and are ported into the valve body through O-ring or gasket seals between valve and subplate.

Subplate mounting is more costly than simply screwing pipe directly into taper pipe threads in the valve body, but this mounting style offers several important advantages:

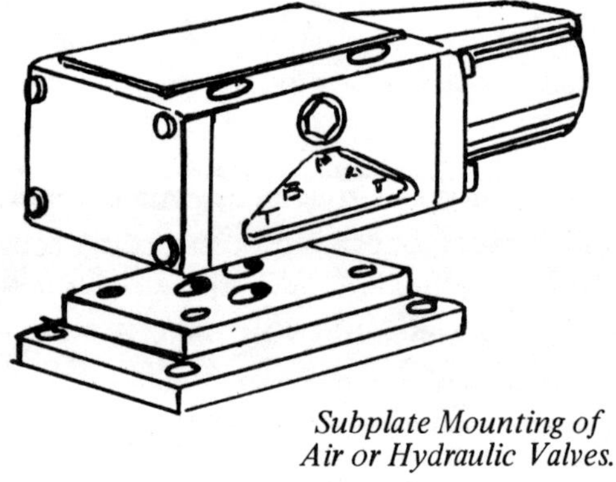
*Subplate Mounting of
Air or Hydraulic Valves.*

(1). The valve body is not distorted by screwing taper pipe threads into it. Leaks around pipe thread connections are eliminated.

(2). In case the valve should fail, it can be quickly replaced, getting the machine back into production in a very short time. Or when troubleshooting a machine, a new valve can be quickly substituted to check faulty valve operation. The additional cost, which is usually only the cost of a subplate, is immediately recovered the first time trouble develops in the fluid circuit.

Caution! Subplates have either taper pipe thread or straight thread connections. Do not ever weld on the subplate either around the connections or to attach the subplate to the machine. Welding may warp the plate, distorting the mounting surface, and causing leaks around the O-ring or gasket seals.

Do not over-pressure subplate mounted valves. Excess pressure may stretch the tie-down bolts, causing the O-ring or gasket seals to leak.

Flange Connected Valves. Subplate mounting becomes somewhat impractical on valve sizes larger than 1½ or 2-inch. Larger valves use bolt-on flanges, square-hole pattern or SAE rectangular pattern. The flanges seal against the valve body with an O-ring. The flanges have a choice of connection threads, taper pipe, straight thread, or plain unthreaded hole for welding to pipe.

The flange also acts as a union, making it relatively easy to disconnect the valve if it should have to be removed.

When welding pipe into a connection flange, be sure to remove the O-ring seal before applying heat. When assembling welded plumbing to a valve, mount the flange loosely on the valve, insert the pipe or tube, then tack the pipe on to the flange with one or two small spots. Then, remove the pipe and tube from the valve, if possible, before running a bead around the pipe, to avoid transfer of excessive heat into the valve body.

Electrical Plug-In Valves. Some manufacturers offer solenoid valves in which the solenoid electrical connections are carried through the valve body into the mounting base to connectors which disconnect when the valve is lifted off the subplate. An important advantage of plug-in connections is that solenoid valves can be replaced even though an electrician may not be available.

To install plug-in valves on existing machines the entire valve and subplate must be replaced.

Indicator Lights. Some manufacturers also offer solenoid valves with built-in pilot light assemblies, usually to replace standard junction box covers.

They are of great value to a troubleshooter in pinpointing a fault when electrical circuit trouble appears. Lights are available either as replacement covers for existing valves or as built-in options on new valves.

As an alternate to pilot lights on each solenoid, a 2-conductor cable can be run from each solenoid to a cluster of pilot lamps on a small panel.

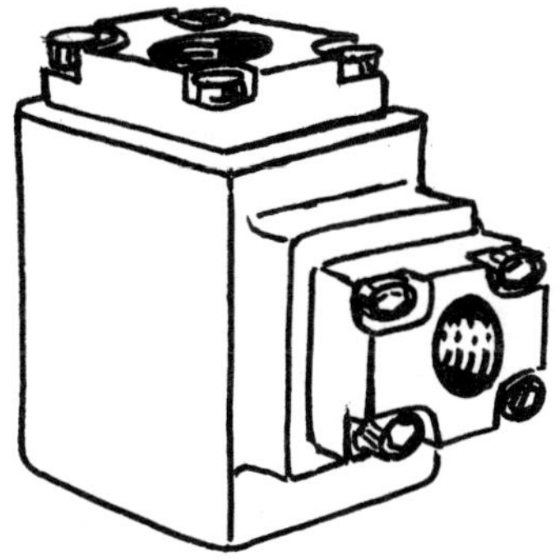

Flange-connected Hydraulic Valves.

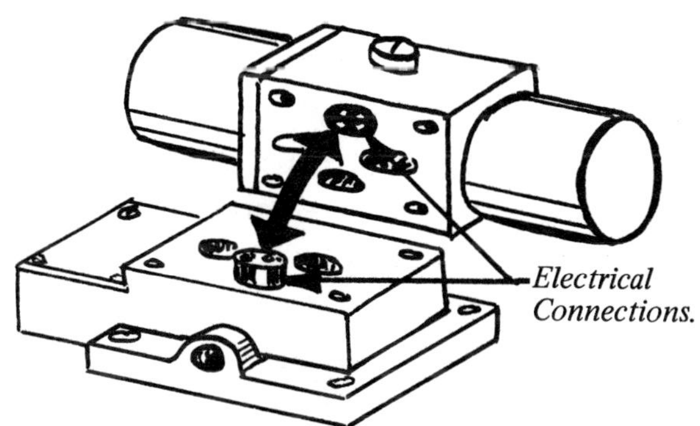

Plug-in Type Solenoid Valves.

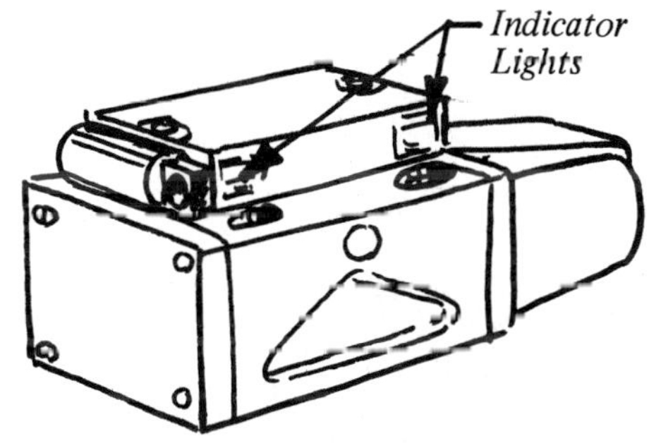

*Indicator Lights Mounted on
Solenoid Valve.*

Pipe Thread Connections. For many years NPT taper pipe threads have been popular as port threads on cylinders, valves, and pumps. A variation of the standard pipe thread, and interchangeable with it, was developed to eliminate or at least to minimize spiral leakage around the crest of the threads. This is the NPTF (National Pipe Thread Fuel) variation. Although this has been a vast improvement, leakage has not been altogether eliminated, especially at high pressures.

At the date of this writing we find many fluid power components still use the NPTF port threads. But the industry has been slowly changing over to the SAE straight thread port which is sealed with an O-ring. But, as industry in the U.S.A. changes over to the S.I. metric international standards, component threads will be metric size straight machine threads or metric straight pipe threads. Incidentally, S.I. metric pipe threads are the former BSP British Standard Pipe threads which have been adopted for international use. Information on S.I. metric port threads is on Pages 220 and 221. But for the present, as long as NPTF port threads are used, these precautions should be observed:

(1). Thread Form. The NPTF instead of NPT thread form should be used on all components. Most manufacturers, even those manufacturing low pressure air components, have changed to this thread. When making your own threads always use NPTF taps and dies. Use sharp tools and keep them well lubricated. It is much better to cut threads on a machine than by hand. Carelessly or inaccurately cut threads are almost certain to leak, even NPTF threads.

(2). Tightening torque. Do not use unnecessary torque when screwing pipe fittings into the body of a component. The wedging action of taper threads may distort the body and cause spools to bind. Over-tightening can also split cast iron bodies. Distortion caused by over-tightening is usually not permanent. If the pipe is slightly loosened the distortion may disappear. When using a thread sealant, friction is reduced and it is easy to over-tighten without being aware of it.

(3). Thread Sealant. We recommend a sealant for threads on hydraulic components. One of the best sealants is Teflon in paste form. Do not use Teflon tape; if not correctly applied, thin threads of tape may be cut off and get into the fluid stream.

(4). Unions. Add unions at every place where it may be necessary to remove a component for repair or replacement. For example on the pump inlet where it may be necessary to replace the pump or replace a suction strainer. Straight thread joints for steel tubing or hose act as their own union and can be easily disconnected when needed.

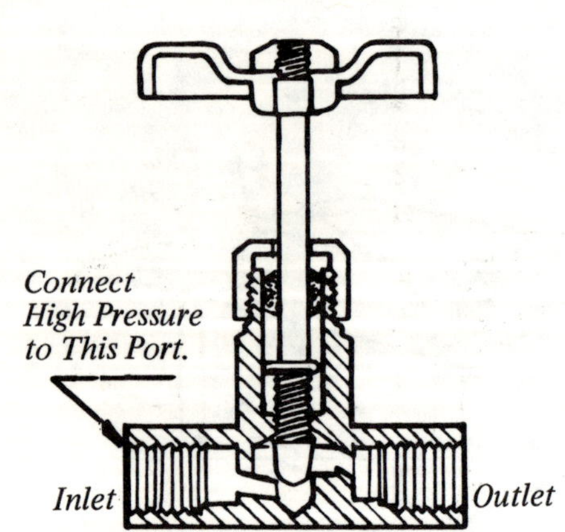

Connect High Pressure to Side
Which Goes to Bottom of
Internal Orifice.

(5). Gauging. Plan ahead for future convenience in troubleshooting. Add tees in the plumbing at every point where it would be helpful to install a pressure gauge when troubleshooting. Plug-in pressure taps can be installed at these points to accept a plug-in pressure gauge. When repairing a machine this is a good time to add gauging points for future use.

Refer to Volume 1 "Industrial Fluid Power" for more information on various threads and plumbing materials.

Needle Valve Orientation. Needle valves are marked for preferred direction of flow. An arrow will be stamped on the body or one port will be marked as the inlet. Either port can be used as the inlet, but when the marked inlet is used, line pressure is removed from the stem seals when the valve is closed. This not only removes unnecessary strain on the seals, it eliminates leakage around the stem if the seals are well worn, and on some models, the stem seals can be replaced when the valve is closed without removing it from the line.

4-Way Directional Valve Orientation. When installing a replacement 4-way valve of a different brand, it might be necessary to interchange wiring to the solenoid coils, or interchange cylinder port connections to get the same direction of cylinder movement with the same direction of valve shift, as there could be a variation of construction, on both air and hydraulic valves, between manufacturers.

In the case of double solenoid valves, interchanging wiring between solenoids would do the job but this would not work on single solenoid valves.

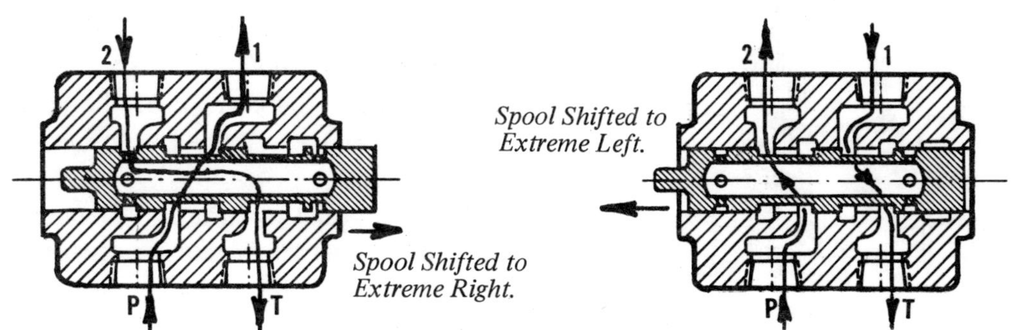

Hollow Spool Valve Having One Flow Return Path Through Center of Spool.

Orientation of 4-Way Valve for Highest Flow Capacity. Some brands of 4-way valves, because of the way they are built, will handle a greater volume of flow from one cylinder port than from the other. On new jobs, or even on existing installations, system performance will be improved if the cylinder port with the greater flow capacity is connected to the blind end of the hydraulic cylinder.

All 4-way valves are constructed so as to join the tank flows from opposite ends of the spool into one common tank port. Some valves, as the one illustrated, use a hollow spool for combining tank flows. Other valves use a cored passage in the body to combine the flows. The advantage to a hollow core spool is that the valve is smaller and lighter, but flow loss through the center of the spool is greater than flow loss across one of the grooves.

In the valve pictured, in the left view, tank flow from cylinder Port 2 passes through the center of the spool and is more restricted than flow from cylinder Port 1 across the spool groove shown in the right view. The proper way to use a valve of this kind is to connect cylinder Port 2 to the rod end of the cylinder.

Most valves available at this time have a cored passage rather than spool center porting, but it may be worthwhile to check the installation drawing for your valve to see if there is any advantage to be gained by connecting it in a certain way.

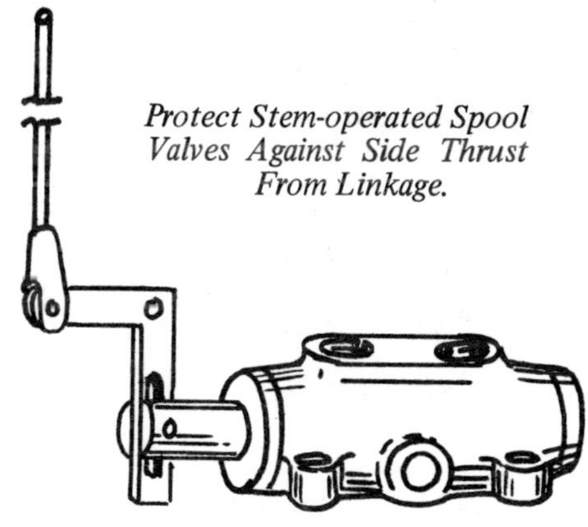

Protect Stem-operated Spool Valves Against Side Thrust From Linkage.

Linkage to Stem-Operated Control Valves. Control valves which are operated remotely through cables and levers must be protected against side thrust on the spool imposed by the linkage. Continued side thrust over a period of time will enlarge the clearance between body and spool, resulting in increased leakage in the valve, or will prematurely wear out the spool end seal.

If levers are attached to the clevis on the end of the spool, they should have slots long enough to take up longitudianl movement as the levers swing in an arc.

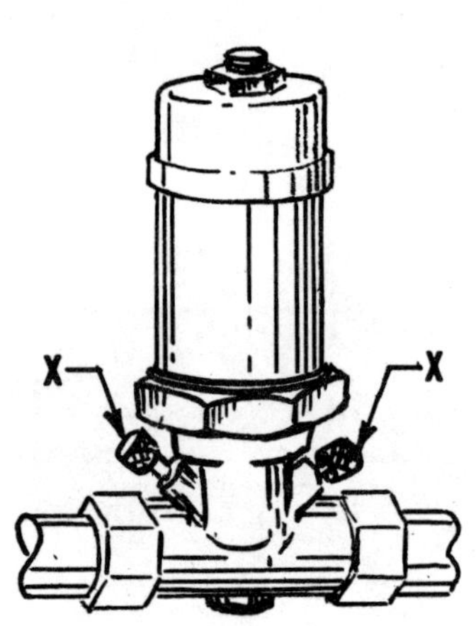

This Type of Solenoid Valve May Have to be Mounted With Solenoid Vertically Upward to Work Properly.

Solenoid 2-Way Valves. Solenoid shut-off valves having the appearance of this illustration are usually of the type known as "solenoid controlled, pilot-operated valves". The solenoid opens and closes a very small control orifice, but the power for opening and closing the main poppet is taken from the fluid line which is being controlled, and is under the control of the solenoid orifice. These valves are described in Volume 1 "Industrial Fluid Power".

The mounting position — horizontal or vertical — may be important. Manufacturers literature should be consulted for mounting requirements.

If there is any doubt, mount with solenoid extending vertically upward. Some valves depend on the weight of the magnetic plunger to return it to its seat after the solenoid has been de-energized. These valves, if mounted in a horizontal position would not work.

Some of these valves have one or two external screw adjustments at locations marked "x". These are probably orifice adjustments to adapt for proper opening and closing speeds on fluids of various viscosities. If there is only one adjustment, it may be a twist-type manual override to manually open the valve in case of an electrical power failure.

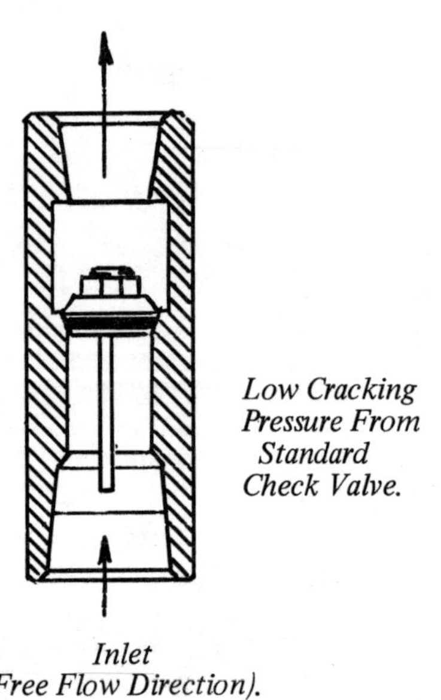

Low Cracking Pressure From Standard Check Valve.

Inlet (Free Flow Direction).

Low Cracking Pressure. A check valve may be mounted in any position for most applications in air or hydraulic service.

There are, however, certain applications in which a very low cracking pressure is essential for the operation. A check valve designed for air may crack at 1 to 3 PSI in the free flow direction. This is too high to be desirable in 20 PSI air instrumentation circuits, and certainly in vacuum applications on which a cracking pressure equal to a few inches of water pressure is desirable.

If a swing or flapper type check valve is not available the spring can be removed from a standard air check valve. The body should be mounted in a vertical position with the weight of the poppet holding it on its seat.

Check valves with soft seal, like an O-ring, seal more tightly than those with metal-to-metal seal.

SOLENOID 4-WAY VALVES

Energizing Both Solenoids at the Same Time. If both solenoids of a double solenoid valve are accidentally energized at the same time, this may or may not be a potential burn-out situation depending on the construction of the valve.

Part A of this illustration shows a valve of the direct-acting type. Both solenoids are attached to opposite ends of the same spool. If both solenoid coils are energized at the same time, one of the armatures will seat and its coil current will drop to the "holding" value. The other armature will not be able to seat and its coil will continue to draw inrush current which may be up to 5 times its normal continuous rating. It will burn out in a short time. Safeguards should be included in the electrical circuit to make it impossible to energize a solenoid if the opposite solenoid is in an energized condition.

Part B shows a pilot-operated type of double solenoid valve. It uses a small direct-acting solenoid valve mounted "piggy-back" on top of the main body to control the movement of the main spool. The piggy-back valve is the same kind of double solenoid valve described in Part A, and the same precautions must be taken to prevent accidentally energizing both solenoid coils at the same time.

Part C shows another kind of double solenoid valve in which a "piggy-back" solenoid valve is mounted on top of the main body but this valve has a different action. Each solenoid operates miniature orifices in the head. Closure of one armature does not prevent the other armature from seating. Therefore, both solenoids can be energized at the same time with no harm to the valve. The solenoid which happens to be energized first will take over control of the main valve spool.

Part D is another kind of construction in which each solenoid is entirely independent of the opposite solenoid. Each one operates its individual orifice and no harm will be done to the valve if both solenoids should accidentally be energized at the same time.

The service diagram for the solenoid valve being used will reveal whether the two solenoids are yoked to a common spool, in which case special safeguards may be necessary in the electrical circuit

A *Direct-acting solenoid valve in which both solenoids are yoked to a common spool.*

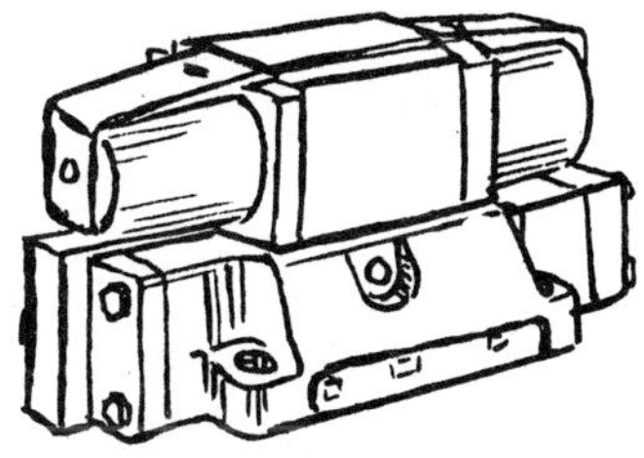

B *Pilot-operated solenoid valve. Both solenoids are yoked to a common spool as in Part A.*

C *Pilot-operated valve does not use a common spool for the solenoids; they are independent of each other.*

D *Pilot-operated valve in which completely independent solenoid operators are used.*

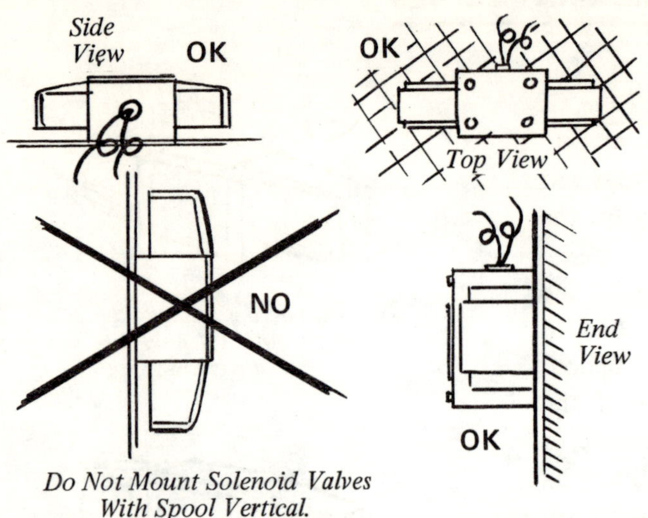

Do Not Mount Solenoid Valves With Spool Vertical.

Mounting Position of Solenoid Valves.

Mounting Position of Solenoid Valves. All spool-type valves should be mounted with their spools in a horizontal position, and this is especially important with solenoid valves. This reduces the danger of the spool drifting out of neutral position, unexpectedly starting the machine and causing an accident.

Most modern solenoid valves which have 2-position spools, with no return or centering springs have a detent to keep the spool from drifting due to machine vibration or other causes. Those valves which do have springs on the main spool are safer but even these valves could possibly drift under a combination of spool weight, machine vibration, and excess flow if the spool operates in a vertical direction.

This does not preclude mounting a spool-type valve on a vertical panel provided the spool is kept in a horizontal operating position.

Self-Shift of Spool-Type Valves. On spool-type valves, especially 2-position models which do not have return or centering springs, the spool can drift out of position when fluid starts to flow across its grooves. Any kind of spool-type valve is subject to self-shifting forces, produced by one or a combination of operating conditions, but the greatest danger is with 2-position double solenoid or pilot-operated hydraulic valves, if they are shifted with a momentary electrical or pilot pressure signal. A circuit using a double solenoid valve, for example, which causes a cylinder to reciprocate, may work reliably for 999 cycles, then fail on the 1000th cycle as the cylinder, after reaching the end of its stroke appears to "bounce" back a short distance and stop. Since the cylinder has moved off the return limit switch, the circuit is in a stalled condition. Operation can only be resumed after the solenoid is pulsed with its manual override button or if the limit switch is actuated by hand.

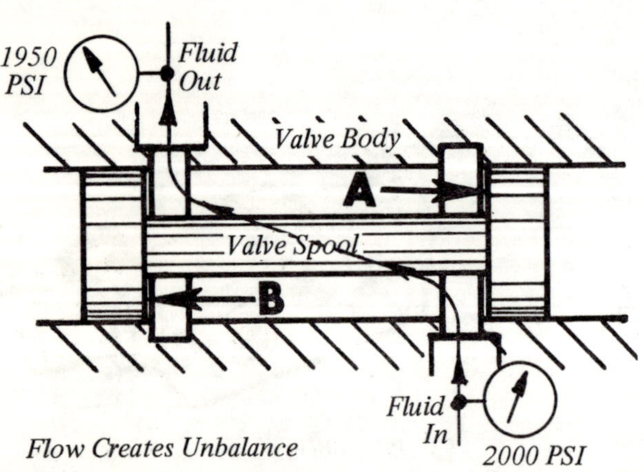

Flow Creates Unbalance

The spool is well cushioned by hydraulic oil and cannot "bounce" back toward its neutral or crossover position. Actually, it is pushed back by unbalanced pressures working on equal areas on opposite sides of the spool grooves. Unbalance of pressure is created in the following manner.

All spool-type valves have equal opposing areas on opposite sides of each spool groove, and will remain in a perfectly balanced condition no matter how high the inlet pressure. But the spool will only remain balanced while the fluid is static (not moving). It will become unbalanced and subject to possible drift as soon as fluid starts moving). It will become unbalanced and subject to possible drift as soon as fluid starts to flow across the spool grooves. The amount of drifting force created varies with the flow resistance across the spool and to the square inch exposure area on each side of the spool grooves. Drifting force is created as shown in the illustration on the preceding page.

In this diagram we will assume that fluid (hydraulic oil) is entering the inlet port under a pressure of 2000 PSI. When fluid starts to flow, pressure in the outlet stream may drop to 1950 PSI, for example, due to flow resistance across the groove. A drifting force in Direction A is produced by 50 PSI working on the exposure area on the sides of the groove. If this drifting force is sufficient, the spool would actually move a short distance to the right.

Other factors such as machine vibration and spool weight, if valve spool is working in a vertical direction, might also add to the drifting force.

Remedies for Self-Shift. The tendency for a spool to drift or "bounce" after shifting can be reduced or eliminated by proper attention to these points:

(1). The electrical circuit, instead of delivering only a momentary pulse to shift the spool, should be designed to keep current on the solenoid coil throughout the entire stroke of a cylinder, even though a 2-position valve will shift and remain shifted on a momentary electrical impulse.

On new designs, current can be maintained on valve solenoids with holding relays. On existing machines, if impractical to alter the electrical circuit, a mechanical camming arrangement can sometimes be installed as shown in the next diagram. In many cases this will not require a major change in the electrical circuit or wiring.

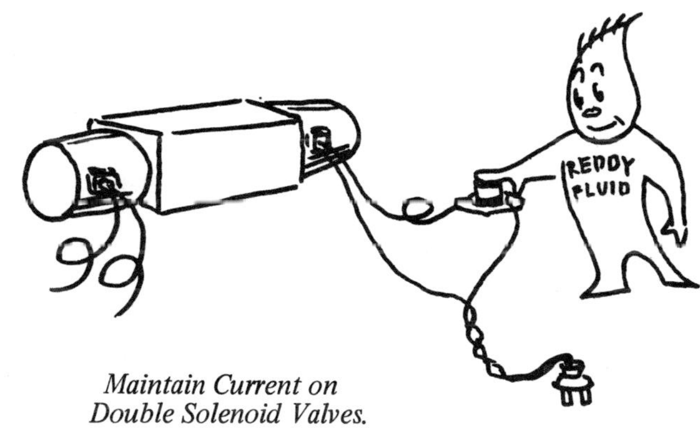

*Maintain Current on
Double Solenoid Valves.*

(2). To maintain holding current on a solenoid throughout the cylinder stroke, the spring return electrical limit switches normally used can be replaced, or the existing switches can sometimes be modified, to have maintained action. In maintained action the switch does not spring back to a normal position; it must be actuated in both directions. This is "toggle" action. The two limit switches normally used, one at each end of the cylinder stroke, can be replaced with one "toggle-type" switch actuated by one or more cams attached to a moving member of the machine. Movable cams can be ad-

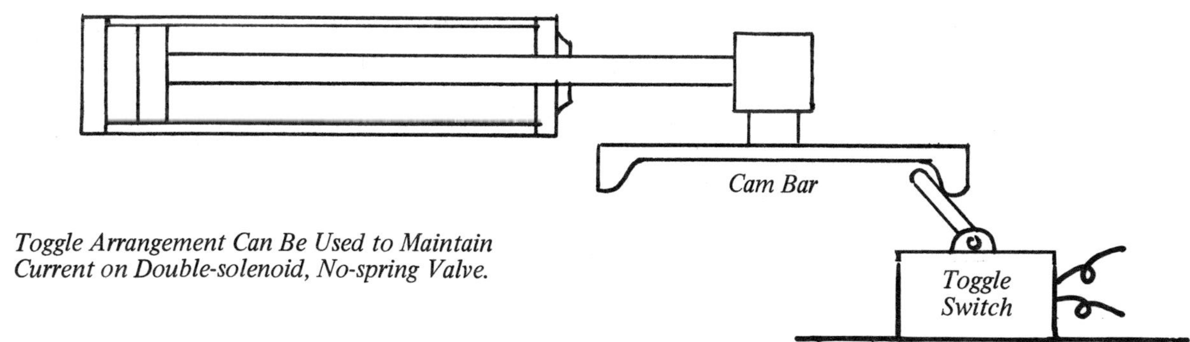

*Toggle Arrangement Can Be Used to Maintain
Current on Double-solenoid, No-spring Valve.*

justed for exact cylinder stroke. If the cams could override the switch in event of an electrical failure, the cams should be designed so the switch is not damaged.

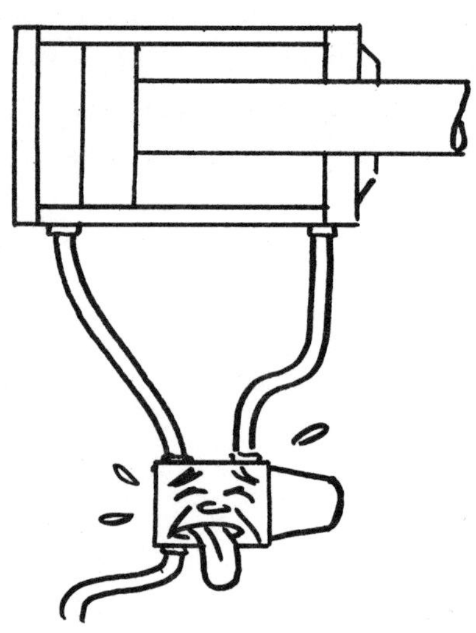

*Excessive Flow May Create
High Pressure Loss and
Spool Unbalance.*

(3). Do not overload the valve with a higher flow than its catalog rating. Unbalance increases with flow volume. And don't forget that return flow from the blind end of a cylinder is greater (on a single-end-rod cylinder) than flow from the rod end. It can be up to twice the pump flow on 2:1 ratio cylinders. If you can't arrange the electrical circuit to maintain current on the solenoid coils, be conservative in your choice of valve size. Stay well below the manufacturers flow rating.

(4). Mount all spool-type valves with spool in a horizontal position. If absolutely necessary to mount with spool vertical, mount the end up which is the direction drift would be produced by excessive flow. In the earlier illustration this would be with the right end up.

(5). When selecting a valve model, choose one which has detents on the main spool. We believe ball and spring detents will outlast friction-type detents.

(6). On pilot-operated solenoid valves, use an external solenoid drain line to tank. Do not combine solenoid drain with main tank return line. On subplate mounted valves, connect external drain line to Port Y on the subplate, and be sure to plug the internal drain passage. Back pressure surges in the tank return line, if internal draining is used, could back up into the solenoid spool chamber and in some situations could tend to self-shift the solenoid spool. These surges will also cause an additional load on solenoid coils and will produce overheating and possible burn-out.

The self-shift effect described above occurs more often on 2-position solenoid or pilot-operated valves, but can be a threat on manually operated spool valves resulting in a momentary "kick" on the handle, or in extreme cases could actually shift the spool.

Self-shifting is a greater problem on hydraulic valves but can be a problem on air valves operating on flows above catalog rating.

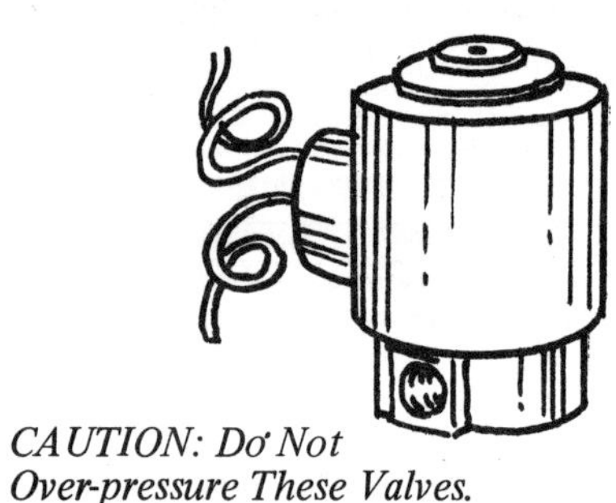

*CAUTION: Do Not
Over-pressure These Valves.*

Effect of Over-Pressure on Poppet-Type Solenoid Valves. This is a common type of miniature poppet valve used by itself in 2-way or 3-way service or as a solenoid operator on larger 3-way and 4-way air valves.

The valve works by poppet action. The poppet spring must be able to hold the valve orifice closed against inlet pressure when the solenoid coil is de-energized. Available in many orifice diameters and pressure ratings.

Warning! It is dangerous to operate this kind of valve on line pressure higher than its rating. Excessive pressure working against the poppet could overcome the poppet spring and cause the valve to shift without being energized. This has been the cause of industrial accidents in the past.

Drain Lines. Many hydraulic valves have external drain ports which must be drained to an area of atmospheric pressure, usually the system reservoir. These drain ports should never be plugged unless there is an alternate way of draining inside the valve to a tank return line. On certain kinds of valves, listed below, an external drain line to tank is mandatory; there is no way to discharge the drain oil to reservoir inside the valve. In all cases a separate drain line to reservoir is better than teeing the drain flow into the main tank return line.

On 4-way solenoid valves of the pilot-operated type, drain flow from the solenoid "piggy-back" solenoid valve can either be discharged to reservoir inside the valve by connecting it to the main tank port, or the flow can be piped to reservoir through an external line connected to Port Y on the subplate. When draining from Port Y, be sure the internal drain passage is blocked.

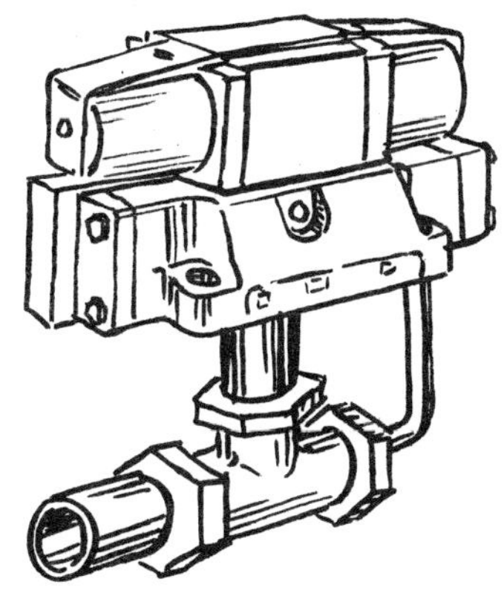

Solenoid drain should not be teed into tank return line.

The advantage of a separate drain line for the solenoid "piggy-back" valve is to isolate this valve from pressure spikes which may be generated by momentary high flow in the tank line. If these pressure spikes are transmitted up into the solenoid drain cavities an extra load is placed on the solenoid each time it is energized. At high cycle rates this may overheat and burn out a solenoid coil.

Drain lines on components prevent surge pressures from entering valve end caps, spring chambers, component cases, and seal pockets. Sequence, pressure reducing, by-pass, and counterbalance valves must be externally drained because their pressure setting must be referenced to atmospheric pressure. While it is possible to connect their drain ports into the main tank return line, a separate drain line back to reservoir is preferred because this will isolate them from pressure surges which may occur in the main tank return line caused by valve shifting, large-rod cylinders retracting, etc. Pressure surges in drain lines would cause components to work improperly. If necessary to tee component drain ports into the main tank return line, the line should be as large as practical to minimize pressure transients. However, drain lines from all components may be combined into one return line back to the reservoir.

Keep Solenoid Coils Covered. Dust covers should be kept on solenoid coils. These not only protect windings against water and oil splash, but keep dirt out of the solenoid armature. If a particle of dirt should lodge under the armature and keep it from fully seating, coil current is increased and the coil will burn out in a matter of time. The cover, if a separate item, should be kept chained to the valve so if it is removed it cannot be lost.

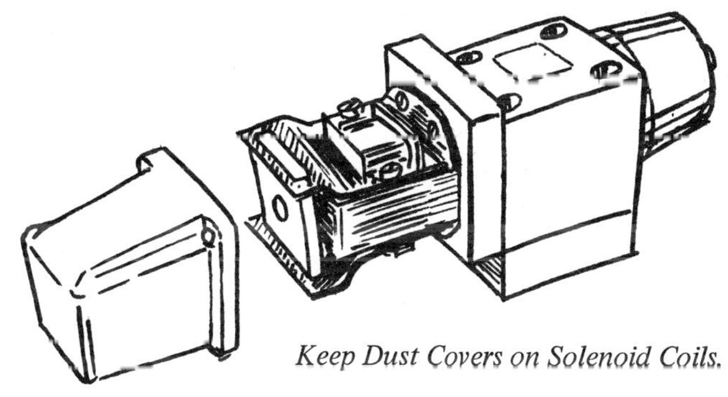

Keep Dust Covers on Solenoid Coils.

This warning applies to older valves in the field. Most modern valves are arranged so the solenoid cannot be operated unless the cover is in place. If valves are mounted inside a dust-tight cabinet, solenoid covers are not needed, and if they can be removed, or if the valves can be purchased without them, heat dissipation from the solenoids will be improved.

Solenoid valves with these spools may not shift out of center position, when energized, unless there is at least 50 to 100 PSI pilot pressure available, either internally or externally.

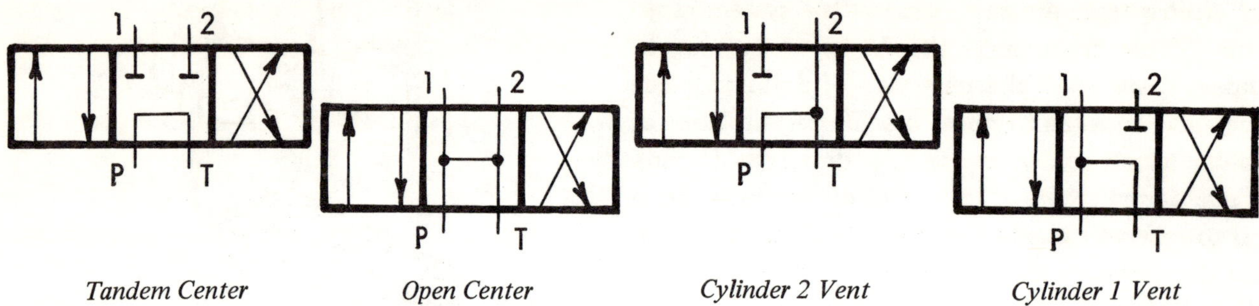

<table>
<tr><td>Tandem Center</td><td>Open Center</td><td>Cylinder 2 Vent</td><td>Cylinder 1 Vent</td></tr>
</table>

Failure of Solenoid Valve to Shift. On pilot-operated solenoid valves which have one of the spools shown above, and which use inlet hydraulic pressure as pilot pressure to shift the main spool, the spool will not shift out of center position, even if a solenoid is energized, if there is not at least 75 PSI on the main inlet port when both solenoids are de-energized. The above spools have the pressure port vented to the tank port in center position of the spool and are subject to this kind of trouble.

The sectional view below shows a valve with tandem center spool. It is connected to use inlet pressure for piloting the main spool (under control of the solenoids). Pressure from the main inlet is taken up through an internally drilled hole to the solenoid valve, and is diverted to one end or the other of the main spool as the solenoids are energized.

To determine the reason the valve will not shift when a solenoid is energized, use the manual

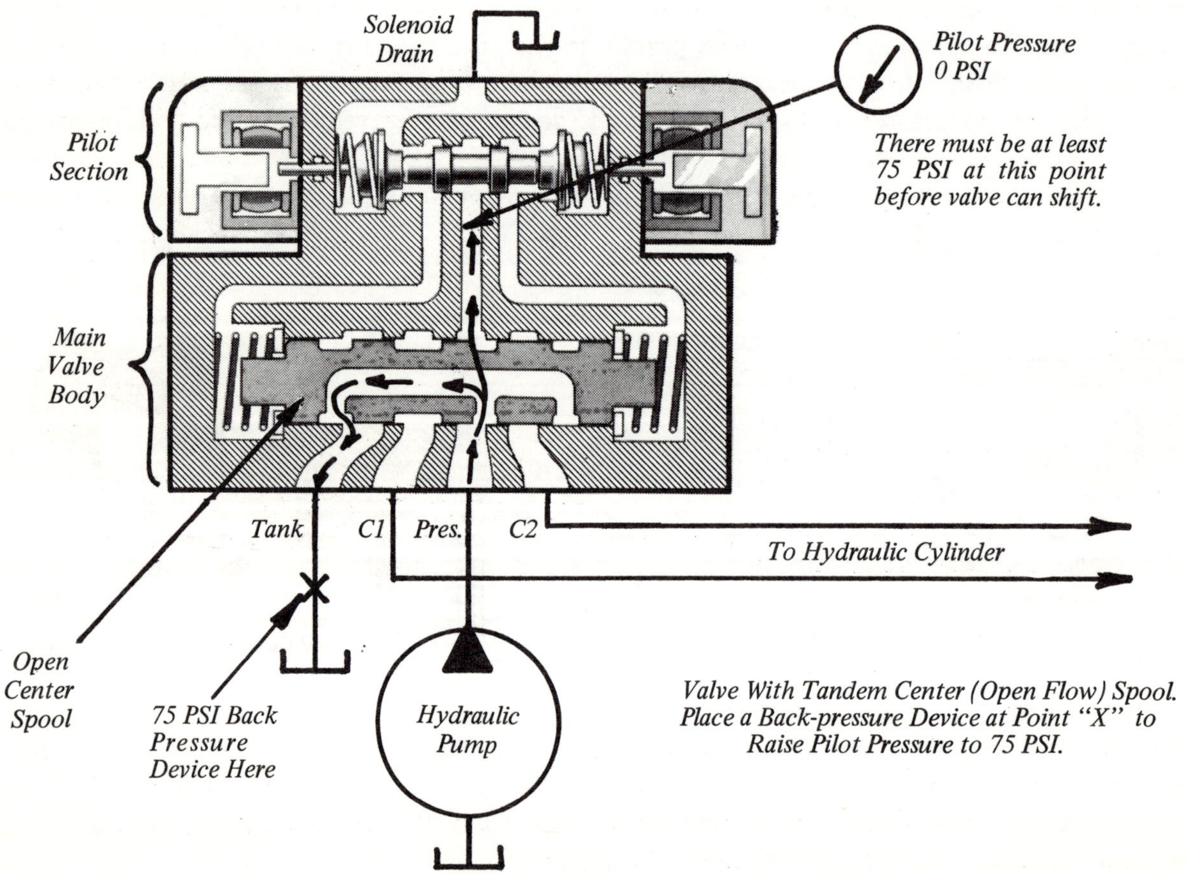

Valve With Tandem Center (Open Flow) Spool. Place a Back-pressure Device at Point "X" to Raise Pilot Pressure to 75 PSI.

override pins on the solenoid covers to manually shift the solenoid spool. If the problem is electrical, the machine should start up. If the problem is insufficient inlet pressure, the machine will not start.

Test for insufficient pilot pressure by connecting a low range pressure gauge directly at the inlet port of the valve. Do not try to shift the valve while the low range gauge is installed; if the main spool should shift out of center position the gauge would be ruined.

If inlet pressure is less than 50 to 75 PSI it must be increased. Perhaps the simplest way to do this is to convert the valve (if necessary) to external draining and run a drain line from Port Y on the subplate to tank. Be sure to plug the internal drain line. Then add a restriction (plug valve or low pressure non-adjustable relief valve) in the tank return line. Adjust this restriction, with valve spool in center position, until pressure at the inlet port is increased to 75 PSI or slightly higher.

A better solution to this problem of lack of pilot pressure may be to add a back pressure valve in the pressure inlet port, and this is described in Chapter 7, "Industrial Fluid Power" Volume 2.

Solenoid Coil Burn-Out. Perhaps occasionally a solenoid coil may burn out because of a defect in its manufacture. But usually the trouble can be traced to some abnormal condition either in the operating conditions of the machine on which the valve is installed, or to unusual environmental conditions. This becomes evident if a burn-out should occur more than once at the same coil location.

Checklist for Burn-Out of A-C Valves. Burn-out is more common on valves with A-C coils than on those with D-C coils because of high inrush current. Until the armature on an A-C solenoid can pull in and close the air gap in the magnetic loop, the inrush current may be up to 5 times the steady state or holding current after the armature seats. On a D-C solenoid, inrush and holding currents are approximately the same. The use of valves with D-C solenoids will eliminate the inrush current problem.

(1). Voltage Too High. Voltage rating is usually marked on the coil. Compare this with measured voltage furnished from the electrical circuit. Operating voltage should not exceed 10% of rated coil voltage.

(2). Voltage Too Low. Operating voltage should be no more than 10% below rated coil voltage. Low voltage reduces solenoid force. It may continue to draw inrush current without being able to pull in.

A test for low voltage can be made by measuring voltage directly on the coil wires or terminals while the solenoid is energized and with the armature blocked open so it will draw full inrush current. Energize solenoid briefly, just long enough to read the voltage. Also take a no-load voltage reading with the solenoid disconnected from the feed wires. A difference of more than 5% between these two readings indicates excessive resistance in the wiring or insufficient volt-ampere capacity in the control transformer if one is used.

Consult Check List For Cause of Solenoid Coil Burn-out.

(3). Wrong Frequency. Operation of a 60 Hz coil on 50 Hz causes the coil to draw excessive current and to overheat. Operation of a 50 Hz coil on 60 Hz causes the coil to draw less than rated current. It may burn out from inability to pull in.

Check Line Voltage

(4). Too Rapid Cycling. Since inrush current of an A-C solenoid may be up to 5 times the steady state current, a standard A-C coil on an air gap solenoid assembly may overheat and burn out if required to cycle too frequently. The extra heat generated during inrush periods cannot be dissipated as rapidly as necessary. The gradual build-up of heat inside the coil winding may, in time, damage the coil insulation.

High cycling applications can be roughly defined as those where the solenoid is energized more than 5 times per minute. Oil immersed solenoid assemblies should be used on valves for those applications.

If instrumentation is available, a thermocouple can be placed on the surface of the winding of two identical valves of the type to be used. One valve can be cycled for several hours at the proposed cycle rate and the other valve can be continuously energized. A difference of more than a few degrees in surface temperature between the two coils indicates a need for molded coils, high temperature coils, or oil immersed solenoid assemblies.

(5). High Electrical Transients. If the current for the solenoids is taken directly from a power line serving large inductive devices such as electric motors, the starting of these motors may generate high voltage transients which may break down insulation of solenoid coils. A "thyrector" should be placed across each valve coil to short circuit these transients. Thyrectors are available at electrical supply houses.

(6). Dead End Service. Fluid circulating through a solenoid valve carries away electrical heat from the coils. Some valves depend on fluid flow to keep excessive heat from accumulating and if they are used on dead end service where the solenoid remains energized for long periods without fluid flow, the coil may burn out from accumulated heat, possibly in combination with other problems.

(7). Ambient Temperature Abnormally High. Coil insulation may be damaged, allowing one layer of wire to short to the next layer. A heat shield or baffle may give some protection from radiated heat. The best protection against heat conducted through metal surfaces or from high temperature surrounding air is to use valves with high temperature coils, molded coils, or oil immersed solenoids.

(8). Ambient Temperature Abnormally Low. Oil becomes more viscous, possibly overloading valve capacity. Mechanical parts of the valve or solenoid structure may distort, causing the valve spool to stick and burn out the coil. Change to an oil more suitable for the low temperature.

(9). Atmospheric Moisture. High humidity, especially with frequently changing ambient temperature, may corrode metal parts of the solenoid structure, causing the armature to drag or the spool to stick. Humidity also tends to deteriorate standard solenoid coils, causing shorts in the winding.

Change to molded coils or oil immersed solenoids. Keep solenoid protective covers tightly in place, and perhaps seal the electrical conduit openings after the wiring is installed.

(10). Excessive Flow. Pressure drop through the spool of a direct-acting solenoid valve caused by fluid flow creates a force unbalance which may impose an excessive load on the solenoid. This effect becomes more pronounced when using fluids of higher specific gravity or higher viscosity.

(11). Dirt in Oil or Atmosphere. A small grain of dirt lodging under the solenoid armature may keep it from fully seating against the core. This will cause coil current to remain above normal. Be sure to keep solenoid dust covers tightly sealed against atmospheric dust.

Small dirt particles in the oil may lodge on the surface of the spool, glued there by "varnish" circulating in the oil, or varnish deposited on the spool may cause it to drag. Varnish forms in systems where the oil is allowed to run too hot. Heat accelerates unwanted chemical reactions which form the varnish. Reduce oil temperature with a heat exchanger.

(12). Overlap in Energization. On those double solenoid valves where both solenoids are yoked to opposite ends of the same spool, if both solenoids are energized at the same time, one of them will burn out in a short time.

The electrical circuit should be examined to be sure the operator, through accident, cannot energize both solenoids simultaneously, even briefly. A relay with sticking contacts or one which does not instantly release could be responsible for a momentary overlap of the solenoids on each cycle, and this would lead to an eventual burn-out.

To prevent the possibility of both solenoids being accidentally energized at the same time, interlocking circuitry can be employed between two push-buttons or between a pushbutton and a limit switch as shown here.

The pushbutton and the limit switch each have an extra set of normally closed (N. C.) contacts. These are wired in series with the active circuit on the opposite switch , making it impossible to have current on both coils at once.

This principle can be applied to wiring between any number of switch or relay contacts of any type.

An example of coil burn-out from energizing both solenoids at the same time is a press with momentary pushbutton or foot switch start. When the press has closed and a return limit switch has been actuated, both solenoids may have current on them if the operator is still standing on the foot switch. The interlocking circuit can be incorporated into the electric circuit to prevent this.

(13). Spool Sticking. Solenoid burn-out may be caused by sticking of the spool in a valve. This may prevent the solenoid armature from fully seating. The probable causes for spool sticking are summarized a little later in this chapter.

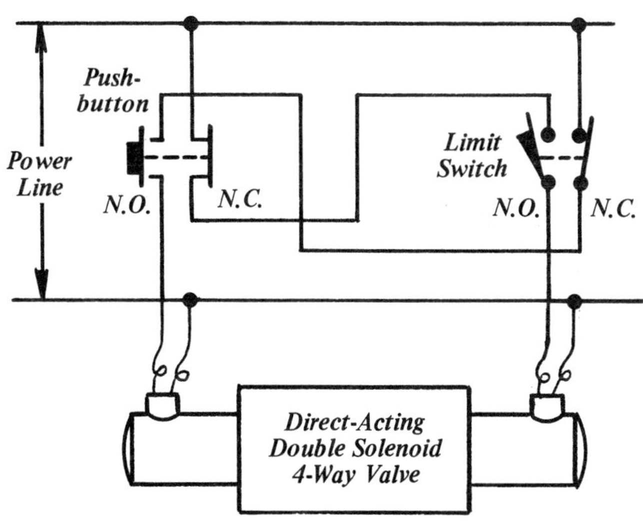

Safety Interlock Circuit to Prevent Accidentally Energizing Both Solenoids at Once on Direct-Acting Solenoid Valves.

Electric Circuit Should be Designed So Improper Operation of Machine Cannot Burn Out Solenoid Valves.

VALVE SIZE SELECTION FOR AIR AND HYDRAULIC SYSTEMS

First, Choose Size of Directional Valve.

The choice of valve size for replacement is no problem, providing the system has been operating satisfactorily. Use the same port size as the original valve. However, this is a good opportunity to slightly increase machine speed by using a larger size replacement.

On new machines the major decision is size of the 4-way directional control valve. Usually all other valving on the machine will work well if they are the same size.

The matter of selecting a suitable size is a matter of experience and good judgement. The valve should be large enough to keep flow loss low, yet not so large that it will be overly expensive.

Please remember that general rules given in this section are no more than approximations which should work well on average applications. Take them as a starting point, then use your judgement on whether valve size could be smaller or should be larger than for an average application.

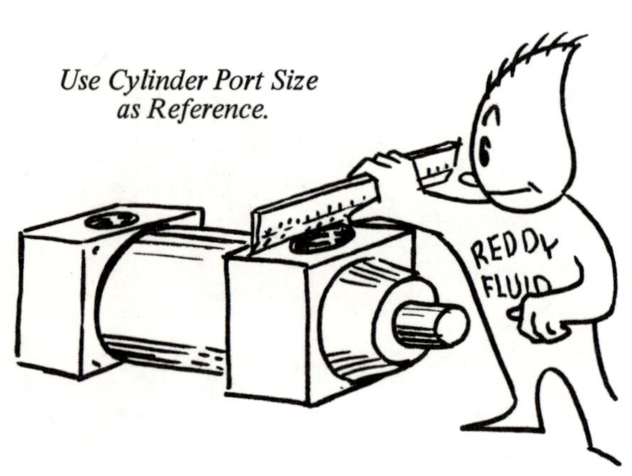

*Use Cylinder Port Size
as Reference.*

Compressed Air 4-Way Valves. A good general rule is to choose valve size the same as the ports on the air cylinder which it will operate. But remember that connection size of a valve may not fairly represent its flow capacity. Small size valves often have smaller internal orifices than the connection size, so match the orifice size, not the connection size.

The above rule is for operation of one cylinder. If more than one cylinder will receive air through the valve, its flow rating will have to be increased in proportion to the number and size of cylinders to be operated.

This table shows the recommended valve size which will usually give adequate flow to operate one air cylinder. The next larger size may be used on applications where high cylinder speed is required.

Cylinder Bore (I. D.)	1½	1¾	2	2½	3	3¼	4	5	6	8
Recom. Valve Size	1/4	1/4	1/4	1/4	1/4	3/8	3/8	1/2	1/2	3/4

Miniature air valves are available for operation of small cylinders in the range of 1/2 to 1½" bores. These valves have connection sizes from 10-32, through 1/8" NPT to 1/4" NPT, but have small diameter internal orifices.

Oversizing of Air Valves. On some applications it may be advantageous to oversize the 4-way valve. These are some of the benefits of oversizing:

<u>(1)</u>. Increased Cylinder Speed. Maximum speed of an air cylinder is related to circuit flow resistance. At the same inlet air pressure and working against the same load, increasing valve size will decrease circuit resistance and permit the cylinder to move at a faster speed.

<u>(2)</u>. By reducing flow resistance with a larger valve, the cylinder will be capable of moving a heavier load when working at the same inlet air pressure.

<u>(3)</u>. Although, while in motion, cylinder speed and/or load capacity will be increased with a larger size valve, the maximum force exerted at stall will be the same regardless of valve size.

<u>(4)</u>. Of course a larger valve is more expensive and may require more shifting effort on a manually operated valve, and more force to shift a cam operated valve.

Air-Over-Oil Systems. These combination systems are described in Volume 2 "Industrial Fluid Power". Shop air supplies the power. Speed control is done in the oil circuit. Since these systems are inherently slower than straight air systems it is important to use large components to obtain a satisfactory speed. These general rules may help in selecting valve size:

<u>(1)</u>. Valve Size in Oil Section. Because very low pressure is available for pushing oil through the system, oil velocity must be kept very low to reduce

Speed Control in Air-over-oil System.

flow loss. Piping and valves should be sized to keep flow velocity at no higher than 2 to 4 feet per second instead of the 15 to 30 feet per second allowed in a high pressure hydraulic system. This would require plumbing at least two pipe sizes larger than ordinarily used in a straight hydraulic system.

Most of the restriction in the oil part of the system is from flow control valves used for speed control. Orifices in flow control valves, both in the free flow and the controlled flow direction, are much smaller than the valve connection size. Therefore, flow control valves should be about two sizes larger than the pipe size to which they are connected. They should be bushed down to mate with connecting pipe. For example, where on a straight hydraulic system 1/4'' NPT pipe would be used, the pipe size for oil flow in an air-over-oil system should be 1/2'' and the flow control valves 1'' size.

Check valves and needle valves should be oversized by the same amount. Eliminate all possible valving from the oil part of the system.

<u>(2)</u>. Valve Size in Air Section. Selection of valve size in the air section can be made in the same way as for a straight air system — by using the cylinder ports as a guide, following the rules given on Page 56.

Selecting Hydraulic Valve Size. The size of a 4-way hydraulic valve does not have the same effect on cylinder speed as the size of an air valve has on air cylinder speed. Hydraulic cylinder speed is determined primarily by the rate of flow from the pump. Using an undersize valve saves money on original cost but can cause the cylinder to lose speed if it restricts the oil enough to cause pump pressure to rise to relief valve setting and spill a part of the oil to tank. Using a proper size or an oversize valve may cost more but reduces power loss and heat generation, and increases system efficiency.

To select a valve size which will have sufficient capacity to keep flow losses to a moderate level yet not be too costly, we recommend first selecting a piping size for the system, then using 4-way and pressure control valves of the same pipe size or flow capacity.

Piping size for the system can either be based on a rule-of-thumb using pump pressure port size as a guide or in larger systems can be selected on the basis of flow velocity and maximum operating pressure in the manner to be described.

Rule-of-Thumb for Piping Size. On small hydraulic systems of less than 15 HP consisting of one cylinder or hydraulic motor, piping in pressure and return lines can be the same size as the pump pressure port. Directional and pressure control valves can also have the same pipe size or flow capacity.

Important! The pump suction line should be at least two, and up to four pipe sizes larger. Flow velocity must be kept low to avoid cavitating the pump.

Piping Size Based on Flow Velocity. On larger systems we recommend piping size be based on flow velocity. Since pressure loss is directly related to flow velocity, the maximum operating pressure of the system should be considered. For example, a piping loss of 100 PSI is a 20% loss of power on a 500 PSI system but only a 2% loss of power on a 5000 PSI system. Therefore, flow velocity should be kept much lower on a low pressure system to prevent power losses from becoming excessive. These are the flow velocities we recommend for general use:

Location	For Petroleum Oil	For Fire Resistant Fluids
Pump suction lines	2 to 4 feet per second	1½ to 3 feet per second
Pressure lines up to 500 PSI	10 to 15 feet per second	7 to 10 feet per second
Pressure lines 500 to 3000 PSI	15 to 20 feet per second	10 to 15 feet per second
Pressure lines over 3000 PSI	25 feet per second	20 feet per second
Oil lines in air-over-oil system	4 feet per second	2½ feet per second

Flow velocity in a pipe is related to its inside area. For lower velocity use pipe with larger area. Pipe and tubing inside areas are on Page 219. See Appendix of Volume 1 "Industrial Fluid Power" for a more detailed treatment of pipe size selection. Formulae for calculating flow velocity are:

V (flow velocity, ft. per second) = GPM (flow from pump) x 0.3208 ÷ A (pipe inside area, sq. inches)

A (pipe inside area, sq. inches) = GPM (flow from pump) x 0.3208 ÷ V (flow velocity, ft. per second)

Selecting Pipe Size. This is part of the pipe pressure and flow rating chart on Page 218 of the appendix. Pressure ratings are shown in bold face type for safety factors of 2 to 6. We recommend a minimum of 4 for most hydraulic systems. Totally shockless systems such as hydrostatic transmissions sometimes use a safety factor of 2.

GPM flow ratings for flow velocities from 10 to 30 feet per second are shown in italic type. For example, the GPM rating of 1/2" NPT Schedule 40 pipe at 15 feet per second is shown as 14.0. For flow velocities not listed, the chart can be interpolated. For example, on the same pipe, GPM rating at 4 feet per second would be 1/5th its rating at 20 feet per second, or 3.8 GPM (19 ÷ 5). Flow and pressure ratings for steel tubing are also shown in the appendix.

(See Complete Table on Page 218)

	Pipe Schedule →	40	80	160
1/2" NPT PIPE 40,000 PSI Tensile	PSI @ S.F. = 6	1730	2330	2980
	PSI @ S.F. = 5	2080	2800	3580
	PSI @ S.F. = 4	2600	3500	4480
	PSI @ S.F. = 3	3460	4670	5970
	PSI @ S.F. = 2	5190	7000	8950
	GPM @ 10 f/s	*9.50*	*7.20*	*5.33*
	GPM @ 15 f/s	*14.0*	*11.0*	*8.00*
	GPM @ 20 f/s	*19.0*	*14.0*	*10.6*
	GPM @ 30 f/s	*29.0*	*22.0*	*16.0*
3/4" NPT PIPE 40,000 PSI Tensile	PSI @ S.F. = 6	1430	1950	2780
	PSI @ S.F. = 5	1720	2350	3340
	PSI @ S.F. = 4	2150	2930	4170
	PSI @ S.F. = 3	2870	3910	5560
	PSI @ S.F. = 2	4300	5870	8340
	GPM @ 10 f/s	*16.0*	*15.0*	*9.33*
	GPM @ 15 f/s	*25.0*	*22.0*	*14.0*
	GPM @ 20 f/s	*33.0*	*30.0*	*18.6*
	GPM @ 30 f/s	*50.0*	*44.0*	*28.0*

See Volume 1 "Industrial Fluid Power", Chapter 1 for detailed information on plumbing materials. See Appendix A in the same book for calculation of plumbing sizes by flow velocity method.

Oversizing a 4-way directional valve to one size larger than would be indicated by the selection methods given on the preceding page may bring some benefits and some disadvantages:

<u>(1)</u>. A larger 4-way valve will not increase cylinder speed, since speed is related only to pump flow rate and to cylinder piston area.

<u>(2)</u>. A larger valve, with its lower flow loss, would reduce heating in the oil and would increase the overall efficiency of the system.

<u>(3)</u>. A larger valve, even though reducing flow loss, would not increase the tonnage of the system when the cylinder stalls against the work.

Consider Advantages of Larger Valve.

<u>(4)</u>. However, a larger valve by reducing flow loss, would increase the system load capacity *while the cylinder is in motion.*

<u>(5)</u>. A larger valve costs more, so the extra cost would have to be weighed against other factors such as reduction of heat, increase in efficiency, and greater load capacity on a *moving* load.

Valve Size for a Vacuum System. There are basically two kinds of vacuum applications: "flowing" vacuum and "static" vacuum systems. In a flowing vacuum system there is a continuous flow of air at less than atmospheric pressure. A vacuum conveyor, in which powdered or granular material is "sucked up" by a nozzle and moved through a pipe, is an example of a flowing vacuum system. In a static vacuum system the vacuum is used for gripping, as with a suction cup. There is little movement of air through such a system.

Choice of valve size depends on the kind of system to be operated. Flowing vacuum systems may require a larger valve than a static system. Valve must be large enough to have a very low flow loss since these systems usually operate on 2 to 5 PSI vacuum, and any loss through a control valve will seriously affect the proper operation. All but strictly necessary valving should be excluded. As a rule, directional control valving should be sized as large as practical and certainly at least 2 to 4 pipe sizes larger than would be used on a compressed air system.

Static vacuum systems are much less critical of valve flow capacity. For example on this vacuum cup application, the only flow through the control valve is the small volume of air trapped in the connecting lines. A relatively small 3-way valve, 1/4" size, should be quite adequate to handle this kind of application up to a vacuum pump capacity of 1½ HP or even more, unless connecting lines are unusually long or unusually large in diameter.

Use a relatively small control valve on static vacuum applications like this one.

Leakage in the valve may be an important factor on static vacuum systems. Unless valve seals are leaktight, it may be difficult or impossible to pull a high vacuum. Spool-type air valves can be used on vacuum if they have O-ring or rubber cup seals between all ports. Metal-to-metal seals between ports, such as in lapped spool or rotary shear types, may be all right on some applications but may have too much leakage for high vacuum applications. Make sure the valve you select is rated for vacuum service.

When connecting a 3-way valve for use on vacuum, remember that atmospheric is the highest pressure in the circuit, so it should be connected to the valve inlet port. The vacuum pump, the lowest pressure in the circuit should be connected to a cylinder port.

Valve Size for Fire-Resistant Fluids. All other hydraulic fluids such as water, water/oil emulsions, soluble oil and water, and phosphate ester have a higher specific gravity than petroleum oil. At the same velocity, through the same size pipe, they have a higher pressure loss. To hold the flow losses through a system to the same level as for petroleum oil, various components must be larger in size.

Pipe size for plumbing can be selected by flow velocity as described on Page 58, using the column in the table for fire resistant fluids.

Suction strainers should have 25% greater mesh area than those for petroleum oil. In fact, instead of 100-mesh stainless steel wire as used for oil, mesh of 74 wires per inch should be used. This is very important, to avoid cavitating the pump. Micronic filters in pressure and return lines should also be oversized at least 25%.

Directional control and pressure control valves should be one size larger than for petroleum oil. The same size valves can be used but will produce a higher power loss.

Conversion of a system from petroleum oil to one of the heavier fluids is quite an involved project if done properly. Simply flushing the system and replacing component seals with those compatible with the new fluid may give a system which will operate, but not necessarily satisfactorily. The fluid may overheat, the pump may cavitate, the electric motor may be overloaded, and the power output will be less than for the same system with petroleum oil. Comparable efficiency and power output can only be obtained by rebuilding almost the entire system.

MISCELLANEOUS VALVE PROBLEMS

Valve application problems described in this section are those not previously covered in this chapter. See also information on Pages 208 to 210 in Chapter 12.

Valve Spool Sticking. This may cause solenoids to burn out, to fail to shift, or to fail to return to center position when de-energized. It may be caused by one or a combination of the following:

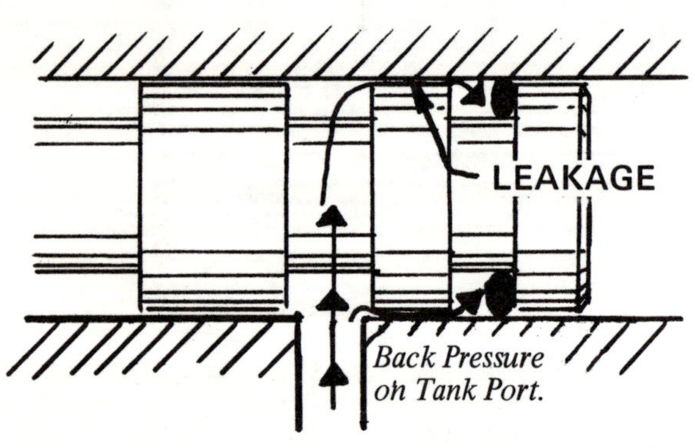

High Pressure May Cause O-ring Drag.

(1). Pressure Lock on O-Ring Seals. Most hydraulic valves have an O-ring on each end of the spool to prevent leakage to the outside. The two outside spool grooves are tank return passages. They join together inside the valve and come out as the main tank port. Any back pressure which appears in the tank return line will find its way into the O-ring grooves through spool leakage. Pressure in the O-ring groove loads the O-ring against the sides of the bore and causes extra spool friction. The higher the back pressure the greater the spool drag. A certain amount of drag may not be harmful on manually operated valves, but on a direct-operated solenoid valve it can cause the solenoid to overheat and eventually to burn out. Most hydraulic valves have a lower pressure rating on the tank port than on other ports. If tank port rating is exceeded the valve may malfunction.

Air valves with O-ring seals *between* each spool groove are subject to O-ring spool drag on any or all of the O-rings. Pressure on all ports must be kept within rating, usually 200 PSI, rarely higher.

(2). Spool Side Drift. In any spool valve there is a tendency for pressure to get on one side of the spool and cause it to drift sideways and make metal-to-metal contact on the opposite side of the bore. At low pressures this effect does not cause real trouble because of the oil film between spool and bore. But at pressures higher than 3000 PSI the oil film may break down and cause the spool to "freeze" in the bore. On solenoid valves, the centering or return springs may not be able to return the spool to normal position when the solenoids are de-energized. Some valves will stick at high pressure if left for sustained periods (say for more than 5 minutes) in shifted position under full pressure. Read valve manufacturers instructions and never operate a valve at higher than its rated pressure.

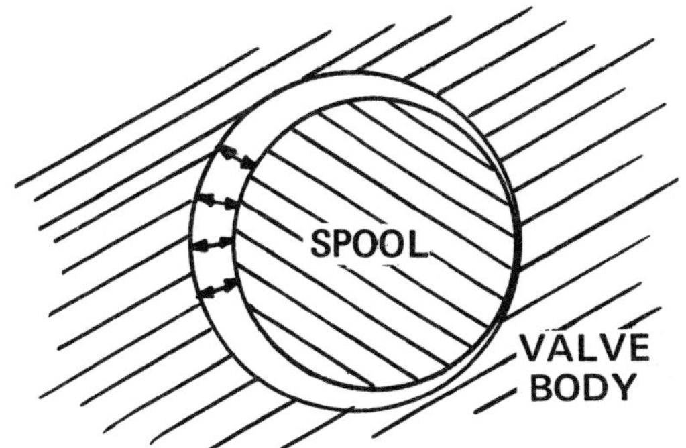

Spool Drifts to Side of Bore on Sustained Actuation.

Valve manufacturers minimize spool sticking by machining tiny grooves around the spool lands. This tends to equalize pressure around the circumference of the spool and in many situations will prevent spool freeze.

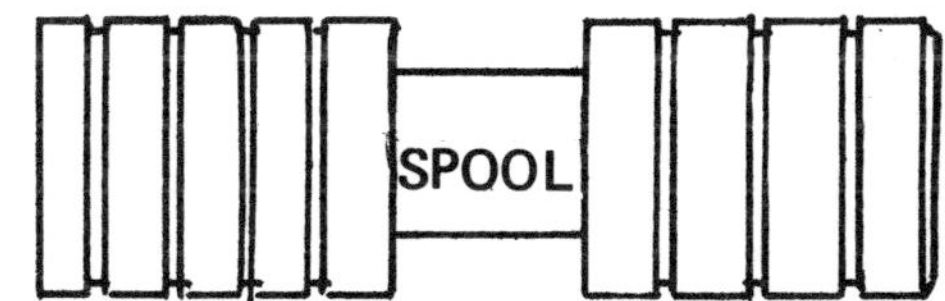

Tiny Grooves .010 Wide and Deep Minimize Spool Lock Caused by Side Drift.

Some manufacturers offer double solenoid valves with a pressure centering mechanism, instead of springs, to return the spool to center when the solenoids are de-energized. This provides a much more powerful centering force than can be obtained with springs, and will in most cases solve the problem of spool sticking. Pressure centered models should be used, if available, for operation above 3000 PSI. Spool sticking seems to be aggravated by microscopic dirt, called silt, in the oil, and may disappear a short time after micronic filters are added to the system.

(3). Valve Body Distortion. Valve bodies with taper pipe thread ports may distort if pipe fittings are over-tightened, causing the spool to bind. This may be a temporary distortion which will disappear if fittings are slightly loosened. Use Teflon paste thread sealant so fittings will not have to be over-tightened to make a fluid-tight seal. See more information on Page 41.

(4). Seals. If synthetic rubber seals on the valve spool are not compatible with the fluid, the rubber may swell and bind the spool. If system is using any fluid other than petroleum base hydraulic oil, even detergent motor oil, check this as a possible source of trouble.

(5). Mounting Surface. Valve body may be distorted by mounting on an uneven surface, causing the spool to bind. This type of distortion is usually only temporary and in many cases can be relieved by slightly loosening one mounting bolt or shimming under one mounting foot. See additional information on Page 41.

Subplates or manifolds for valve mounting must also be correctly mounted. Twisting or distortion of the subplate may cause the valve body to distort, and may produce an oil leak between valve and subplate. Caution! Do not weld pipe into a subplate and do not weld the subplate to a mounting surface. Heat may cause warping and spool bind. See more information on Page 42.

(6). Silting. Micronic particles in the oil can accumulate around valve spools even at low pressure and cause spool sticking. This is more noticeable after spring offset or spring centered valve spools are kept in a shifted position under full pressure for a long time. The obvious remedy is to install micronic filters in the pressure or return lines, or to replace elements in existing filters. Do not install micronic filters in the pump suction line; this may cavitate the pump.

(7). Spool Lapping. Sometimes spool sticking develops for no reason that can be determined. Aging of cast iron bodies is one possible cause. If all other remedies fail, sticking must be relieved by removing the valve from the machine and lapping the spool in the body using a very fine abrasive, and cleaning the valve thoroughly before re-installing. Do not remove any more metal than necessary; excessive clearance can produce excessive leakage across the spool.

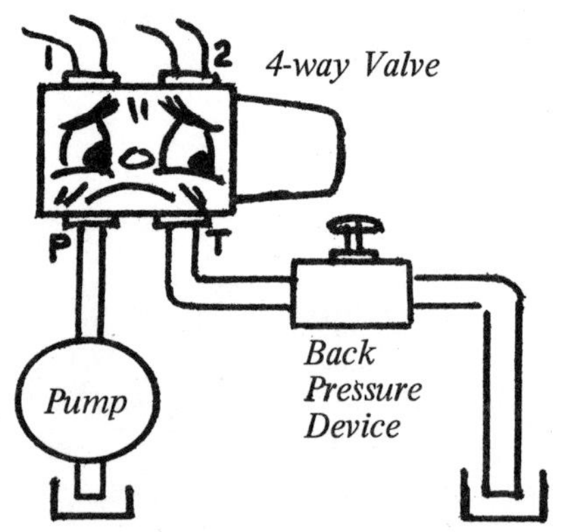

Be Careful About Adding Restrictions in the Tank Return Line.

Excessive Back Pressure. Directional and pressure control valves cannot tolerate excessive back pressure in lines which are drained to tank. Back pressure in drain lines is caused by the flow of oil through resistance in the line. It can be reduced by increasing the diameter or shortening the length of drain lines. Some of the possible effects of back pressure are:

(1). On some valves, particularly spool-type directional valves, seals may be dislodged or washed out.

(2). Back pressure can actually cause the spool to bind by pressurizing spool end seals. See Item (1), Page 60.

(3). On solenoid valves, back pressure reflected into the solenoid drain line places a momentary extra resistance against the solenoid armature. In turn, this will cause average solenoid current to be higher than normal, and may cause the solenoid coil to burn out. See information on Pages 60 and 61.

To prevent this extra load on the solenoid, the drain port (not the main tank port) on pilot-operated solenoid valves can be externally drained to tank by running a drain line from Port Y on the subplate directly to tank. In this case the internal solenoid drain passage must be blocked.

(4). On hydraulic motors and pumps of the piston type, excessive back pressure in case drain lines will cause the shaft seal to blow out.

(5). On some kinds of manually operated 4-way valves, a back pressure surge in the tank return line may cause the handle to "kick".

(6). May cause premature actuation of a sequence valve or pressure switch. On a sequence valve a surge would be only momentary, but on a pressure switch this could trigger the remainder of the electrical cycle and could be disastrous.

(7). Some 4-way air valves will not tolerate back pressure caused by placing needle valves in exhaust ports for cylinder speed control. This may unseat internal seals. Flow control valves placed in connecting lines between cylinder and 4-way valve must be used. As a rule, exhaust ports on either air or hydraulic valves should not be plugged unless permitted by the valve manufacturer.

Spool Bounce. The apparent "bounce" of a spool-type 4-way hydraulic valve back to center position was described on Page 48. This effect occurs when a hydraulic cylinder is being reciprocated by short duration electrical signals to a double solenoid, 4-way, 2-position valve and when the signals are obtained from limit switches at each end of the stroke. When a malfunction occurs, the cylinder starts its return stroke, moves about an inch, just far enough to back off of the limit switch, then stops. The circuit remains "dead" until manually started with an operator pushbutton or by using the manual overrides on the solenoid valve.

Actually, the valve spool does not "bounce", but it does move to its center or closed crossover position under the influence of pressure unbalance created by hydraulic flow across the spool grooves.

Adding Bank Valves in Series. When connecting two bank valves together or when adding a second bank valve in series with an existing one to obtain additional branch circuits, several problems may have to be dealt with:

(1). Relief Valve. If each bank valve has a built-in relief valve, the two relief valves would be placed in series if one handle in each bank valve were to be operated at the same time.

For example, in this illustration, if each relief valve was set for 1000 PSI, the total load on the pump from both circuits could go to 2000 PSI.

The solution in this case is to install a separate relief valve at the pump and set it to the maximum pressure for

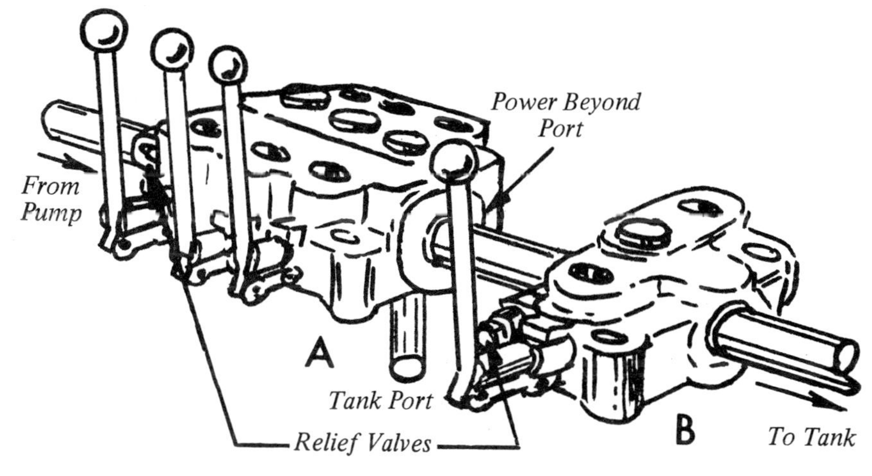

When Connecting Two Bank Valves in Series, the Upstream Valve Should Have "Power Beyond" Feature.

the machine. Screw down both relief valves on the bank valves to their maximum settings.

(2). Power Beyond Feature. When two bank valves are connected in series, the upstream bank must have the optional feature "power beyond". Without this feature, when a handle of the downstream valve was shifted, load pressure would be reflected upstream into the case of the first valve. This would have several effects: (a), one or more of the handles in the upstream valve would shift by themselves; (b), reflected pressure, if high enough, would burst the casting of the upstream valve; (c), load pressure from the downstream valve would add to the setting of the upstream relief valve as described in Part (1) above. This would remove relief valve protection from sections in the upstream valve if a handle should be shifted while a downstream section was working.

Power beyond option on the upstream valve provides two outlet ports located on the end or bottom. One of these ports, the return or tank port must always be connected directly to reservoir. The other outlet port, the power beyond port, is shown coming out of the end of the first section. It is connected to the inlet port of the downstream bank. With power beyond, both sections operate as though they were all in the same bank.

If a second bank valve is to be added to an existing bank valve, order the new valve with power beyond feature and install it upstream of the existing valve, next to the pump.

When the upstream bank has the power beyond option, the relief valves built into both banks can be used without adding a separate pump relief valve as described in Part (1).

For a detailed description of bank valve operation, refer to "Industrial Fluid Power" Volume 1, Chapter 4, and Volume 2, Chapter 7.

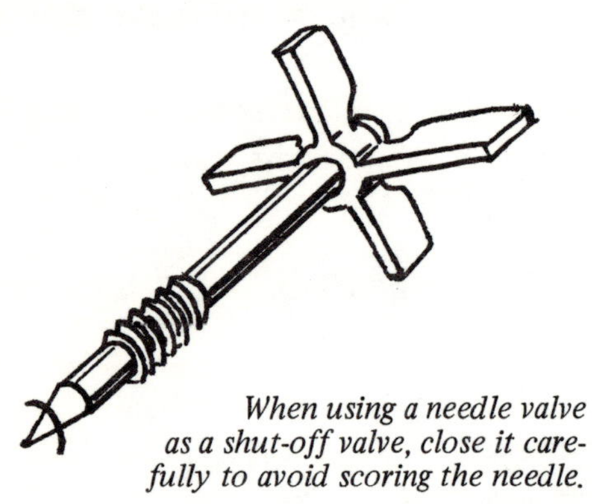

When using a needle valve as a shut-off valve, close it carefully to avoid scoring the needle.

Needle Valves for Shut-Off Service. Needle valves have a high pressure rating, 5000 to 10,000 PSI, and are quite inexpensive. They make good shut-off valves for pressure gauges and similar applications where they are operated only occasionally. However, the needle will soon become scored and will not make a leaktight seal if closed repeatedly with unnecessary torque.

Water Service for Hydraulic Valves. Spool-type hydraulic valves for use with petroleum hydraulic oil do not work well on water and water base fluids. Consider these problems:

(1). Wiredrawing. The wiredrawing effect is the wearing away of metal molecules in orifices and sharp edges on spool valves by the molecules of high specific gravity fluids traveling at high velocity. Water, water base, and synthetic fluids have a higher specific gravity than oil and produce more wiredrawing. Valves used in those systems should be oversized to keep velocity low. Poppet-type valves are less subject to wiredrawing than are spool-type valves.

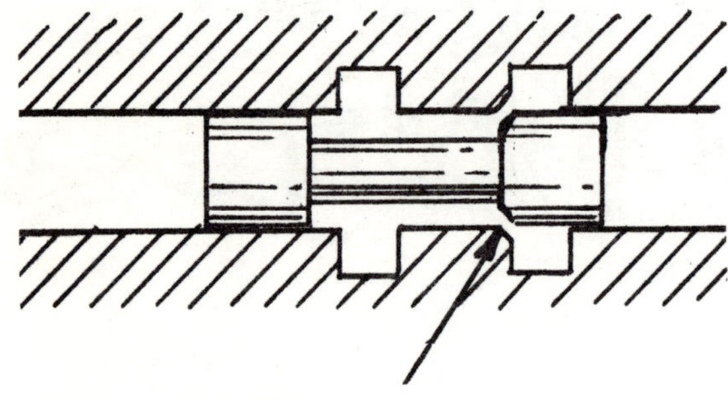

High Velocity Water Wears Away Sharp Edges.

(2). Rusting. Oil hydraulic valves constructed of iron and steel are subject to rusting when used on water and water base fluids. The fluid should contain rust inhibitors. If available, valves constructed with brass bodies and stainless steel spools will give longer life on water hydraulics.

(3). High Leakage. Ordinary spool-type valves for hydraulic oil may have a high leakage rate when used on water or other low viscosity fluids. Poppet-type valves are preferred for water because they are less subject to wiredrawing, they do not need lubrication, and they make a leaktight seal.

(4). Short Life. Standard oil hydraulic spool-type valves have a short life on water and water base fluids because of wiredrawing, lack of lubrication by the fluid, and their tendency to rust even though the fluid may contain a rust inhibitor.

Chapter 3

Hydraulic Pumps & Motors

Pumps and Motors, Hydraulic. The hydraulic pumps and motors used for power generation and transfer in industrial and mobile fluid power systems are the positive displacement type. They include basic gear, vane, and piston pumps with several variations of each. A positive displacement pump is one in which the working elements make a tight sliding fit, to prevent or minimize backward slip of the pumped oil. Therefore, on each revolution of the shaft, a fixed amount of oil will be pumped which is proportional to the displacement of the working elements and to the shaft RPM. A positive displacement pump is distinct from an impeller pump in which the impeller does not make metal-to-metal seal with the housing, and internal slippage is too great for high pressure to be developed. Energy imparted to the fluid by an impeller pump is proportional to fluid mass times velocity squared. This pump, also called a centrifugal pump, can move a large volume of fluid but cannot develop pressure high enough to be useful in a hydraulic system.

The term "positive displacement" means that regardless of conditions of load or speed, the pump will always displace the same cubic volume on each shaft revolution. Pump ratings are usually given in gallons per minute (GPM) which is computed by multiplying cubic inch per revolution (C.I.R.) displacement times shaft speed in RPM divided by 231 (conversion factor from cubic inches to gallons).

Material in this section applies only to positive displacement pumps. It consists of suggestions for correct installation, mounting, and operation of pumps. Please refer to Volume 1 "Industrial Fluid Power" for a complete discussion of pump types, how they work, drive power required, and circuits in which they are used.

Hydraulic pumps and motors are usually the components which are most sensitive to damage from dirt, heat, water, overloading, and mechanical abuse. They should be selected, installed, and maintained with great care. They have a definite life expectancy. If the proper model for each job is selected and if a planned inspection and maintenance schedule is followed, their life will be extended and there will be fewer breakdowns of the system on which they are installed.

INSTALLATION OF HYDRAULIC PUMPS AND MOTORS

Shaft Alignment. If a hydraulic pump and electric motor or engine are to be joined shaft-to-shaft, the two shafts should be very carefully aligned, up and down, sideways, and angularly. Then they should be joined with a flexible coupling to take up any remaining mis-alignment. A solid coupling should not be used. Any remaining mis-alignment would quickly wear out the pump bearings.

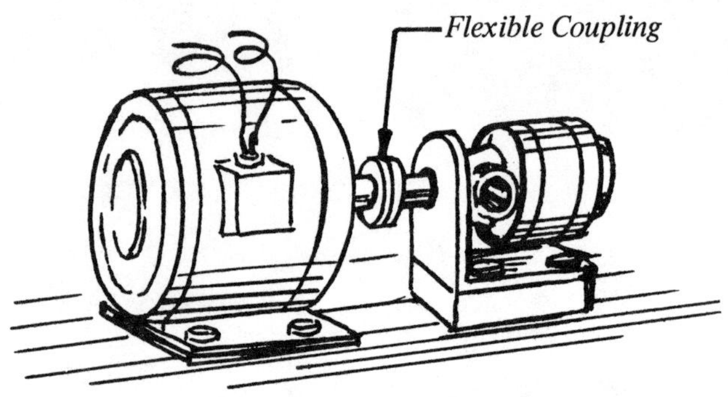

*Use a Flexible Coupling for Direct Drive
of Hydraulic Pump.*

*Leave Foot Bracket
Undisturbed When Replacing Pump.*

When the two shafts have been as accurately aligned as possible, a good practice is to drill and pin both units to the mounting base to prevent mis-alignment if mounting bolts should become loosened.

Pump and motor should be mounted on a common baseplate which is sufficiently strong that it will not distort under the twisting that will develop between the two when load is applied. If the baseplate is not sufficiently rigid, the two units may become mis-aligned when load is applied.

When replacing a foot mounted hydraulic pump with one exactly like the original, do not replace or disturb its mounting bracket unless this bracket has been damaged. The new pump should fit the old bracket and should be in perfect alignment with the motor.

When replacing the electric motor, the new motor will have to be aligned with the pump. Manufacturing tolerances on shaft height make it necessary to repeat the entire alignment procedure.

Precision gauges, if available, can be used to check alignment. On some couplings such as the popular roller chain coupling, alignment can be accurately checked without gauges. After the motor and pump have been aligned as accurately as possible by measurement, the chain is installed on the coupling. Move the chain back and forth sideways to be sure it is free from any bind. Rotate the shaft about 30° and check again for bind. Do this all the way around. There must not be the slightest bind at any angular position.

(1). Angular Mis-Alignment. A small amount of angular mis-alignment can be handled by a flexible coupling, but any detectable error should be corrected without relying on the coupling.

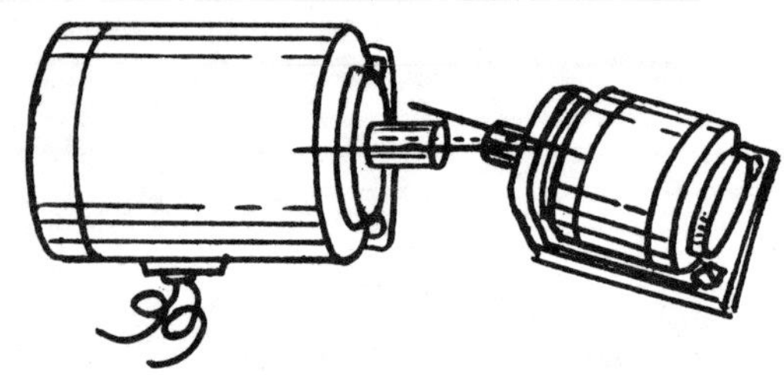

Angular Mis-alignment of Motor and Pump

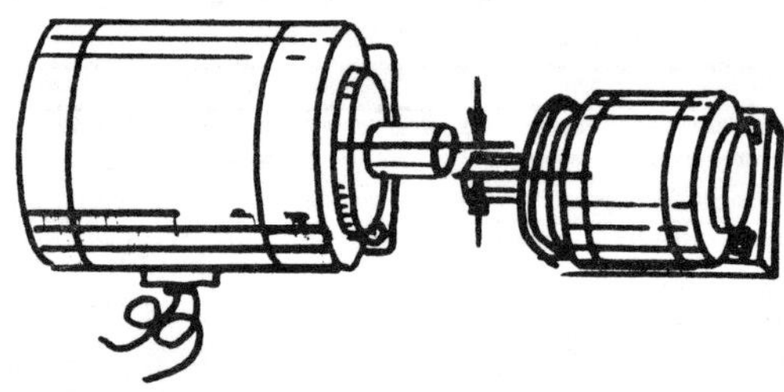

Off-set Mis-alignment of Motor and Pump

(2). Off-Set Mis-Alignment. Off-set mis-alignment can be up and down, side to side, or at any angle. Few couplings can correct more than a very small amount of off-set mis-alignment. This kind of mis-alignment can easily be detected with the test described above.

All alignment errors, whether or not corrected by a flexible coupling, are undesirable. They place an extra load on the coupling causing it to fail prematurely, and place a side load on pump bearings, reducing their life expectancy.

Side Drive With Belt, Chain, or Gear. Any kind of side drive produces a side load on pump shaft and bearings. Two undesirable results are produced: (a), a bending of the shaft itself which in extreme cases can fatigue the shaft and cause it to break; (b), side load on shaft bearings reduces life expectancy. When a shaft bearing fails, this often leads to total failure of the pump as the working elements supported on the shaft may dig into their mating parts inside the pump.

Some pumps are not rated for side drive. Others have inboard or outboard extra bearings specifically for the purpose of accepting a side load. Before using side drive, investigate the pump ratings to see how much the rated bearing life will be reduced. If necessary to side drive, observe these rules:

(1). Mount the pulley or gear as close as possible to the front face of the pump case, and install with hub facing *away* from pump. This will minimize bending and flexing of the shaft.

(2). Choose pulleys or gears with as large a diameter as practical. The larger the diameter of the one on the pump shaft, the less the side thrust. Small diameter pulleys or gears cause higher bending stresses on the shaft and more load on pump bearings. Pumps which are not recommended for side drive, if so used, should be operated at a reduced pressure level.

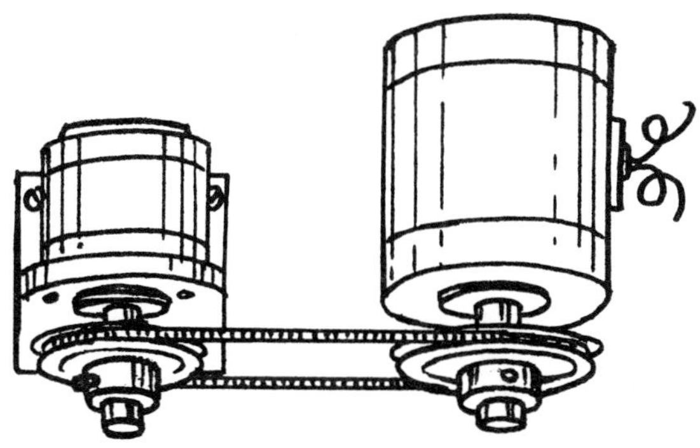

Some Pumps Are Designed for Side Drive.

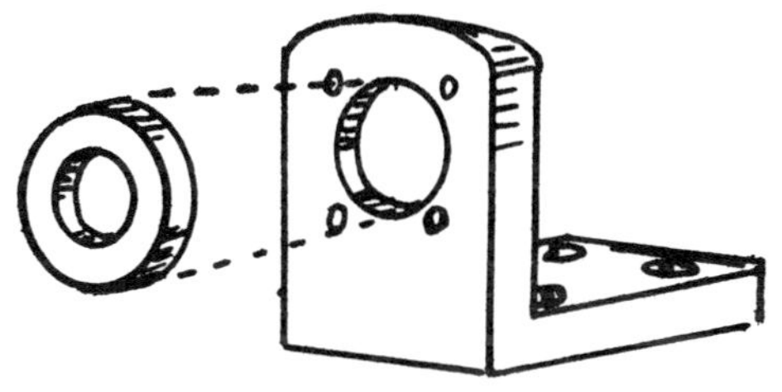

An Outboard Bearing May be Added to the Mounting Bracket.

Adding Outboard Bearings. When it is necessary to side drive a pump which has not been designed for this kind of drive, the user can install an extra bearing in the mounting bracket. A press-fit cavity can be bored or turned in the bracket. Machining must be done with precision so the cavity will be concentric with the pump shaft or the pilot shoulder.

Neutralizing Side Load. On those pumps in which the pumping element is unbalanced, a heavy side load is placed on shaft bearings, proportional to internal hydraulic pressure at which the pump is operating. These pumps include all gear pumps and their variations — external gear, internal gear, gerotor, etc., and also include single lobe vane pumps. If these pumps can be mounted with outlet port

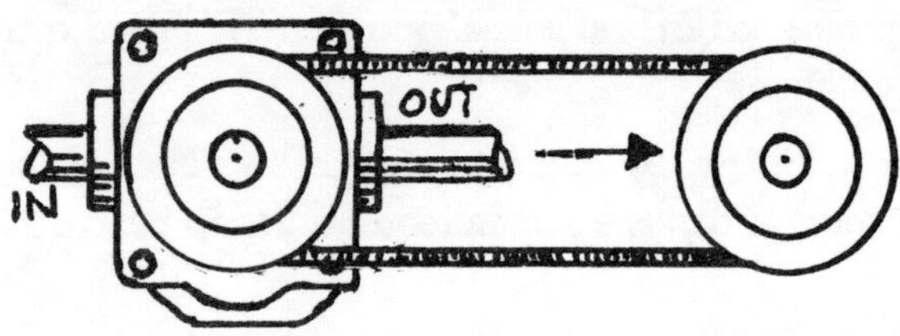

To minimize bearing load, mount pump with outlet port facing drive motor.

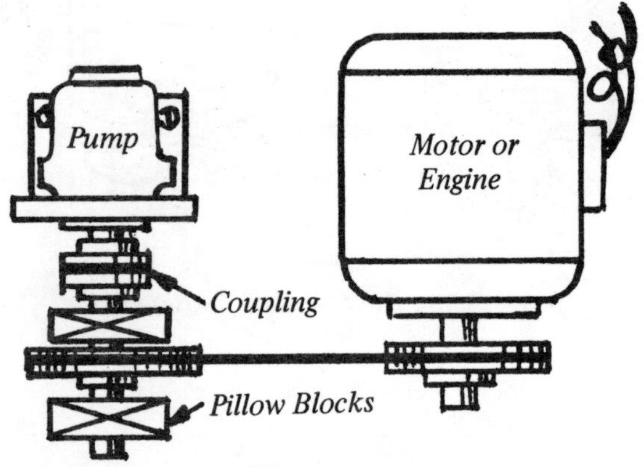

Jackshaft Drive Eliminates Side Load on Pump.

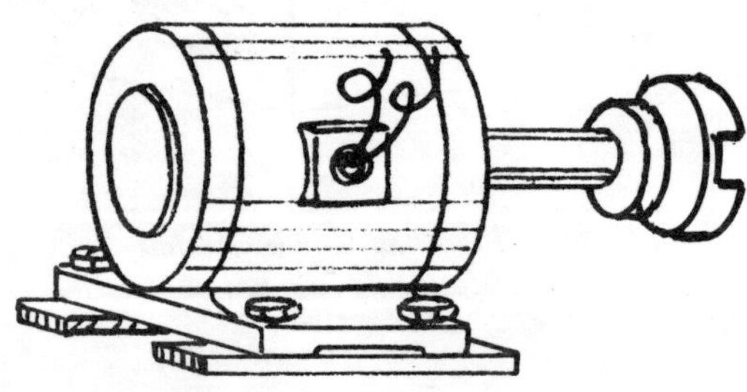

Mount Motor on Cleats, Then Weld Cleats to Mounting Surface.

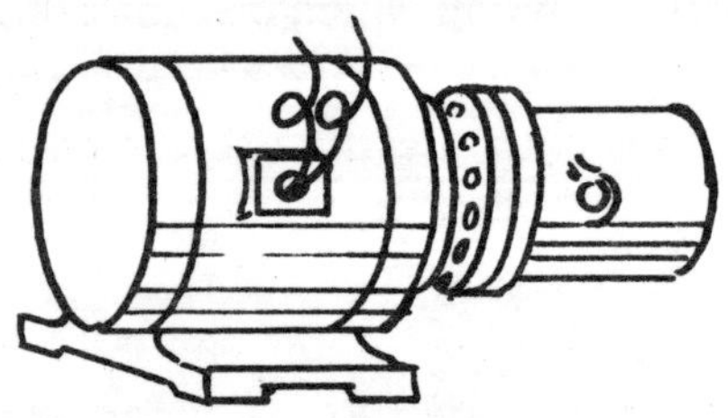

Integral Pump Mounting on Motor Frame.

facing toward the drive motor, the side load on shaft bearings produced by internal pressure will be in opposition to external side load caused by the pulley side drive, Bearing loads will be reduced during those periods when the pump is running loaded, and pump life expectancy will be extended.

Jackshaft Drive. When side drive is used, whether or not the pump was designed to accept side drive, the best drive arrangement is to mount the pulley or gear on a jackshaft which is supported by pillow block bearings. Then align the pump with the jackshaft and couple the pump to it with a flexible coupling. To take up slack in the belts, an adjustable idler puller or a standard sliding motor base may be used.

Electric Motor Mounting. During the original assembly of a hydraulic pumping unit, alignment between pump and motor can be facilitated by using metal cleats under the motor feet.

First, bolt and pin the motor to the cleats. It can then be moved around for easier alignment. Put shims either under the motor or the pump as needed. Then, when alignment has been completed and checked, the cleats can be tack welded to the mounting surface.

Motor-Mounted Pumps. There are several important advantages to using a motor with a flanged end bell and mounting the pump on the motor flange. It solves pump alignment problems; it makes a more compact assembly; it saves separate mounting of the pump; and most important, it adds stability to pumps which, due to centrifugal unbalance, tend to run with a lot of vibration. They will run more smoothly at higher speeds.

One important disadvantage to using end bell motors is that they are not as readily available as standard motors. On important jobs it is a good idea to keep a spare motor in plant stock.

Driving Two Pumps From One Motor or Engine. The following illustrations show several suggested arrangements:

(1). One pump may be coupled to each end of a double-shaft electric motor. Note: One pump must be for clockwise (CW) rotation, the other for counter-clockwise (CCW) rotation. Flexible couplings should be used on each motor shaft.

Double-shaft electric motors are usually available from jobbers stock. For emergency replacement of a burned out motor, if one is not available from stock, some motor jobbers either have a machine shop or have arrangements with a local machine shop to add a shaft extension to a single-shaft motor. This modification, when expertly done, has given performance equal to the original motor. However, it is practical only for integral horsepower motors of, say, 5 HP or more.

(2). The two pumps can be driven from a double sheave on the motor. Pumps can be driven from jackshafts if practical. Pumps with shaft facing in the same direction as the motor must be of the same rotation. Pumps facing the opposite way must be of the opposite rotation.

Idler pulley must be provided for take-up of belt slack, or, pumps can be mounted on adjustable slides. For this arrangement, hose connections must be used.

(3). Many other mounting arrangements can be worked out for driving multiple pumps. This diagram illustrates a direct drive to one main pump, with one or more auxiliary pumps being direct-driven from a jackshaft which is belt driven from the motor.

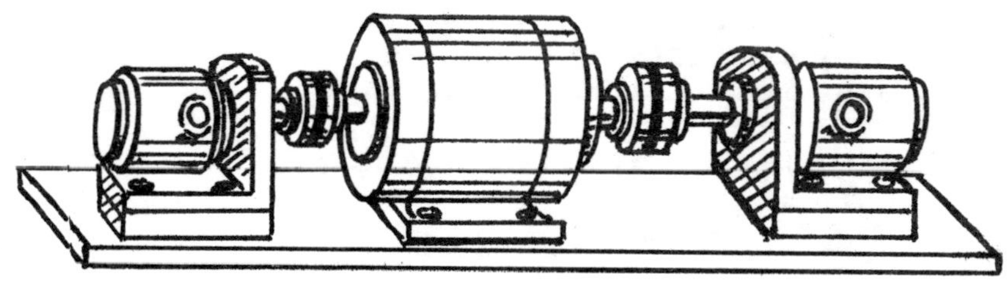

One Pump Must be for CW, the Other for CCW Rotation.

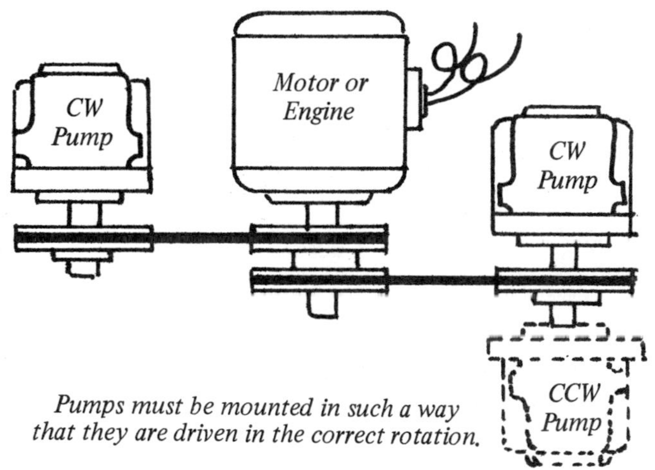

Pumps must be mounted in such a way that they are driven in the correct rotation.

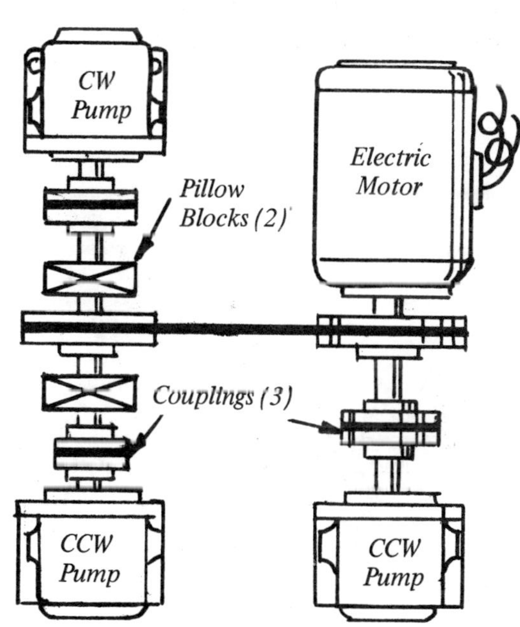

Driving Multiple Pumps

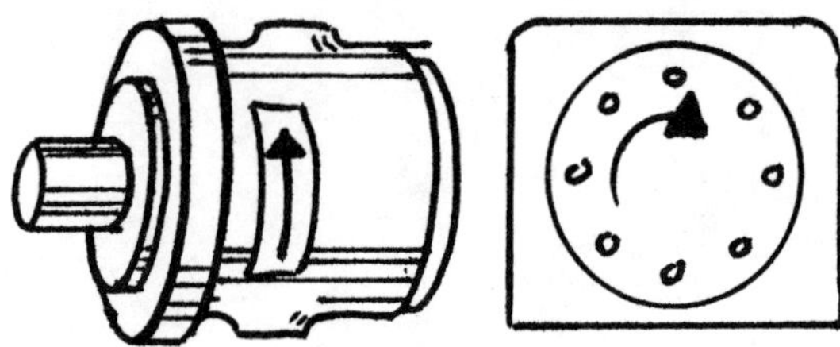

Examine Pump for Markings Indicating Proper Direction of Rotation.

A Clockwise Gear Pump Shows Its Outlet When Placed In This Position.

Direction of Rotation. Direction of shaft rotation is of prime importance. When first installing the pump, examine it for markings indicating correct rotation. Standard practice in the fluid power industry is to specify rotation as clockwise (CW) or counter-clockwise (CCW) when looking at the shaft end. This is opposite to the usual practice for specifying electric motor rotation. When ordering pumps, be sure to clearly state direction of rotation from a specific direction of viewing.

Uni-directional pumps are built for one direction of rotation only, and this will be marked either by an arrow cast into the pump case, on the end cap, or on the nameplate.

In some cases a gear-type pump may be marked "Inlet" and "Outlet". Proper direction of rotation can easily be discerned by a visual inspection. If the pump is placed in the position shown, with its shaft nearer the top, a CW pump will show its outlet port. A CCW pump placed in the same position will show its inlet port.

Another way to check rotation is to squirt a little oil into the outlet port and rotate the shaft by hand. Bubbles will come from the outlet port when the shaft is turned in the right direction. Or, if the gears can be viewed through the side outlet port, proper rotation is when the two gears rotate toward each other. Running a pump under pressure in the wrong direction will blow out the shaft seal.

Jog Motor to Test for Rotation.

Checking Direction of Rotation. When installing a pump, back off the pressure relief valve, then jog the electric motor very briefly to see which way the pump shaft will turn.

If the motor and pump shafts are enclosed and cannot be seen, crack the fitting on the pump outlet port. When jogged, oil should come out this port. If reservoir oil can be seen, observe the oil while jogging. If rotation is wrong, air bubbles will come to the surface from the suction line.

On hydraulic pumping assemblies moved to a new location and re-wired, always test for correct rotation before operating. On 3-phase electric lines, if the wires happen to be re-connected in a different order from the original connection, motor rotation will be backward. Single-phase motors, once connected correctly, will not change rotation when moved to a different location.

Reversing Electric Motor Rotation. To reverse rotation of a 3-phase electric motor, reverse any two of the three wires either on the line side or the load side of the starter or switch box or in the motor junction box.

A single-phase motor cannot be reversed by simply reversing connections to the power line. Polarity of the starting winding must be reversed in relation to the running winding. This must be done in the motor junction box, and instructions will be found inside the box on how to make this change. Once a single-phase motor has been connected for the correct rotation, it will always run in this rotation even when moved to another location and wired to a different power line.

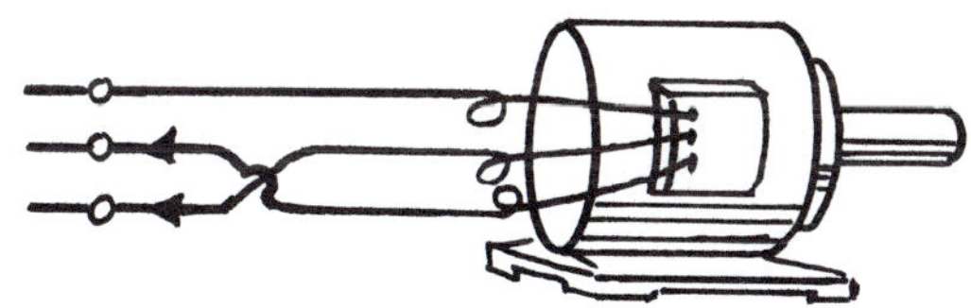

Interchange Two Line (or Load) Wires To Reverse a 3-phase Electric Motor.

Note: A single-phase motor can be reversed if both windings are brought out to a reversing switch, but rotation cannot be reversed while it is running; it must be brought to a stop before it will start in the other rotation.

Reversing an Engine-Driven Pump. Since most engines cannot be reversed, and if the rotation is wrong for the pump to be used, correct rotation is obtained by mounting the pump on the other side of the drive belt (location shown in dotted lines).

Other alternatives, sometimes possible, are to mount the pump on the front rather than behind the engine, or to belt drive it from the main power shaft instead of the power take-off. Perhaps it can be driven from one side or the other of a jackshaft. If side-driven with a set of gears, rotation will be opposite to that with belt drive.

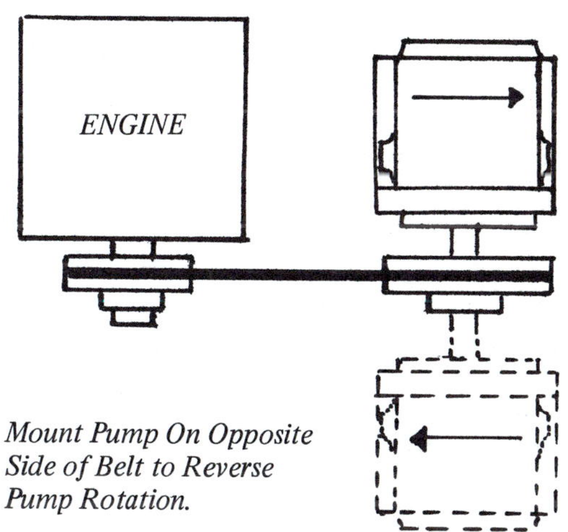

Mount Pump On Opposite Side of Belt to Reverse Pump Rotation.

Reversing Pump Rotation. Direction of rotation of most vane pumps can be changed in the field. Because of the wide variation in vane pump construction the pump service sheet should be consulted. But the following rule generally applies:

If the pump is of balanced vane construction (has two lobes on the cam as illustrated), rotation is reversed if the cam can be re-positioned 90° from its original position. This is accomplished by various means according to the pump brand. In some brands, the complete cartridge including vanes, cam ring, and side plates can be slipped out and either turned over or rotated 90°. In

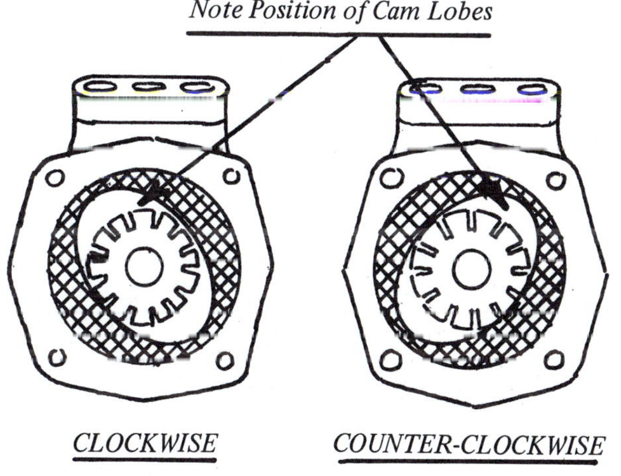

Vane Pump Rotation is Reversed if the Cam Ring Can Be Re-positioned 90° From Its Original Position.

other pumps, only the cam ring needs to be re-oriented either by turning it over, or by rotating it. Often the presence of dowel pins will determine whether the cam ring must be turned over or rotated. But in any case, the new position of the cam lobes must be 90° from the original position.

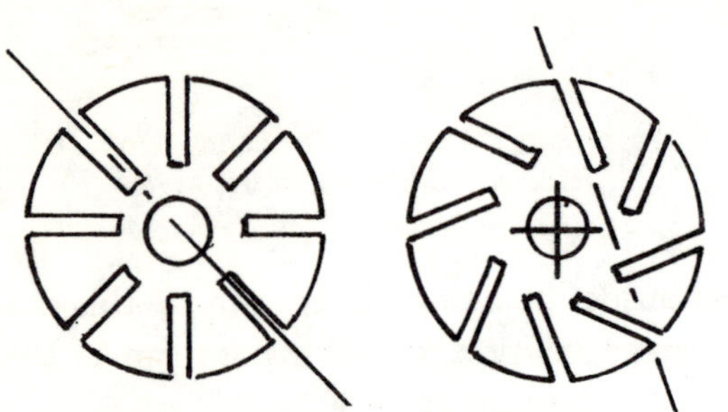

If Slots Are Not On a True Radius, Rotor Must Also Be Turned Over.

If the vane slots are cut on a true radius as in the left view, rotation can usually be changed by not turning the rotor over, only rotating or flopping the cam ring over, whatever is appropriate for the particular pump. But if the slots are not cut on a true radius as in the right view, then not only must the cam ring be rotated 90°, the rotor itself must also be turned over.

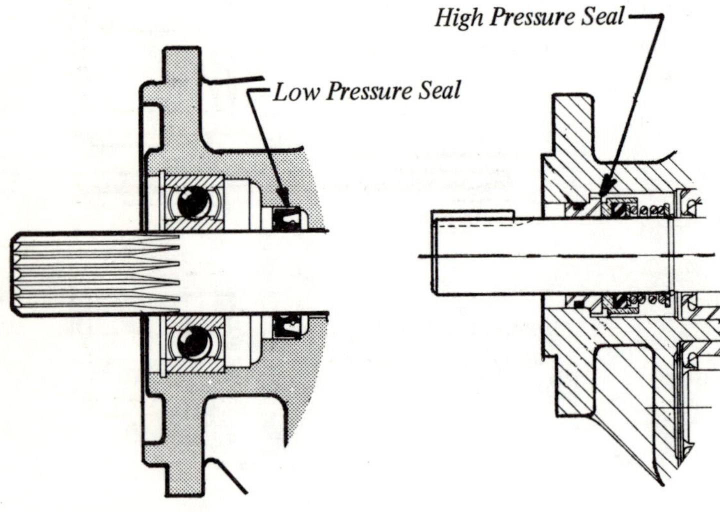

Low and High Pressure Shaft Seals.

Gear Pumps. Gear-type pumps have internal drain holes venting the space behind the shaft seal to the inlet port. On most of them there is no way to change them in the field to the opposite rotation. Those pumps which use a standard cup seal, shown on the left, can tolerate only about 15 PSI behind the seal before it blows out. If they are run in the wrong rotation under pressure, the shaft seal will blow out.

Bi-rotaional gear pumps, those which can be run in either rotation, may use a face-type seal, shown on the right, and the seal will stand full outlet pressure without blowing out.

Note that reversing shaft rotation on most pumps also changes the former inlet port to the new outlet port, etc.

Calculation of Side Load on Shaft. Many pump catalogs show the amount of shaft side load, in pounds, for which the pump is rated. Side load can be calculated with this formula:

$$F = [HP \times 63024] \div [RPM \times R]$$

in which F is the developed side load, in pounds; HP is the driving power to the pump; RPM is pump shaft speed; R is pitch radius, in inches, of the sheave.

From the formula we note that side load decreases as the radius of the sheave increases. Choose sheaves with as large a diameter as practical.

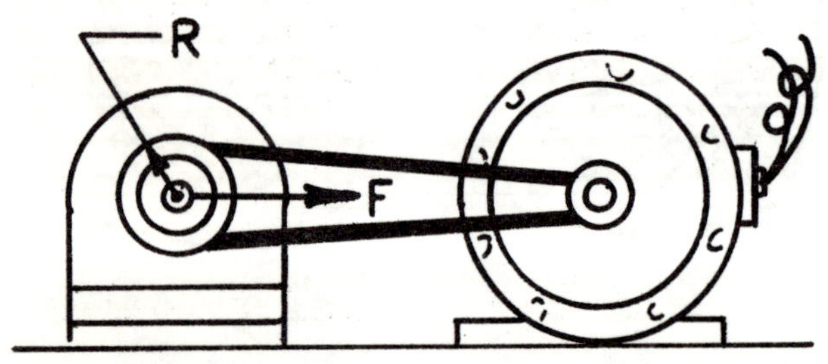

Use Formula To Figure Side Loading On Pump.

PUMP PROTECTION WITH A PRESSURE RELIEF VALVE

Why a Relief Valve is Necessary. No one would think of operating a steam boiler without a pop-off safety valve because there would be no limit to how high the pressure could go. A hydraulic system is similar. Without a relief valve, unless the pump is badly worn and the internal leakage is high, there is no safe limit on how high the pump pressure could go as load increased. The pressure would continue to rise until the pump or one of the other components burst, or until the limit of input HP was reached, in which case the entire system, including the prime mover would stall. The bursting of a pump, valve, cylinder, or fitting is a safety hazard to personnel and could be very costly.

Without a pump relief valve, dangerously high pressures could be generated when a cylinder piston reaches the end of its stroke and stalls, when too great a load is placed against the system, or momentarily during transit of the spool in a 4-way valve.

Precautionary measures when assembling a hydraulic system are to be sure there is a relief valve across the pump outlet line, that it has not been installed backward (inlet and outlet reversed), and that its adjustment has been backed down to minimum setting. In all common relief valves, unscrewing the adjustment knob reduces the cracking pressure.

The illustration shows a simple, direct-acting relief valve. The poppet is held on its seat by an adjustable spring. This retains all the oil in the pump line unless the system pressure, due to an increase in load, should rise high enough to compress the spring. This would open an escape path for part or all of the oil to flow to tank, and would prevent any further rise in pressure.

Relief valves, both the direct-acting and the pilot-operated types, are covered in more detail in both Volumes 1 and 2 "Industrial Fluid Power".

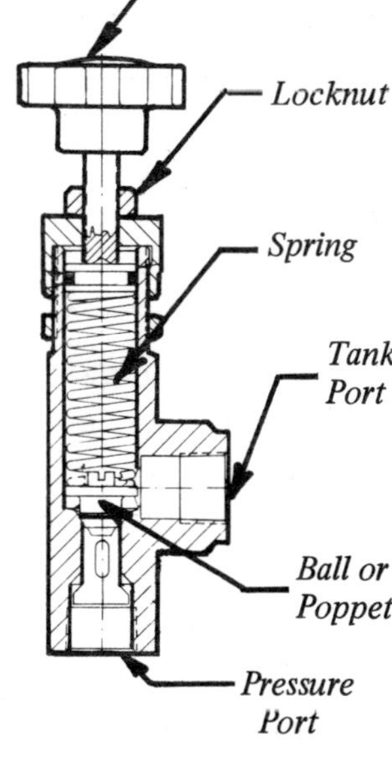

Simple Pressure Relief Valve Using Soft Seat Poppet and Adjustable Spring.

The Relief Valve "Knob Twister" at His Nefarious Work.

Relief Valve Tampering. A problem which plagues maintenance people is having unauthorized persons screwing down the adjustment of a relief valve in an attempt to improve machine performance. If the adjustment has been set correctly, it will seldom, if ever, need to be changed.

Usually, the temptation to screw down the relief valve comes when the hydraulic machine seems not to be performing as it should. It operates more slowly than it should, or is not producing as much force as it should, or is not working at all. In an attempt to improve machine performance, a non-technical machine operator may experiment with all adjustments he can find, and the entire machine may have to be re-adjusted later.

As a matter of fact, the relief valve is seldom at fault. The problem is usually in the pump circuit — a worn-out pump, a dirty pump suction strainer or one of the problems described below, or in the troubleshooting section of this book or in Volume 1 "Industrial Fluid Power".

Several precautionary measures can be taken to prevent relief valve tampering. A locking cover is available for some models; the relief valve itself can be hidden in the tank or a concealed location; the valve body can be cross drilled and a pin pressed in to limit adjustment beyond a maximum safe setting; or a relief valve model can be used which must be removed from the line to be adjusted.

System Failure Caused by the Relief Valve. If a hydraulically operated machine, which previously has been working normally, should develop a malfunction, a little analysis may turn up some possible clues to the source of the trouble. For example:

(1). If there is a complete or almost complete loss of system pressure, the relief valve could be at fault even though this is unlikely. Possible causes of relief valve failure are breakage or binding of the spring in a direct-acting relief valve; seizing of the spool in a pilot-operated relief valve, perhaps caused either by expansion due to overheated oil, or caused by accumulation of silt in the oil. To determine whether there is a complete loss of pressure, use a pressure gauge to make sure.

(2). If there is only a partial loss of pressure, the relief valve can almost certainly be ruled out as the cause. Sometimes, screwing down the relief valve to a higher setting may improve machine operation temporarily, but this is symptomatic of other trouble. Suspect such faults as leaking or blown cylinder piston seals; a badly worn or damaged pump; overheated oil which has become too thin to be pumped effectively; pump cavitation from dirty strainer; or air leaks into the pump from low oil level, leaks around a worn pump shaft seal, etc.

Fortunately, a relief valve is usually a "fail-safe" device. A malfunction in the relief valve causes a loss of pressure, usually a complete loss, and only in rare cases does it cause a rise in system pressure.

Installation of a Pressure Relief Valve. To protect the pump, a relief valve must be teed into the pump pressure line at a point where there is no valving or component between it and the pump. Accidental closing of a shut-off valve, or spool sticking in a diverter valve in the line between pump and relief could isolate the pump from its protection. However, the use of free flow check valves between pump and relief is permissible but be certain they are connected correctly with inlet toward the pump.

The relief valve does not have to be physically located close to the pump, as long as the line connecting it to the pump is free of valving which could accidentally close it off.

Keep in mind that pressure at the pump is not necessarily related to the adjustment of the relief valve; it is directly related to the load resistance placed against the machine. So, a relief valve cannot be set to a certain cracking pressure except when an infinite load is placed on the circuit. This can be done by stalling the cylinder against

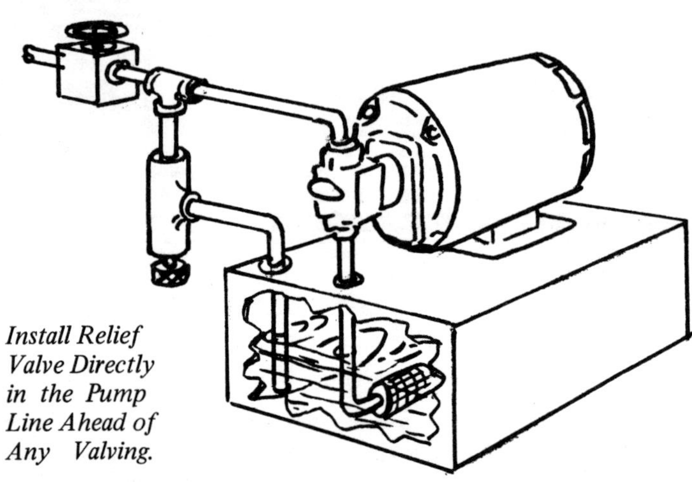

Install Relief Valve Directly in the Pump Line Ahead of Any Valving.

an immovable object, or by temporarily placing a shut-off valve in the pump pressure line, *downstream of the relief valve*, and running the pump across the relief valve for a brief period while the relief valve is adjusted. There are exceptional circuits which maintain relief valve pressure on the pump for the entire machine cycle. These are circuits using a flow control valve in a series-type speed control circuit. Speed control circuits are described in Volumes 1 and 2 "Industrial Fluid Power".

Important! Washers or spacers should not be placed behind the spring in a relief valve to raise the pressure setting. This might prevent complete opening of the valve and its inability to pass the entire pump flow to tank in case of a severe overload.

Be careful of any relief valve which can be over-tightened. This could force the poppet against the seat where it could not open on a pressure overload.

PUMP INLET CONSIDERATIONS

Pump Suction Strainer. We believe that all positive displacement hydraulic pumps should be protected against accidental intake of solid particles. Such particles can destroy the pump by lodging between the working elements; they can accelerate pump wear; they can cause spool valves to stick, poppet valves to leak, cylinder barrels to be scored, etc.

However, the pump suction strainer, if not kept clean, can cavitate and destroy the pump. For this reason, some users prefer to omit the strainer and to allow the pump to pick up unstrained oil from the reservoir. This practice is only safe if the system is kept clean during assembly, and if the reservoir is sealed air-tight and the vent protected with a micronic filter. Even on these systems, a suction strainer should be installed when the system is first put into operation and used for a few weeks to pick up initial contamination, then it can be removed.

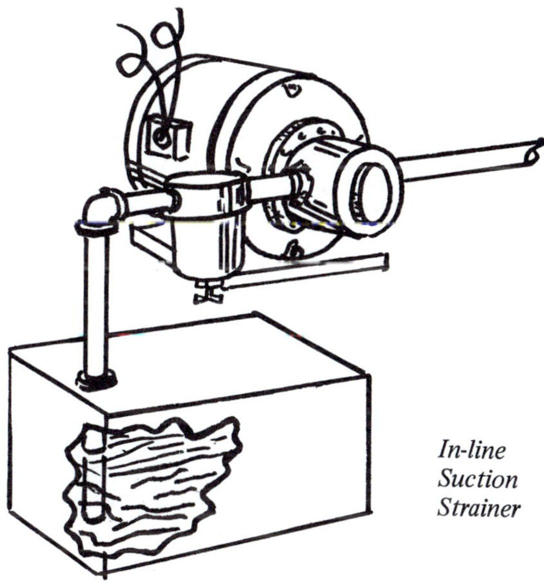

In-line Suction Strainer

The usual practice, when using petroleum hydraulic oil, is to pass pump intake oil through a stainless steel wire mesh strainer of 100 mesh, 100 wires to the inch. This will have about a 150µm rating. On systems where a higher degree of filtration is desirable, a supplementary micronic filter should be added in the pump pressure or tank return line.

The suction strainer may be adequate for gear-type pumps as these are more dirt tolerant than most other kinds, or for systems used very occasionally, but for continuously operating systems using vane or piston pumps, 150µm filtration is not considered sufficient.

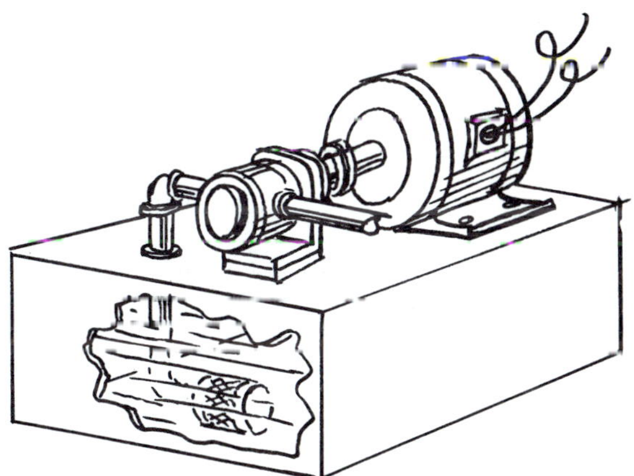

Sump Strainer Buried Under Oil Level.

In-Line Suction Strainers. Refer to top illustration on preceding page. These are easier to install and service than the ones buried under the oil. They are particularly useful on certain mobile hydraulic systems where it may be difficult to obtain access to in-tank strainers. However, they are more expensive and considerably larger than the other type and are available only in limited sizes.

Caution! The Y-strainers commonly used in processing industries for in-line straining are not generally satisfactory for hydraulic oil systems. Pressure drop through them is too high for suction strainers, their pressure rating is too low for pressure line strainers, and their micron filtration rating is too low. Their best application is for straining return line oil before it is discharged into the reservoir.

*Choose Pump Suction Strainers
of Generous Size.*

Sump Strainers. See illustration on preceding page. Since a sump strainer is buried under the oil in the hydraulic reservoir, it does not need a housing. Most hydraulic systems use this type because it is less costly and is available in a wide range of sizes up to 100 GPM. Several strainers can be piped in parallel to obtain virtually any higher capacity.

A suction strainer should be generously sized to reduce flow restriction and danger of cavitating the pump. Some designers purposely oversize the strainer as a margin of safety against pump cavitation if the strainer is infrequently or never serviced. The rule-of-thumb recommends 10 square inches of 100-mesh surface for each 1 GPM flow of petroleum oil of 150 SSU viscosity. Make a generous allowance when using oil of higher viscosity. Flow resistance is approximately proportional to viscosity in centistokes (not SSU). Straining area should be increased in proportion to the increase in centistoke viscosity. A conversion chart SSU to centistokes will be found in the appendix.

We recommend sizing a suction strainer by the rule-of-thumb for straining area, then selecting a size as much larger as practical. If it should be impractical to use as large a strainer as recommended, then a model with 60-mesh instead of 100-mesh wire should be selected to reduce flow resistance.

Suction Strainers for Heavier Fluids. Petroleum hydraulic oil has a specific gravity of about 0.9. All other common hydraulic fluids, including water, water base, water/oil emulsions, and synthetic fluids have a higher specific gravity. They require more power to push them through the same system and they produce higher flow losses. Sometimes larger piping, filters, and valves must be used to produce the same output HP on a given input HP, and to hold down oil heating. When using these heavier fluids, choose a strainer with more square inches of surface than would be used for petroleum oil, or select one with a more open mesh, 60 mesh instead of 100 mesh.

SIZING A PUMP SUCTION STRAINER

Allow At Least 10 Square Inches Straining Area per GPM Flow.

*(This is for 100-mesh strainers and for oil of 150 SSU at 100° F oil;
With heavier oil or finer wire mesh, allow more area).*

Valve Plate Piston Pumps. Most susceptible to micronic dirt in the oil are those piston pumps which have bronze piston shoes riding against a rotating or non-rotating cam plate. Most hydrostatic transmission pumps and motors are of this type. A small grain of dirt which happens to lodge between a piston shoe and the surface on which it rides will soon wear a groove in the sliding members. This will cause internal leakage (called slippage) which will reduce pump efficiency. Over a period of time, as the grooves wear deeper, the pump will become useless when fluid pressure in the grooves causes separation between shoes and cam surface.

A simple wire mesh suction strainer is by no means sufficient for these dirt-sensitive pumps, and transmission manufacturers recommend 3 to 10μm filtration of the system oil either by a high pressure micronic filter in the pump pressure line, or micronic filtration in the case drain lines. Putting micronic filters in the suction inlet of any pump is not recommended unless there is no other practical place to put them, or unless the pump inlet is supercharged to about 5 PSI by another pump. Hydrostatic transmissions usually have a gear-type charge pump driven from the shaft of the main pump. Unless a port is provided for taking the charge pump oil outside the main pump case, running it through a micronic filter, then returning it to the main pump inlet, a micronic filter in the charge pump suction line must be used. This filter should be sized for no more than a 1/2 PSI pressure drop when its element is clean. It should also have an electrical or visual warning device to indicate when the element has become partially restricted by dirt.

PUMP CAVITATION

Cavitation Defined. Our definition of cavitation is any condition at the pump inlet which prevents the pump from taking in a full charge of oil on each revolution of its shaft. The causes are:

<u>(1)</u>. A partial vacuum at the pump inlet because atmospheric pressure is not sufficient to push a full flow of oil to the pump through flow restrictions in the inlet line or in the pump internal passages.

Most probable causes for vacuum cavitation are running the pump too fast, a dirty inlet strainer, too long or too restricted an inlet line, too high a gravity head, a partially collapsed or internally swollen hose to the pump inlet, or the use of an oil with too high a viscosity.

<u>(2)</u>. A partial intake of air at the pump inlet also meets our definition of cavitation, but is technically called aeration.

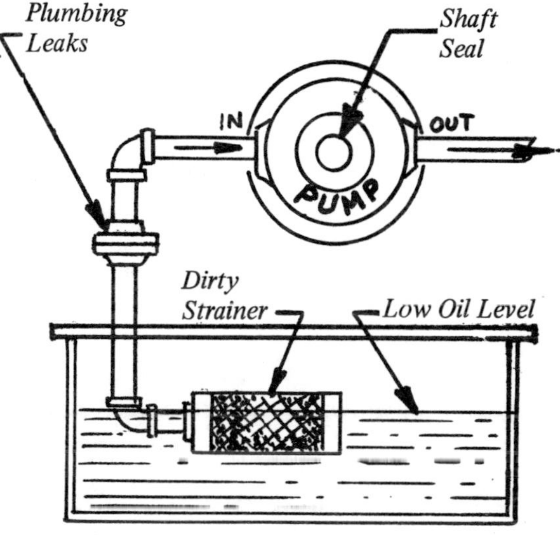

Some Causes of Cavitation.

Most probable causes for air intake are leaky joints in the suction line, low oil level in the reservoir, a worn pump shaft seal, and on certain applications a cylinder rod seal.

Either type of cavitation is destructive to the pump and each cause listed above will be discussed in greater detail.

Symptoms of Cavitation. Cavitation, on a correctly designed system, may gradually develop over a period of time and because it develops gradually, may go unnoticed until it becomes pronounced. The pump becomes noisier. It may run quietly when idling then become extremely noisy as sys-

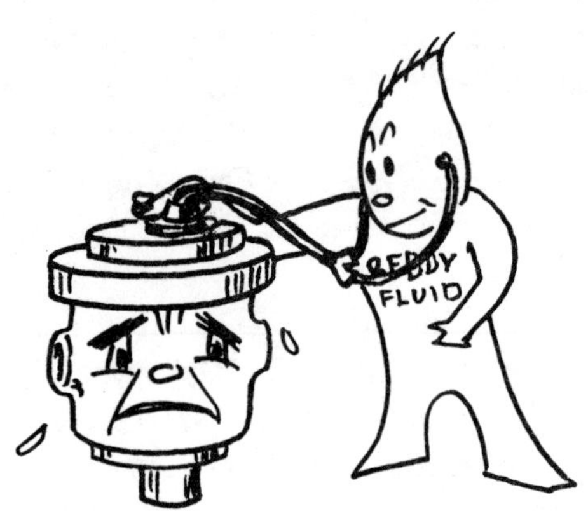

Excessive Noise Indicates Cavitation.

*Cavitation May Cause Pump Overheating
and Excessive Mechanical Wear.*

tem pressure builds up. It may run hot, particularly near the shaft bearing. It may fail to develop sufficient pressure to perform the work. The cylinders will move more slowly, operate erratically, or may stall completely under load. Oil in the reservoir may become "milky".

Effects of Cavitation. Pump bearings will overheat and may be permanently damaged. In normal usage the pump bearings are not only lubricated, but cooled by the flow of oil. When this flow is reduced or cut off, the pump will overheat.

Cavitation deprives working elements in the pump of proper cushioning, causing accelerated mechanical wear. In extreme cases the working elements may not receive adequate lubrication.

Noise produced by a cavitated pump is evidence of increased impact between working parts, and will eventually lead to pump damage.

Troubleshooters Checklist for Cavitation. Some of the causes for pump cavitation have already been mentioned. Those systems which do not receive regular maintenance, including cleaning of the suction strainer, inspection of reservoir oil level, inspection and replacement of pump shaft seal, will eventually develop cavitation, perhaps so gradually that it will not be noticed until damage has been done to the pump. We believe there may be more equipment failures in the field from this than any other single cause.

Although it is normal for pump noise to increase as load pressure builds up, cavitation should be suspected if it becomes greater than previously noted as normal for that particular pump, for that pressure level, and for that speed. If cavitation is suspected, check these conditions:

(1). Suction strainer may have become loaded with dirt. Nearly all hydraulic systems use a pump suction strainer, and if it cannot be seen in the pump inlet line, it is most likely buried under the oil in the reservoir. To test for cavitation caused by a dirty strainer, some users temporarily remove the strainer in the reservoir, or take the element out of the housing on an in-line strainer, then run the pump briefly to see if the noise decreases. Do not run the pump without the protection of the strainer for more than the few minutes needed to make this test.

A wire mesh element can best be cleaned with an air blow gun, blowing from the inside. The use of a solvent may be necessary if there is an accumulation of brown "varnish".

(2). Air may be entering through leaks which have developed in the suction line. Check tightness of all joints in the line. Air can enter around a worn pump shaft seal. Check this by squirting a little oil around the seal while the pump is running. If there is an air leak, the oil will be sucked into the pump, and the pump noise should decrease.

(3). Reservoir oil level may be too low, exposing part of the strainer. Oil level can be observed in the oil level sight gauge. Level fluctuates during the machine cycle while cylinders are operating, and fluctuates more with large rod cylinders. Bring the oil level to the full mark on the gauge when all cylinders are retracted. Single-acting and ram-type cylinders cause more fluctuation than double-acting cylinders. Bring oil level to full mark with all these cylinder retracted. Reservoir should be large enough that suction strainer remains covered 2 to 4 inches at minimum oil level.

(4). The oil may be too viscous. It may have thickened because of cold weather, or someone may have added a more viscous oil than the system can handle. Thickening may sometimes result from mixing of different kinds of oil.

*If Cavitation is Present, Check
These Possible Causes.*

(5). If engine driven, the pump speed may have been increased. This could cavitate a pump which did not cavitate at a lower speed. If the pump is belt-driven, the pulley ratio may have been changed, causing the pump to run too fast.

(6). The oil may be carrying an excessive amount of foam or bubbles in suspension. This can be caused by too small a reservoir, from too small a tank return line, causing the oil to enter the reservoir at excessive velocity. It could be caused by use of an oil which does not have a foam suppressant additive. Draw an oil sample from the reservoir after the system has been running about an hour. Visually inspect the sample for color. Entrained air will give it a milky appearance.

Testing for Cavitation. A simple test for cavitation was suggested above. Vacuum cavitation may be present even with clean suction strainer. The pump inlet can be tested for degree of vacuum by drilling and tapping a 1/4" NPT gauge port in the inlet connection flange and installing a vacuum gauge. Connect the gauge with a short length of hose (rated for vacuum) so machine vibration is not transferred to the gauge. A leaktight plug valve can be installed in the line so the gauge can be installed or removed without stopping the pump. The chart shows maximum safe inlet vacuum that can usually be tolerated by various kinds of pumps:

Gear Pumps	Vane Pumps	Piston Pumps
3 to 5 PSI	2 to 3 PSI	2 PSI
or	or	or
6 to 10" Hg	4 to 6" Hg	4" Hg

Ratings in the above table are *maximum* vacuums when the strainer becomes too dirty for efficient pump operation. In choosing a strainer size select one which will have only about 1/3rd these vacuums when clean and when oil is at normal operating temperature of

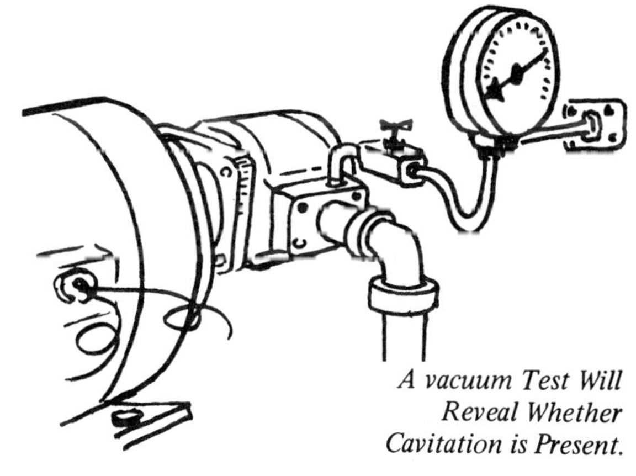

*A vacuum Test Will
Reveal Whether
Cavitation is Present.*

130 to 140° F. When vacuum exceeds these levels the strainer should be removed and cleaned at the first opportunity. Note: this information is for popular pumps which we checked. It is always best to consult the specification sheet for the particular pump used.

Caution! On new designs keep pump speed as low as practical. Pump bearing life will be extended in inverse proportion to the amount speed is reduced. Pumps run at speeds above their rating will almost certainly cavitate. When pumping fluids of higher specific gravity than petroleum oil, shaft speed should be less than the normal catalog rating to avoid cavitation.

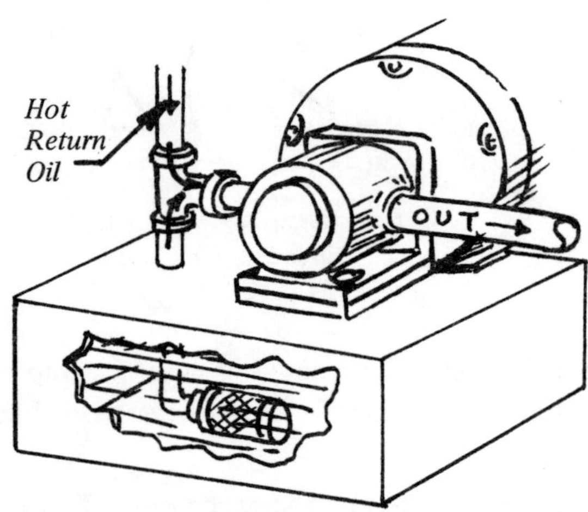

Do Not Tee Return Oil Into the Pump Inlet.

Special Examples of Cavitation. (1). Do not tee tank return oil into the pump suction line. This would feed hot return oil into the pump, and fresh oil from the reservoir could only enter the pump while the cylinder was retracting. This would defeat the beneficial functions of the reservoir: to keep the oil cool, to allow the oil to purge itself of entrained air and dirt.

Tank return oil should enter the reservoir at a point far removed from the suction intake of the pump. The reservoir should have a center baffle plate; return oil should be discharged on the opposite side of the baffle from the pump intake. See reservoir construction in Volume 1 "Industrial Fluid Power".

(2). Where a pump is driven from a vehicle engine, a small oil reservoir is often used in the engine compartment near the pump. But if the reservoir must be larger than can be mounted near the pump, it must be installed elsewhere, often a long distance from the pump with a long pump suction line. This is a critical application. The suction line must be one or two sizes larger than would normally be used if the reservoir were near the pump. It should be sized to keep flow velocity to 2 feet per second or less to avoid cavitating the pump. It will be necessary to bush the line down at the pump to the inlet port size. Be sure to use *vacuum* hose. This hose is especially built for vacuum with a spiral wire inside to prevent collapse.

The suction line must be large enough to handle the pump flow at the maximum engine speed. After the system is installed, a vacuum gauge should be installed temporarily at the pump inlet to make sure the inlet vacuum does not reach the cavitation level.

Long Suction Lines Must be Large in Diameter.

(3). The pump suction line should be carefully designed with as large diameter as practical, with a minimum number of bends, as short as possible, and with the minimum suction head (this is the distance the pump must suck the oil to a higher elevation).

The ideal mounting position is with the pump inlet port down to eliminate bends, as at A, but it is difficult to make provision for withdrawing the suction strainer out of the reservoir.

The usual and more practical mounting position is shown at B with inlet and outlet ports in a horizontal position.

Mounting with inlet port up, as in C, should be avoided. In addition to flow restrictions, the high point in the plumbing makes an air trap. Air may accumulate in this pocket causing erratic operation of the system and perhaps even causing the pump to lose its prime after being stopped.

Flooded Suction. Mounting the reservoir higher than the pump is very good design if practical. Inlet vacuum, due to suction head is eliminated, but vacuum caused by flow of oil is not reduced, so the inlet line still needs to be very large. Flow velocity still needs to be in the 2 to 4 feet per second range.

A rotary plug valve is usually placed in the inlet line to cut off oil drainage when the pump must be removed for servicing. Caution! A lock should be placed on this valve to prevent someone from closing it while the pump is running. This would very quickly destroy the pump.

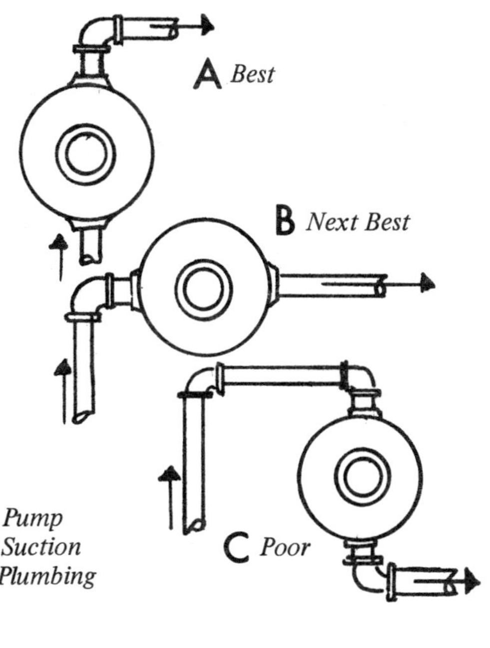

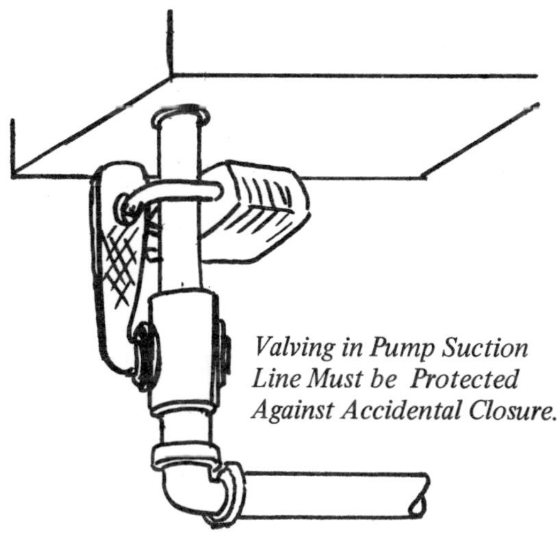
Valving in Pump Suction Line Must be Protected Against Accidental Closure.

PUMP PROBLEMS

Shaft Seal Blows Out. Although a pump shaft seal can become loose and work out of its housing because of a poor fit, replacement with the wrong seal, or improper installation, in most cases the seal blows out because of a build-up of pressure behind it. A visual inspection will usually reveal whether it worked out or was violently blown out. If it was forcibly blown out, check these possible causes.

(1). The pump was being rotated in the wrong direction. Refer to the instructions given earlier in this chapter for determining rotation and checking for correct rotation. Not only is the pump running in the wrong rotation, but:

(2). The inlet and outlet ports are reversed, with the true pump inlet being connected to the pressure line and the true outlet being connected to tank. The inlet port is exposed to a build-up of pressure in the system.

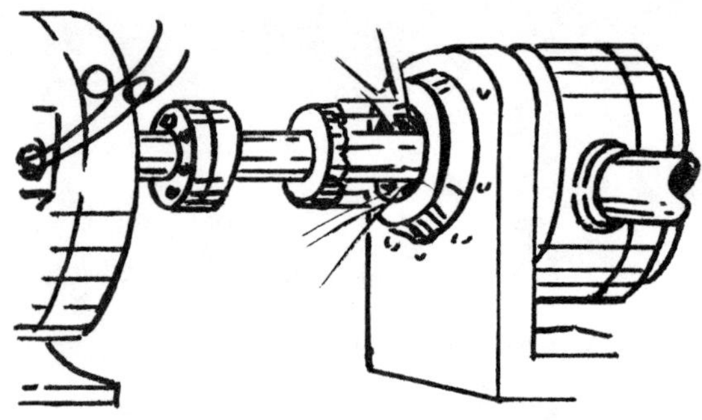

*Wrong Rotation May Cause the Shaft
Seal to Blow Out.*

*Failure to Build Up Pressure May be Due
to One of Several Causes.*

*A Hot Front Bearing May Indicate
Excessive Pump Slippage.*

Pump Fails to Build Pressure. If a system failure appears to be caused by the pump, check these possible trouble areas:

(1). Sheared key or pin on the shaft.

(2). If belt driven, check for proper belt tension. If oil has been spilled on belts, they can slip even though tight.

(3). Check for the common causes of pump failure such as dirty inlet strainer, low oil level in the reservoir, etc.

(4). Open the pump and check for a sheared key or pin holding the drive gear to the shaft. Generally inspect the inside of the pump for broken parts including vanes, springs, or gear teeth. Varnish deposited on internal parts indicates the system has been running much too hot.

(5). If pump seems to work normally when the oil is at room temperature, then pressure falls off as oil temperature rises, this may indicate a badly worn pump, or oil too thin from use of wrong viscosity or from excessive oil temperature. Measure reservoir temperature and add a heat exchanger to the system if necessary. Preferred operating temperature 130 to 140° F. If hotter than 160° F the oil life and the pump life will be materially shortened.

Pump Runs Too Hot. It is normal for the pump to run hotter than the reservoir temperature, especially around the front shaft bearing. Mechanical power loss in the shaft and bearings generates heat. The pumped oil carries the heat back to the reservoir where it can be dissipated. An alert maintenance man will be aware of pump temperature under normal operation, and should notice if the pump seems to be overheating. A heat build-up higher than normal may be caused by one of the following:

(1). Badly worn pump. The increased internal leakage will generate an abnormal amount of heat.

(2). Cavitation. Check causes of cavitation described earlier in this chapter.

(3). An increase in temperature in the entire system due to some cause other than the pump.

Pump Fails to Prime. On a new installation, failure to pick up oil is caused by air trapped in the pump inlet line, in the pump case, or elsewhere in the system. The pump is unable to develop sufficient vacuum to purge this air. In some cases, cracking a fitting on the pressure line may help by allowing trapped air to escape.

Before running the pump, be sure the rotation is correct by briefly jogging with the electric motor. A piston pump must have its case filled with oil. This is important not only for lubrication but many of these pumps draw inlet oil from the case, and they cannot prime from a case full of air. Fill the case through one of the case drain ports. These methods of priming a new pump are suggested:

(1). Some pumps, like gear pumps, will immediately prime. If the pump does not immediately pick up oil, as shown by a slight movement of the pressure gauge, stop the motor and crack a fitting on the pressure line and try again.

(2). If the pump still fails to prime, rotate the pump shaft *by hand* until a few drops come out the cracked fitting. Tighten the fitting. Jogging by hand is more effective than by the motor. Due to the mass of the oil, it cannot respond fast enough to the brief vacuum impulses generated at high speed, but almost any pump will respond to slow shaft rotation. If impractical to rotate pump shaft by hand, jog very briefly with the electric motor, allowing the motor to come to a complete stop before jogging again.

(3). If the above methods do not produce results there are other and more drastic methods. Pressurize the reservoir with shop air through a low range pressure regulator. Be careful to apply very low pressure: if the reservoir is tight, as it should be, 1 PSI will raise the oil 2½ feet. More pressure than this may damage the reservoir.

If the tank is not airtight it should be repaired to make it so. Breather openings will, of course, have to be plugged during tank pressurization.

An easy way is to use an air hose through the filler opening. Use a shop towel to seal the hose in the filler opening. Run the pump while air pressure is applied.

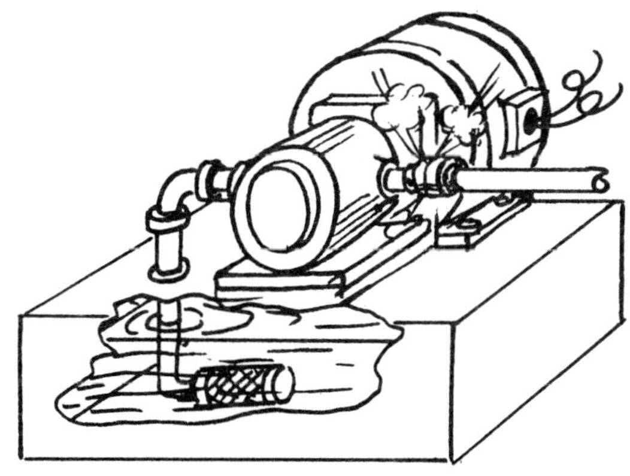

Crack a Fitting on the Pump Outlet to Relieve Trapped Air.

Pour Oil Into Pump Case.

Rotate Pump by Hand With Fitting Cracked on Outlet Line.

A Pressure of 1 or 2 PSI On Top of the Oil Will Help the Pump to Prime.

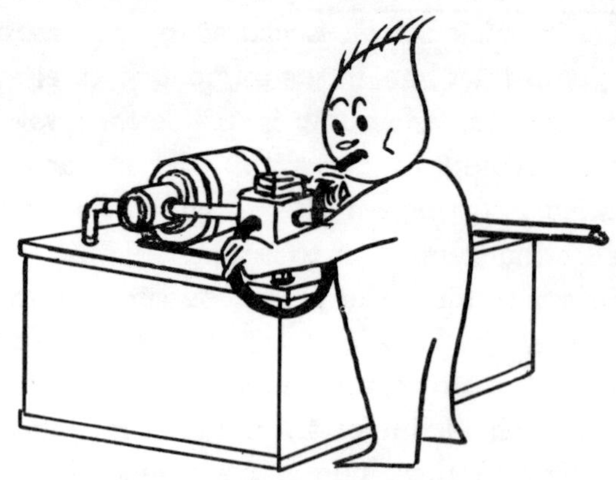

A Small Amount of Suction Often Helps the Pump to Prime.

<u>(4)</u>. A small vacuum pump, or mouth suction can be applied to the pump pressure line through the pressure gauge port after the gauge is removed. Rotate the pump shaft by hand while applying vacuum.

Once a pump has been primed it should never lose its prime unless the plumbing is opened and air is admitted. If the pump does loes its prime, perhaps while stopped overnight, air is entering the system through a leaking fitting, a small crack in the pump case (or other component), or more likely, through a worn pump shaft seal.

Positive Displacement Pumps Are Not Ordinarily Designed to Pump Non-lubricating Fluids.

Oil Pumps on Other Fluids. Positive displacement pumps which include, gear, vane, and piston types and their variations, are designed for use on fluids, like petroleum oil, which lubricate the bearings and provide a lubricating film on internal parts which operate in metal-to-metal contact. They can be used on hydraulic fluids other than oil which provide adequate lubrication, although many times the pump should be operated at less pressure than permissible on oil.

They should never be used on fluids such as water, kerosene, fuel oil, jet fuel, gasoline, etc. because these fluids do not provide the necessary lubricity.

Pump Running Out of Fluid. Positive displacement pumps, if used on fluid transfer or sump pump applications, should be monitored and immediately stopped if they run out of fluid. Cavitation will cause rapid wear, heating, and will very definitely reduce pump life. For fluid transfer applications where very little pressure is required, impeller pumps are usually more satisfactory.

Overspeeding of Pumps. The manufacturers recommendation on top speed should be closely followed. The danger of cavitation increases in proportion to speed increase because of the higher volume of oil which must be drawn into its inlet port by vacuum. If the pump must be operated at high speed, an overhead reservoir will help reduce cavitation, but may not entirely eliminate it. It may be necessary to supercharge the inlet at 5 to 15 PSI with a separate supercharge pump.

Bearing loads on some pumps increase with speed. Bearing life is inversely proportional to RPM. For example a pump run at 3600 RPM will have one-half the bearing life expectancy of the same pump operated at 1800 RPM.

At high speeds there is an additional hazard of mechanical failure from vibration created by centrifugal unbalance and the mass of the rotating parts.

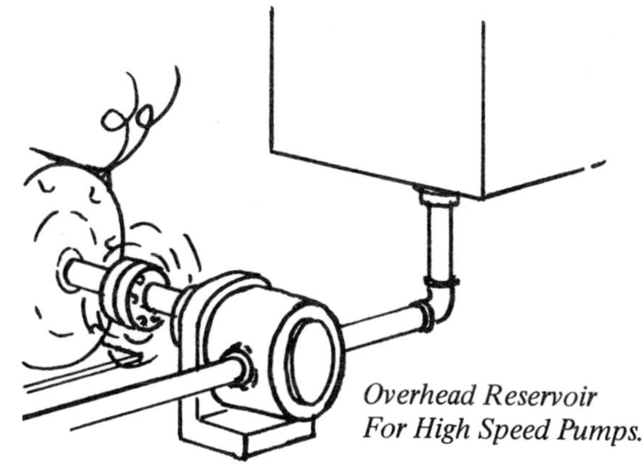

*Overhead Reservoir
For High Speed Pumps.*

Minimum Speed of Pumps. Gear, gerotor, and piston pumps continue to pump as shaft speed is reduced, with pump flow approximately proportional to shaft speed. But at low speeds, at a given pressure, their efficiency decreases. Slippage remains about the same regardless of speed, being primarily proportional to pressure, so at low speeds the slippage GPM becomes a higher ***percentage*** of pumped GPM. For better efficiency, then, these pumps should be operated as close as practical to their rated maximum speed, but not above this speed.

Those vane pumps which depend only on centrifugal force to keep the vanes in contact with the cam ring do not operate efficiently below 1000 RPM. Vane pumps which use springs or hydraulic pressure in addition to centrifugal force do operate reasonably well at low speeds, but their efficiency decreases for the same reason as described for gear pumps.

For any positive displacement pump, the highest horsepower per pound of pump weight is obtained at maximum catalog speed rating. We recommend a minimum speed of 600 to 1000 RPM for any of the common gear, vane, and piston pumps.

Over-Pressuring of Pumps. A positive displacement pump is usually designed with a safety factor of at least 3, so it will not be destroyed by operation at a steady pressure considerably over its maximum catalog rating. But it can be destroyed by brief or transient pressure spikes which can be many times the steady pressure as read on a gauge. These transients are generated by shock during valve shifting, the impact of a cylinder against a positive stop, or by the sudden release of high pressure on a press. An oscilloscope must be used to measure their intensity. So, on systems which are subject to these shocks, the maximum pressure rating of a pump should not be exceeded.

*Over-pressuring a Pump Greatly
Shortens Its Life.*

Aside from the danger of mechanical rupture from high pressure, the effect of higher pressure is to shorten pump bearing life. Life expectancy is increased as the cube of pressure decrease. Reducing pressure from 3000 to 1500 PSI increases bearing life expectancy by 8 times (the cube of 2). However we do not necessarily recommend reducing operating pressure simply to increase pump life because there may be other factors to consider.

Matching the GPM of a Broken Gear Pump. The chart on this page covers gear-type pumps. Please refer to the formulae on Page 223 for replacement information on all kinds of pumps.

When purchasing a replacement for a worn-out or damaged gear pump, an exact replacement, brand and model, may not be available. It then becomes necessary to purchase a pump of a different brand. If the GPM rating of the broken pump is not known, it can be estimated with this chart. Purchase the new pump with as near as possible the same flow rating, considering the shaft speed rating of the new pump. Matching the flow may be quite important. If the new pump has less flow, the system will not operate as fast. If it has more flow the system will be faster but will require more HP. The driving motor or engine may not be able to supply this extra horsepower.

To use the chart, measure the thickness and the pitch diameter of the gears in the old pump. Pitch diameter is found by measuring the distance between the two gear shafts, not by measuring the outside diameter of the gears. Use pitch diameter and gear thickness to determine GPM rating of the old pump at 1200 RPM. The new pump should have very close to the same flow at 1200 RPM.

For GPM at pitch diameters not listed in the chart, use rule-of-thumb that when pitch diameter doubles, GPM increases about 3 times.

Figures in the body of this chart are approximate GPM displacements of a gear pump when running open (0 PSI) at 1200 RPM.

Gear Thickness	Pitch Diameter of Gear			
	1-1/2"	2"	3"	3-5/8"
5/8"	3.9			
3/4	4.6	8.01	GPM Flow	
7/8	5.4	9.39	at 1200 RPM	
1	6.2	10.7		
1-1/4	7.8	13.4	23.8	50
1-1/2	9.4	16.0	28.5	60
1-3/4	11.0	18.7	33.3	70
2	12.6	23.4	38.0	80
2-1/4	- - - -	24.0	42.8	90
2-1/2	- - - -	26.7	47.5	100
2-3/4	- - - -	29.4	52.3	110
3	- - - -	32.0	57.0	120
3-1/4	- - - -	- - - -	61.8	130
3-1/2	- - - -	- - - -	66.5	140
3-3/4	- - - -	- - - -	- - - -	150
4	- - - -	- - - -	- - - -	160

DRIVING HORSEPOWER FOR A HYDRAULIC PUMP

The amount of power for driving a hydraulic pump depends on three factors: (a), the GPM flow which the pump will produce; (b), the PSI pressure which it must put up for the load, and (c), the pump efficiency.

(a). Input HP is in direct proportion to the rated flow. A pump with twice the GPM rating will require twice the input HP (at the same pressure level).

(b). Input HP to a pump is in direct proportion to the pressure it must put up to just balance the load resistance. Twice the HP will be required to balance twice the load resistance. In addition, the pump must put up enough extra pressure to push the pump flow through flow resistance in plumbing and valves. On an average system 25 to 35% additional pressure will take care of flow losses.

(c). Most positive displacement pumps operate, under their normal ratings, at about 85% efficiency.

Some may be slightly more efficient, perhaps 90%; others may be less efficient, perhaps 75 to 80%.

Since GPM flow of a pump is directly proportional to its shaft RPM, increasing pump speed from 1200 to 1800 RPM (a 50% speed increase) will require a 50% increase in input HP (at the same pressure).

The driving power for a pump, therefore, can be computed from the GPM, the PSI, and the efficiency using the fluid HP formula:

Input HP = GPM x PSI ÷ [1714 x 0.85] (for a pump with 85% efficiency)

The approximate driving power is shown in this chart. It has been calculated from the above formula. If pump efficiency is known to be other than 85%, use the actual efficiency in place of 0.85 in the fluid HP formula. Please refer to Page 222 for complete HP chart and more details on input HP requirements.

HORSEPOWER REQUIRED TO DRIVE A HYDRAULIC PUMP

Figures in this chart are the HP needed at the PSI and GPM shown. Pump efficiency is assumed to be 85%. Please see complete HP chart on Page 222 of the Appendix.

GPM	200 PSI	250 PSI	300 PSI	400 PSI	500 PSI	750 PSI	1000 PSI	1250 PSI	1500 PSI	2000 PSI
5	.70	.88	1.05	1.40	1.72	2.57	3.43	4.29	5.15	6.86
10	1.40	1.75	2.10	2.80	3.43	5.15	6.86	8.58	10.3	13.7
15	2.10	2.63	3.15	4.20	5.15	7.72	10.3	12.9	15.4	20.6
20	2.80	3.50	4.20	5.60	6.86	10.3	13.7	17.2	20.6	27.5
25	3.50	4.38	5.25	7.00	8.58	12.9	17.2	21.4	25.7	34.3
30	4.20	5.25	6.30	8.40	10.3	15.4	20.6	25.7	30.9	41.2

Example: Find the driving HP for a 15 GPM pump to produce 1000 PSI, assuming 85% efficiency. The answer, 10.3 HP, is found at the intersection of the vertical 1000 PSI column and the horizontal 15 GPM line. If the pressure were raised to 1500 PSI, a HP of 15.4 would be required.

Rule of 1500. This rule-of-thumb comes in handy for making quick estimates of mechanical HP in terms of fluid flow and pressure. The rule states that 1 drive HP must be available for each combination of PSI and GPM which multiplies to 1500. For example, 1 GPM at 1500 PSI is equivalent to 1 HP. Or, 2 GPM at 750 PSI is also equivalent to 1 HP.

Other examples: 7½ GPM at 1000 PSI = 5 HP; 20 GPM at 3000 PSI = 40 HP. Many examples can be done quickly by mental arithmetic or easily on a scrap of paper.

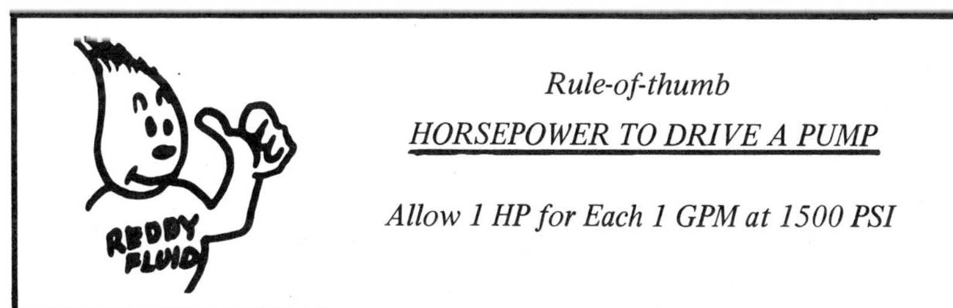

GPM	750 PSI	1000 PSI	1250 PSI
5	2.57	3.43	4.29
10	5.15	6.86	8.58
15	7.72	10.3	12.9
20	10.3	13.7	17.2
25	12.9	17.2	21.4
30	15.4	20.6	25.7

Values Not in the Chart. From the key values listed in the chart, input horsepower for any other operating condition can very easily be estimated.

Example: Find the input horsepower needed for 15 GPM at 2250 PSI.

Solution: Since HP is proportional to pressure, values in two (or more) columns can be added. In this case the values in the 1000 and the 1250 columns can be added: 10.3 + 12.9 = 23.2 HP.

In a similar manner, at higher flows, beyond 30 GPM, values in the same pressure column can be added:

Example: Find the input HP for 40 GPM at 1000 PSI:

Solution: Add values in the 1000 PSI column for 30 and 10 GPM: 20.6 + 6.86 = 27.46 HP.

A more complete chart on input horsepower is on Page 222 of the Appendix.

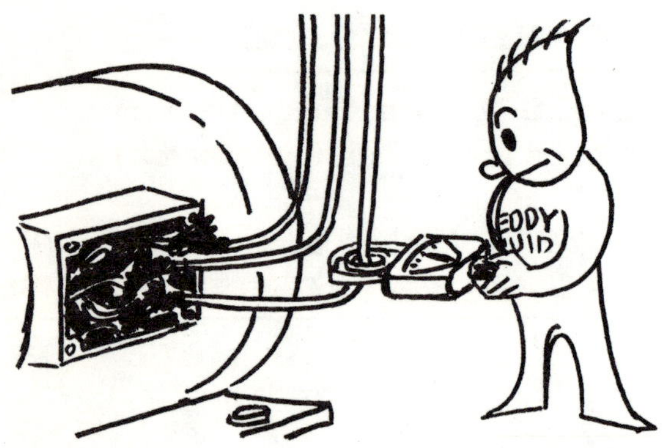

Use a Loop Ammeter for Measuring Motor Current.

Electric Motor Drive. The 3-phase, squirrel cage induction motor with NEMA Design B characteristics is the one more often used for pump drive of positive displacement hydraulic pumps.

It can be overloaded for short periods up to 2 or 3 times its normal full load torque before it will stall. When stopped, it can develop up to 1½ times its full-load running torque for starting.

To determine if an electric motor is running at its full rated load, or above its rated load, the load current can be measured and compared to the current rating on the motor nameplate. Use a loop ammeter. This instrument can be hooked around any one of the three line wires without breaking the insulation or disconnecting the wire.

Motor speed on a 60 Hz power line can be selected from, and is limited to, one of these synchronous speeds: 3600 RPM (1 pair of poles); 1800 RPM (2 pairs of poles); 1200 RPM (3 pairs of poles); or 900 RPM (4 pairs of poles). Actual no-load running speed on an 1800 RPM motor, for example, is about 1770 RPM, and the full-load speed is about 1725 RPM. No speed adjustment is possible in between these values. Motors connected to a 50 Hz line will run at 5/6ths their speed on a 60 Hz line.

The temperature in the windings of an open frame motor, running at normal room temperature, will level off at about a 40°C rise above room temperature. These motors have a service factor of 1.1, which means they can be run continuously at 10% above their nameplate horsepower without the windings being damaged, assuming they are operating in a normal room temperature. The winding temperature will exceed 40°C but will not rise high enough to cause permanent damage. For example, a 10 HP motor can be run at 11 HP under normal room temperature.

Running a motor continuously at 10% above its nameplate rating is not good design practice. Above rated HP the line current increases at a much more rapid rate than HP output. Electric power costs may, for example, be 25% higher for an increased output of 10%.

Most hydraulic systems operate at full power output for only brief periods during the cycle. During the remainder of the cycle they are operating far below their full rating. For this reason it may be good design practice to allow the motor to run slightly overloaded during these brief peak periods.

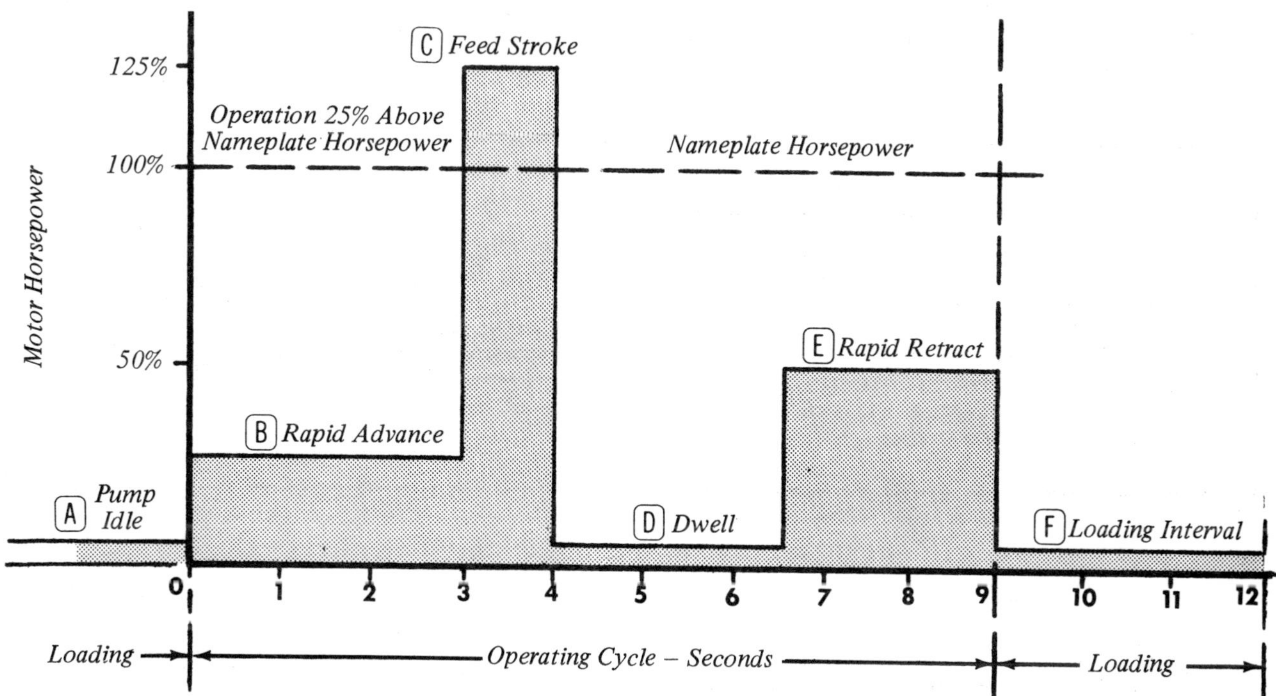

Typical Cycle of a Hydraulic Press Showing HP Usage at Various Stages in One Complete Cycle.

The cycle of a typical hydraulic press is shown in this diagram. The complete cycle time is 12 seconds including reloading time. When the cycle starts, the cylinder moves forward at maximum speed to close the distance to the work. The work is contacted and full tonnage is exerted for 1 second. The cylinder dwells for a period of 3 seconds at full extension before retracting at full speed. A period of 3 seconds is allowed for unloading and reloading before the start of the next cycle.

Some designers use the rule-of-thumb that a 3-phase induction motor can be overloaded to a limited degree and for brief periods during the cycle if the average HP over, say, a half hour, does not exceed the nameplate rating. This graph illustrates the amount of overloading which would usually be permissible according to the rule-of-thumb.

<u>Line A</u>. Before the cycle starts the pump is unloaded. The motor consumes only enough power to run the pump in an unloaded condition.

<u>Line B</u>. At the start of the cycle the cylinder moves in free rapid traverse to close the distance to the load. The motor consumes only enough power to move the cylinder against friction.

<u>Line C</u>. When the cylinder contacts the load, the press produces its maximum force for an interval of 1 second. During this interval the motor is running at 25% overload. This amounts to less than 10% of the cycle time on a 12-second cycle.

<u>Line D</u>. The dwell period in this cycle is assumed to be with the cylinder holding at its full extension and with the pump unloaded. If full tonnage must be held against the load during this dwell period, this would place too long an overload on the motor, unless a variable displacement pump was being used which would be unloaded by its compensator during this period.

<u>Line E.</u> During the cylinder retraction the motor is running at only a fraction of its rated power.

<u>Line F.</u> During the reloading period the motor is allowed to run at low power — only enough to run the pump in an unloaded condition.

It is obvious from this graph that the average power developed by the electric motor is far below its nameplate rating, so there is no danger of accumulating enough heat in the windings to cause them to be damaged.

RULE-OF-THUMB

A 3-phase induction motor can be overloaded to 25% above its name-plate HP rating for brief periods provided the average horsepower does not exceed the nameplate rating. A 25% HP overload might be represented by current about 30 to 40% above nameplate current rating. Current can be measured by a loop ammeter. The overload should not be maintained for longer than 10% of the cycle time on the average.

Be sure the switchgear — switches, breakers, and wiring are designed to handle this current overload.

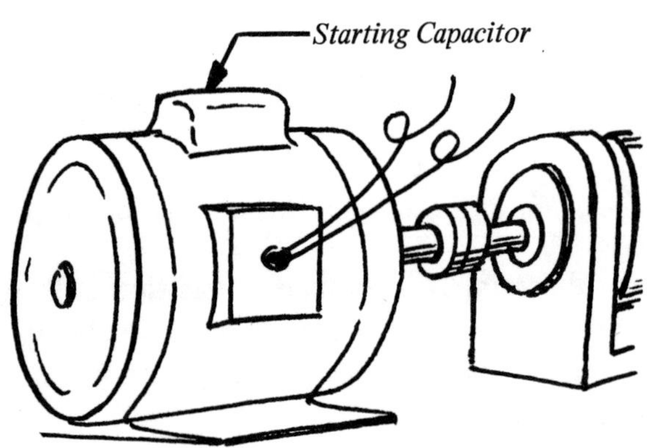

Single-phase Electric Motor

Unload Pump Before Starting Engine.

Single-Phase Electric Motor Drive. There are several kinds of fractional and integral HP single-phase electric motors, but probably the kind most often used to drive hydraulic pumps is the capacitor start induction motor because it has a higher starting torque than some of the other kinds. The starting capacitor is usually mounted on top of the housing.

Single-phase motors have less starting torque than 3-phase motors, so a hydraulic circuit should be designed to remove the hydraulic load against the motor for starting.

Single-phase motors may stall on a 25% torque overload, so they should not be used under conditions which may require in excess of a 10% torque and HP momentary overload.

Engine Drive. Unlike a 3-phase electric motor, an engine does not have the torque reserve to permit it to be overloaded. Circuits must be designed to prevent any torque overload, even momentarily. To obtain maximum HP, speed must be maintained because HP is proportional to speed.

An engine will suffer a loss of HP due to aging, so circuits should be designed with a safety reserve of HP when the engine is new.

HYDRAULIC HAND PUMPS

Single-Acting and Double-Acting Pumps. A single-acting hand pump delivers a flow of oil when the handle is moved in one direction, with a free return. A double-acting pump delivers a flow of oil in both directions of handle movement. A single-acting pump usually has only one piston; a double-acting pump may have two pistons, both connected to the same handle, one on its discharge cycle while the other is on its intake or suction cycle.

One "cycle" of a hand pump is usually defined as a complete forward and return movement of the handle back to the starting position.

"Displacement" of a hand pump is usually specified as (so many) cubic inches of volumetric displacement on a full handle stroke through one cycle.

Mounting Position. Correct mounting position is important with all hand pumps. Manufacturers literature should be consulted if there is any doubt on correct position.

Two factors determine the position in which the pump must be mounted: (a), the location of the air vent on the reservoir, and (b), the shape and routing of the suction intake line inside the reservoir.

(a). The air vent must be in such a position that it will always be above the maximum oil level in the reservoir. On some pumps it may be necessary, when mounting in other than the specified position, to plug the original vent and to drill and tap a new vent in a suitable location above the oil level. On some pumps it may be possible to attach a curved piece of tubing or pipe to the inside terminus of the vent, and run this connection to a position above the oil level.

Again, on some pumps, which have a relatively large reservoir, the vent hole can be sealed if sufficient air space is left for the trapped air to compress and decompress. The usual practice is to fill the reservoir to about one-half to two-thirds of its maximum capacity.

The vent is sometimes sealed or protected with a micronic filter to prevent ingestion of atmospheric dust, water vapor, or other contaminants.

(b). Please refer to these illustrations. The suction tube must be able to pick up oil from the lowest part of the reservoir if the full reservoir capacity is to be utilized.

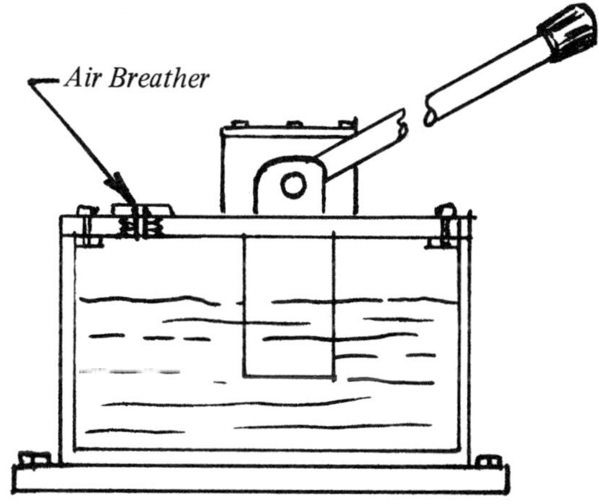

Mount Hand Pump With Air Vent Up.

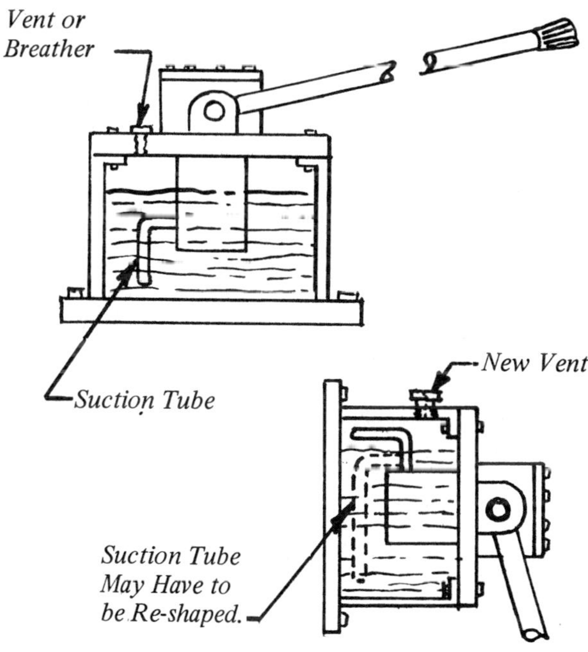

Suction Tube May Have to be Re-shaped.

Usually, the suction tube can be re-shaped, or even replaced, as shown in the illustration on the preceding page, to bring the oil pick-up point near the bottom of the oil.

Ram-type Cylinder

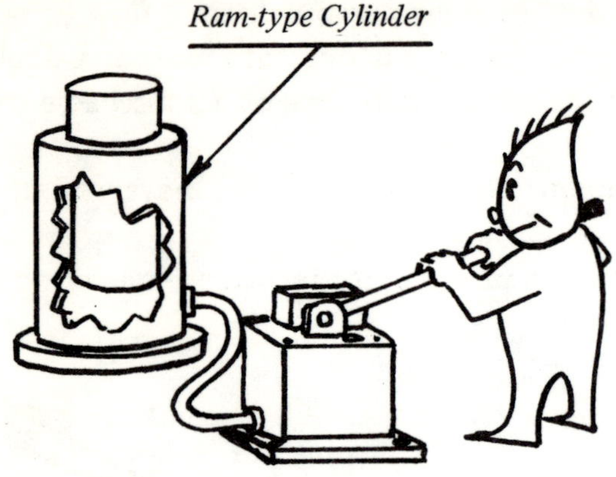

Pump Reservoir Capacity Must be Greater Than Internal Volume of Ram.

Single-acting Cylinder

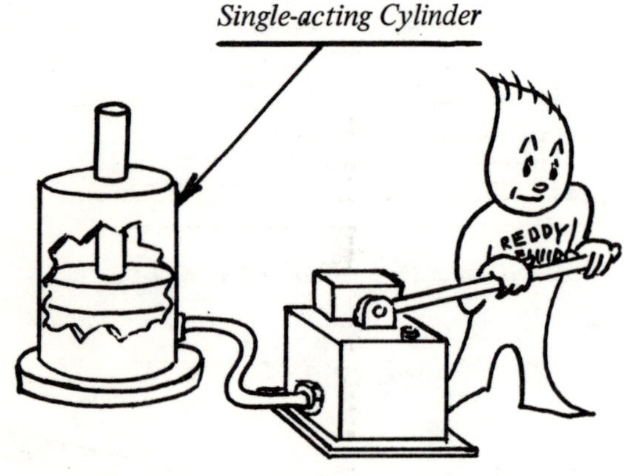

Multiply Piston Area Times Length of Stroke to Find Cubic Inches of Oil Required.

Double-acting Cylinder

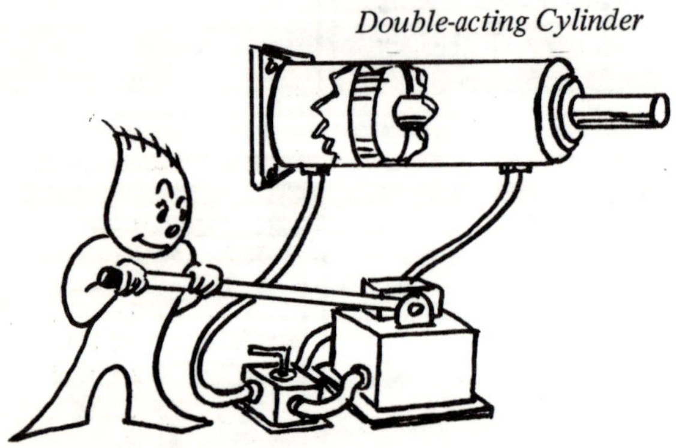

Reservoir Furnishes Oil to Make Up Cylinder Rod Volume on Double-acting Cylinders.

Calculation of Reservoir Capacity. Sometimes there is a choice of reservoir size. The most satisfactory operation will usually be with the largest reservoir which can be accommodated by the available mounting space. Calculate amount of oil required for cylinder operation, as described below, then use a pump with a reservoir capacity at least 50% greater than cylinder displacement.

Ram-Type Cylinder. This is one which works by displacement. The ram is also the piston. To calculate volume of oil required by the ram, take diameter of ram, calculate its square inch area (or refer to table in Appendix), then multiply times the length of stroke. This will give total ram displacement in cubic inches. Convert to gallons by dividing by 231 (the number of cubic inches in 1 gallon).

Example: Find gallons of oil to fill a hydraulic jack (ram) with 6" diameter ram and a 2 foot (24 inch) stroke.

Solution: Area of a 6" ram is 28.27 square inches. Multiply times 24" = 678.48 cubic inches. Divide by 231 = 2.9 gallons. A 4 to 5 gallon reservoir should be used for this ram.

Single-Acting Cylinder. This is a cylinder which has an internal piston and a smaller diameter piston rod. Usually the rod port will be vented to tank to prevent external leakage around the rod seal.

Calculate in the same way as for the ram, using the full piston area.

Double-Acting Cylinder. Delivers power in both directions of movement. A 4-way valve would be used for directional control. This kind of cylinder can use a smaller reservoir because oil returned from the rod end keeps the reservoir level from dropping as rapidly.

Oil required for a double-acting cylinder is equal only to the volume displaced by the piston rod and is calculated as described on the preceding page, using rod square inch area.

Example: Calculate the oil volume displaced from the reservoir when an 8'' cylinder extends 17 inches. Rod diameter is 3½ inches.

Solution: The full piston area is irrelevant. The rod area determines displacement. The area of a 3½'' diameter piston rod is 9.62 square inches. Multiply times stroke: 17'' x 9.62 = 163.54 cubic inches. Divide by 231 = 0.708 gallons. A 1-gallon reservoir should be sufficient.

Adding Reservoir Capacity. Sometimes a pump is not available with sufficient reservoir capacity. An auxiliary reservoir of any size can be added. It must be on the same elevation as the pump reservoir, and must be connected to it with a tube or pipe. For most hand pumps a 1/2'' tube may be sufficient. In most cases it will be necessary to machine a connection in the original reservoir, and this should be as near the bottom of the reservoir as practical.

If a double-acting cylinder is being operated, the tank return oil from the 4-way valve should be returned to the add-on reservoir, to take advantage of the larger oil volume for filtering and air purging.

How to Calculate Number of Strokes. In order to calculate the number of strokes (or cycles) to extend a certain size cylinder through a certain stroke, the cubic inch displacement of the hand pump must be known. If it is not known it can be measured. Pump a number of strokes with oil discharged into a measuring cup. Divide the volume pumped by the number of strokes made.

To find the strokes required to move a cylinder a certain distance, take square inch area of piston, ram, or rod (whichever is applicable), and divide by the pump displacement in cubic inches.

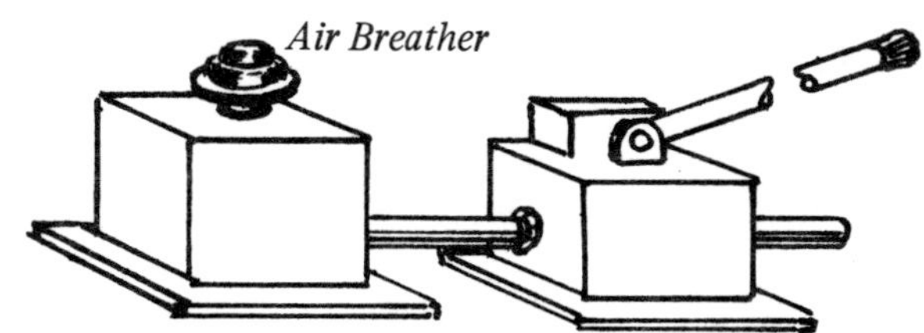

Additional Reservoir Capacity Can be Added to a Hand Pump.

Strokes = Ram Area ÷ Pump Displacement

Example: Calculate the number of strokes required to extend a jacking ram a distance of 3 inches. Ram diameter is 5 inches. Pump displacement is 0.32 cubic inches per stroke.
Solution: Area of a 5'' ram is 19.64 square inches. Cubic inches of oil required = 19.64 x 3 = 58.92. Strokes required = 58.92 ÷ 0.32 = 184.

Example: A single-acting cylinder with 36-inch stroke has a volume capacity of 452.52 cubic inches. How many strokes of a 2.0 cubic inch displacement hand pump are required for the 36-inch stroke?
Solution: 452.52 ÷ 2.0 = 226 strokes, or 226 ÷ 36 = approximately 6 strokes per inch of travel.

Example: A double-acting cylinder has a 3'' bore and a 1'' diameter piston rod. Using a hand pump with 0.3 cubic inch displacement, how many strokes are required for 6'' cylinder extension?
Solution: Area of a 3'' piston is 7.07 square inches. Oil volume required for 6'' stroke = 6 x 7.07 = 42.42 cubic inches. Strokes for a 6'' extension = 42.42 ÷ 0.3 = 142.

HYDRAULIC PUMPS USED AS HYDRAULIC MOTORS

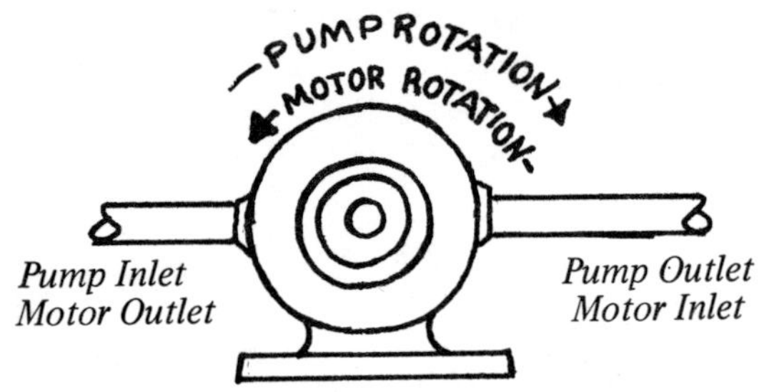

A Pump Becomes a Motor of the Opposite Rotation

If a suitable hydraulic motor is available it should always be used rather than trying to adapt a pump for service as a motor. Some pumps can be adapted in case of emergency but others cannot.

There are so many brands and kinds of pumps that we can only give some general rules. If possible, contact the pump supplier to find what modifications, if any, must be made and what precautions should be observed.

Gear Pumps. Many gear pumps can be used as motors. When so used, the original inlet port should be used as the motor outlet, and the original outlet must be used as the new inlet. Of course this means the rotation as a motor must be opposite to its correct rotation as a pump. Single-rotation pumps have internal passages to drain slippage oil to the lower pressure port which, on a pump is the inlet port, and on a motor is the outlet port.

A pump designed for bi-rotational operation can readily be used as a motor of either rotation or as a reversible motor.

A new pump, because of its tightness, may not start easily as a motor at low pressure. If possible, it should be run in before use.

Caution! A pump designed for single direction rotation should not be used on any application where there is more than 15 PSI back pressure on its outlet port. A high back pressure is often developed on a motor during dynamic braking or while decelerating normally to a stop. This will blow out the shaft seal.

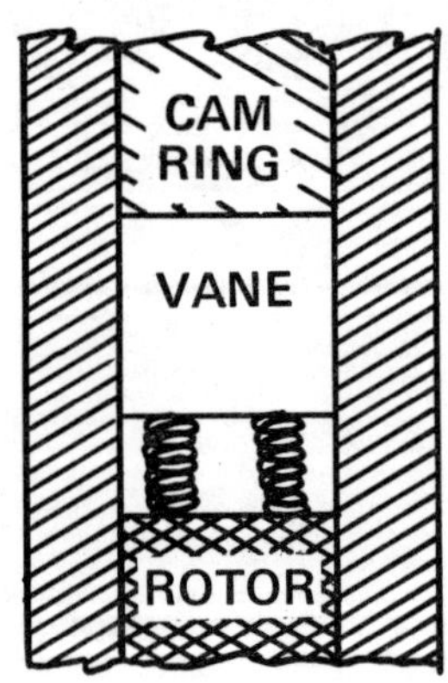

Pumps With Spring-loaded Vanes Will Often Work as Motors.

Vane Pumps. Those vane pumps which have a set of springs under each vane to extend it against the cam ring can ordinarily be used in reverse as hydraulic motors. As with a gear pump, the rotation as a motor is opposite to that as a pump, and the outlet port as a pump becomes the inlet port as a motor.

The same restriction applies as described above for gear pumps, that a vane pump should not be used as a motor in circuits where the back pressure on the outlet port may exceed 15 PSI. This would blow out the shaft seal.

Those vane pumps which use hydraulic pressure from the outlet port under each vane to extend it, or those which simply use centrifugal force to extend the vanes, will not work as motors because they are not self-starting. When stopped, the vanes collapse. Incoming oil can by-pass them without producing starting torque. Usually they cannot be modified in the field. Again, we suggest consulting the pump manufacturer to see if the pump can be used as a motor.

Gerotor Pumps. These are a modification of an internal gear pump. When using them as motors, the same restrictions apply as described above for gear pumps.

__Piston Pumps.__ Those pumps which have a fixed or rotating valve plate can often be used as motors. If they are bi-rotational as pumps they can be used as reversible motors. If they are restricted to single-rotation as pumps, they can only be used as motors in the opposite rotation.

Those piston pumps which pick up inlet oil from the case cannot be used in the reverse rotation as motors. Another exception is a hydrostatic transmission pump which has a built-in charge pump on the main shaft. The main pump may be reversible, but cannot be run in reverse because of the uni-rotational nature of the charge pump which is usually a gear or gerotor type.

Those piston pumps of the check valve type, using an inlet check and an outlet check valve for each piston can never be used as motors; oil cannot flow in reverse through check valves.

__Storage of Hydraulic Pumps.__ An unmounted pump which is to be stored for a considerable time should have its case filled with oil and its ports plugged. Ordinary hydraulic oil can be used, or a rust preventive oil may be better if a kind is selected which is compatible with the rubber seals (shaft seal, etc.) in the pump. Leave a small air space for the oil to expand if the pump should be exposed to heat. The shaft and other external machined surfaces should be coated with grease. If the pump is left in storage for several years, the shaft seal should be replaced before putting it in service.

SAE FRONT FLANGE SIZES FOR HYDRAULIC PUMPS AND MOTORS

This information is only for identification of size. Please see a more complete table on Page 225 of the appendix. Complete dimensional data can be found in SAE Handbook Supplement 39 entitled "Construction and Industrial Machinery" available from the Society of Automotive Engineers.

2-BOLT FLANGE

SAE Size	HP Rating	"Y" Dimension	Hole Diameter
A	10	4-3/16"	7/16"
B	25	5-3/4	9/16
C	50	7-1/8	11/16
D	100	9	13/16
E	200	12-1/2	1-1/16
F	300	13-25/32	1-1/16

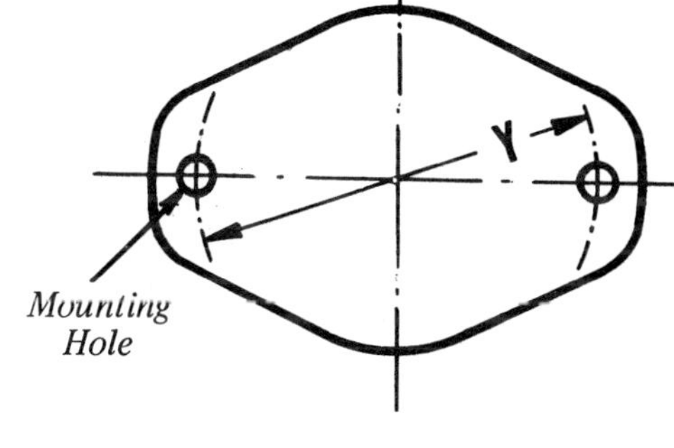

4-BOLT FLANGE

SAE Size	HP Rating	"X" Dimension	Hole Diameter
B	25	5"	9/16"
C	50	6-3/8	9/16
D	100	9	13/16
E	200	12-1/2	13/16
F	300	13-25/32	1-1/16

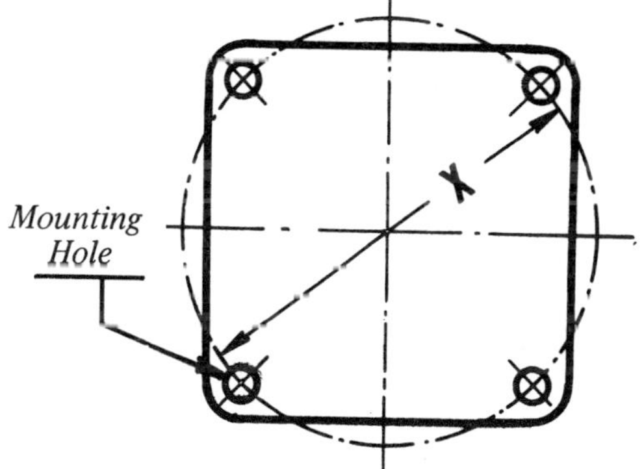

Chapter 4

Hydraulic Oil Reservoirs

The reservoir must be considered as an important part of the total hydraulic system. Unless it is sized adequately, constructed properly, and installed correctly, the operation, the life expectancy, and reliability of the entire system will be adversely affected. Its important functions are to cool the oil, purge it of trapped air, water, and dirt, and to provide a reserve supply of oil to take care of the rise and fall of the oil level as cylinders retract and extend. Also to supply oil for make-up of leakage or spillage.

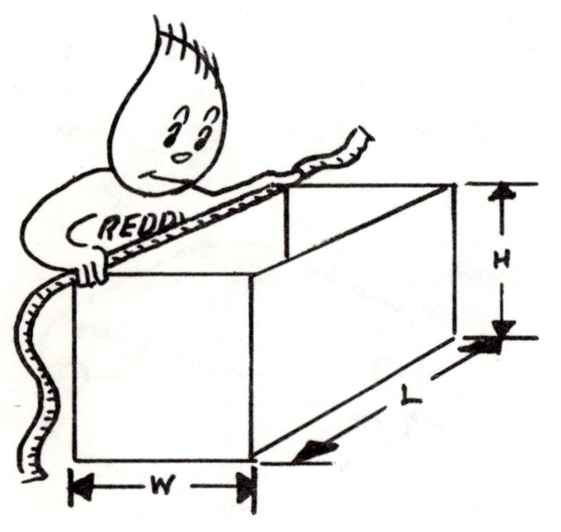

Multiply L x W x H to Calculate Volume.

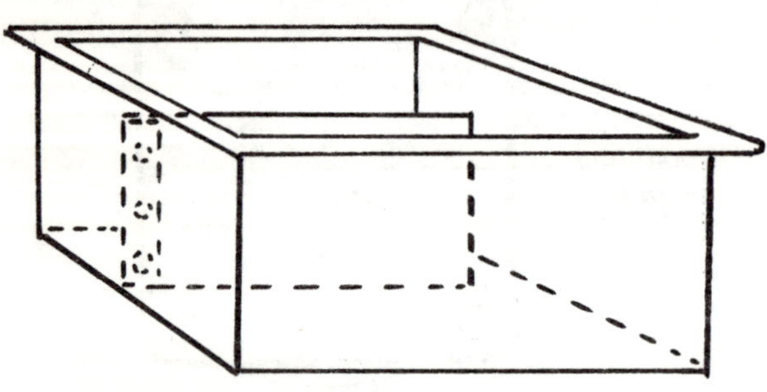

A Vertical Baffle Should be Located About in the Center of the Tank, With Ample Oil Flow Space Around it.

TIPS ON RESERVOIR CONSTRUCTION

Gallon Capacity. The absolute minimum gallon capacity in any hydraulic system is that amount to cover the pump suction strainer to a depth of 2 to 4 inches when all cylinders are in their extended positions. This may be sufficient on systems operated occasionally, with long resting periods during which the pump is not producing pressure.

But for industrial systems a rule-of-thumb is to make the gallon capacity at least twice the gallon rate of circulation per minute, or up to 3 or 4 times if practical. For a pump flow of 20 GPM the recommended reservoir size is at least 40 gallons, etc.

To calculate the capacity of a tank, multiply length x width x height, in inches. This gives capacity in cubic inches. To convert to gallons, divide by 231.

Baffle Placement. A vertical baffle plate should divide the reservoir into two approximately equal compartments. The baffle height should extend slightly above maximum oil level. At one end a generous size notch should be cut to connect the two compartments. The pump should pick up suction oil from one compartment and return flow should be discharged into the other compartment.

The purpose of the baffle is to separate the point of oil discharge from the point of pick-up, to allow the oil to remain longer in the reservoir for more effective cooling, and to allow dirt to settle and air to be purged. The baffle causes hot return oil to circulate in contact with the outside wall surfaces for better heat transfer. No more than one baffle should be used. Two or more baffles will cause the oil to move at a higher velocity through the reservoir and will reduce the ability to settle out dirt and to purge air.

Important! Be sure to allow a large passageway around the end of the baffle so oil transfer from one side to the other will not be restricted. The main passageway should be at one end and should have about 1 square inch opening for every 3 GPM pump oil flow. It may be helpful to have a small opening at the opposite end, or to have a few small holes drilled in the baffle at the opposite end. This will prevent stagnation of oil in the corners of the reservoir opposite to the main baffle opening.

Air-Tightness. A serious effort should be made to have the reservoir air-tight. Gaskets should be used under all covers, lids, and removable plates. Sealing flanges should be used around all pipes entering the reservoir through clearance holes in the cover.

Cover gaskets can be made from almost any gasket material which is compatible with the fluid being used. Strips can be cut from 1/16"

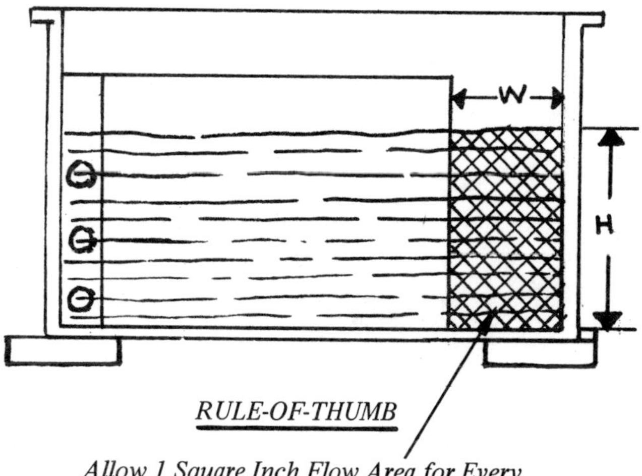

RULE-OF-THUMB

Allow 1 Square Inch Flow Area for Every 3 GPM of Flow.

Use Air-tight Seals on Reservoirs.

Sealing Flange

sheets and cemented to the tank surface with ordinary gasket cement. Hold-down bolts on the lid should be placed close enough to one another to keep the lid flat and assure an air-tight seal. Surfaces to be sealed with a gasket must be reasonably flat. If they are not, it may be necessary to use a thicker gasket, possibly of synthetic rubber. Do not apply gasket cement to the under surface of the lid. The lid may, at some time, have to be removed.

Sealing flanges to fit all standard pipe outside diameters can be purchased from most stores which sell hydraulic components. Rubber seal rings are furnished which fit snugly around the pipe.

Drop Lines. Main oil discharge lines inside the reservoir should terminate well below the minimum oil level, actually very close to the bottom of the reservoir. To reduce return flow velocity, the drop line diameter can be made one or two sizes larger than used in the hydraulic system. Reducing flow velocity reduces the turbulence set up in the tank by return oil.

To maximize oil cooling, the discharge stream can be directed against a side of the reservoir wall or into one of the corners where oil may tend to stagnate. If the construction permits, a 90° elbow at the

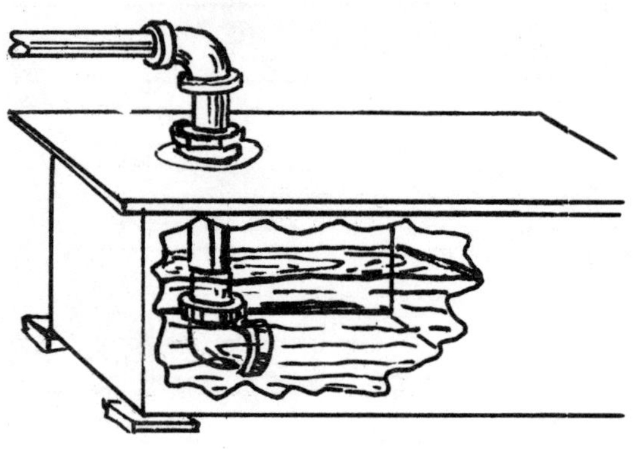

Carry Drop Lines to Bottom of Tank.

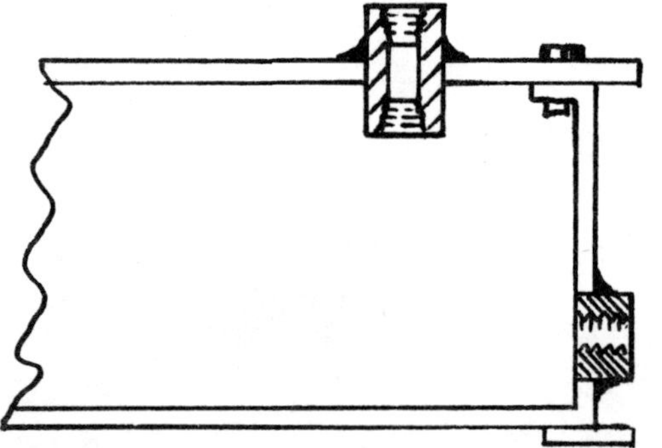

Weld Pipe Couplings Into Tank for Connections.

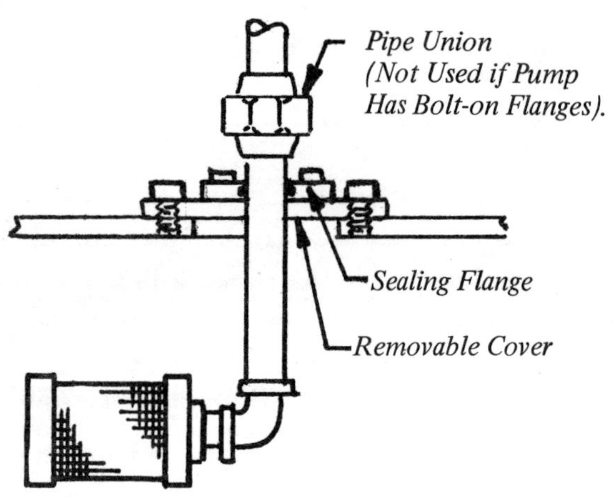

Suction Strainer Must be Removable for Cleaning.

bottom of the drop line will serve. External drains from components such as sequence, counterbalance, reducing, and flow control valves can be combined externally and run via a separate drain line to the reservoir without being combined with the main tank return flow. This is to prevent pressure spikes generated in the main return line from backing up into components and causing false operation. The drop line for drain flow can discharge either on top or underneath the oil level. Discharging on top of the oil level does not cause excessive turbulence because the oil volume and velocity are low, and it does prevent siphoning of oil out of the reservoir if an external drain connection on a component is opened.

Drop Line and Drain Connections. An effective way to carry permanent connections through the lid or a side wall of a reservoir is to use a black iron pipe coupling. Drill a clearance hole, insert the coupling and weld around it. This gives a pipe thread connection both on the inside and outside of the reservoir. If a thread is needed on only one side, a pipe coupling can be sawed in half, and one of the halves can be welded over a hole which is the same diameter as the I. D. of the coupling.

Clean-Out Plates. A means should be provided for getting to the inside of a hydraulic reservoir. On reservoirs of less than 30-gallon capacity, the lid is usually removable and this access may be acceptable. On larger tanks the lid is often welded to the sides or is a part of a wrap-around construction and cannot be removed. A clean-out opening should be provided on one or both ends of the tank, fitted with a gasketed cover plate. Cover plate bolts must be spaced closely to prevent leaks. These clean-out openings should be placed to give access to both sides of the center baffle.

Removal of Pump Suction Strainer. Some means should be provided for easily and quickly removing the suction strainer for cleaning or replacement without opening the tank. Usually the best arrangement is to install it through a clearance hole in the lid. A gasketed sealing flange should be used to make an air-tight seal.

Also, the plumbing should be designed to be easily disconnected. If the pump has pipe thread ports, a pipe union should be included in the pump suction line. If the pump has port pads for bolt-on flanges, these will serve as a union. All joints in the suction line must be air-tight to prevent pump cavitation.

Reservoir Air Breather. For most hydraulic systems we recommend the use of an air breather rated for 10μm filtration. On those reservoirs serving a single-acting cylinder (ram), the reservoir level rises and falls a considerable distance on each cylinder stroke. A large volume of outside air is drawn in each time the cylinder extends. Over a period of time a large volume of outside air is breathed in and out, and if the air is not well filtered, a large volume of dirt particles will enter the reservoir and mix with the oil. Good quality 10μm air breathers are available, or, a 10μm oil filter can be used for air filtering.

If there is debris floating in the air, such as cotton lint in a cotton gin area, the tank breather should be protected by forming a 10 or 20-mesh piece of screen wire around it to prevent debris from quickly smothering the air intake.

Oil Level Gauging. A sight glass (or glasses) should be used rather than a dipstick for measuring oil level. A dipstick will introduce contamination into the oil, and an air-tight seal cannot be maintained around a dipstick. Also, a dipstick may not be replaced after its use.

A composite sight gauge is available in various lengths to show both high and low levels. It is installed and sealed in drilled holes. Some of these gauges also contain a thermometer.

On large tanks a suitable composite gauge may not be available, and it may be necessary to use two small sight glass windows for observing high and low oil levels.

Various kinds of level gauges are available from suppliers who sell hydraulic components.

Filler Caps. Filler caps should be chained to the reservoir to keep them captive. Filler cap assemblies should have a wire mesh strainer for straining out "nuts and bolts" as oil is poured into the reservoir. However, the preferred construction is with an air-tight (non-vented) filler cap and a separate air breather hole protected with a 10μm filter.

Measuring Inlet Vacuum. Proper pump inlet conditions can be checked with a vacuum gauge installed at the pump inlet. A vacuum reading can be recorded when the suction filter is clean. Increased vacuum will develop as the suction filter becomes clogged with dirt or at start-up

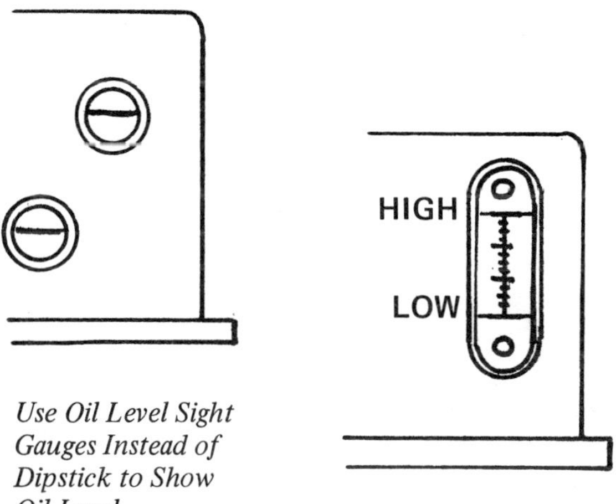

Use Oil Level Sight Gauges Instead of Dipstick to Show Oil Level.

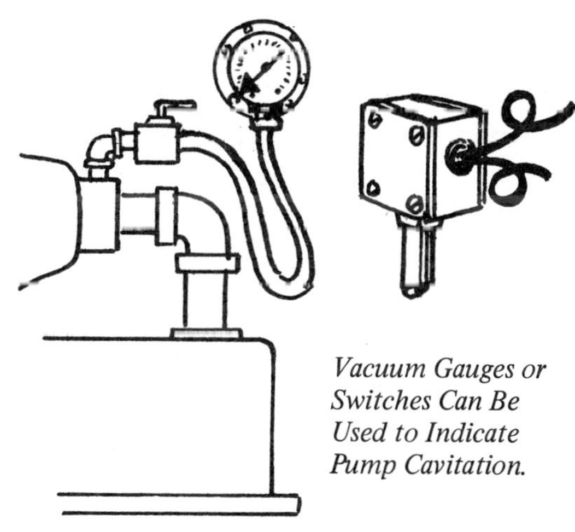

Vacuum Gauges or Switches Can Be Used to Indicate Pump Cavitation.

Vacuum-indicating Suction Strainer.

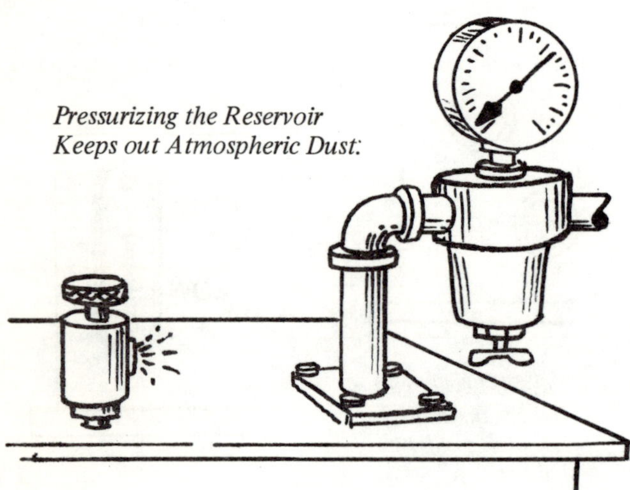

Pressurizing the Reservoir Keeps out Atmospheric Dust:

when the oil is cold. To prevent false readings, the vacuum gauge should be mounted on a separate structure, away from mechanical vibration, and connected to the pump by a length of hose. To preserve the gauge, it can be cut off when readings are not being taken, by a cut-off valve installed in the gauge line.

Instead of a vacuum gauge, a vacuum switch can be used to operate a warning light or sound an alarm when high vacuum develops. Or it can be connected to shut down the electric motor.

Indicating Filters. The use of suction strainers with vacuum indicators or electric switches to actuate when vacuum reaches a certain level are now used on most industrial hydraulic systems. These strainers are mounted in the suction line outside the tank. Electrical switches or mechanical indicators are optional, to be specified when ordering the strainer.

Pressurized Reservoir. In very dirty locations such as mills and foundries, atmospheric dust can be effectively excluded from the system by maintaining a low pressure inside the reservoir. But for this to be practical, the reservoir must be virtually air-tight.

The breather is replaced by a special low range precision pressure regulator operating from the shop air line. It must be accurately adjustable down to 1 PSI. A 1/4" size regulator should be adequate for almost any size reservoir.

To protect the tank against accidental over-pressure, a low pressure non-adjustable air relief valve should be connected to any point on the tank above oil level. It can be teed in with the pressure regulator in the former breather port. It must be set to a very low cracking pressure. An air check valve with cracking pressure of 1 to 3 PSI makes a very good non-adjustable relief valve. It must be connected with its free flow direction toward atmosphere.

Caution! Tanks, especially large ones, can be damaged by internal air pressure. Be very careful to keep the internal pressure very low. It only takes a very low pressure above surrounding atmospheric to keep dust from entering.

Drilling Holes in Reservoir. All holes into the reservoir should be drilled prior to bolting on the cover so drill chips can be cleaned out. Where it is essential to drill a hole into an assembled reservoir, reach in through a clean-out opening and hold a cup or pan to catch chips as the drill breaks through. If this is not possible, and if the tank is air-tight, put a low pressure, 1 or 2 PSI inside the reservoir to blow the fine chips out as the drill comes through. Large chips will be prevented from getting into the system by the pump suction strainer.

Cleanliness in Assembly. Weld splatter and slag should be removed from the inside surfaces of the reservoir by chipping. Sandblasting is effective but some users will not permit sandblasting inside the tank because of the danger of sand and dust being left in the tank. Great care should be used in thoroughly cleaning a tank which has been sandblasted. Pipe, tubing, and hose used for plumbing should be left with protective end plugs in place until actual assembly. Each piece of pipe or tubing should be de-burred after cutting, then blown out with an air hose. Some users insist that each piece of tubing or pipe be washed out with solvent after forming and before assembly. Sophisticated systems built to exacting specifications may have to be flushed after assembly before adding the permanent hydraulic fluid.

Painting. Hydraulic reservoirs constructed of steel sheet or plate should be painted on the inside. Rust will form on unpainted steel surfaces from water which condenses in the oil due to atmospheric temperature changes. Rust particles will flake off in the oil and will cause damage to the system if they are small enough to pass through the pump suction strainer.

The inside of the reservoir should be coated with a paint which is compatible with the fluid used. For petroleum oil, use a rust resistant primer. For additional protection use at least one coat of oil base enamel. Check with your paint supplier to be sure the primer and enamel are compatible with petroleum fluid.

Be very careful of paints to be used with synthetic fire resistant fluids. Ordinary paint is not compatible. For all fire resistant fluids, get paint recommendation on proper paint from the supplier of the fluid.

The outside of the reservoir is often not painted. Paint slightly reduces heat transfer to atmosphere. If reservoir or components are painted, get paint recommendations as stated above.

When changing from hydraulic oil to a synthetic fluid like phosphate ester, it is very important that all traces of the old paint be stripped, particularly from the inside of the reservoir, but from all outside surfaces of the reservoir and components. Phosphate ester is positively not compatible with ordinary paint. These surfaces must be re-painted with a compatible paint.

Magnets in the Reservoir. Magnets installed in the reservoir, often as part of the suction strainer assembly, are very effective in picking up iron and steel particles. Magnets can be installed at any place in the reservoir below the oil level as long as they can be easily removed for cleaning through an access plate, breather opening or filler port without removing the tank lid.

Electric Motor Mounting. Usually the top of the reservoir is constructed of hot rolled steel plate no more than 1/8" thick. This material is not sufficiently sturdy for direct mounting of electric motor and pump. A torque or twist is created between the motor and pump under load and this will flex a thin mounting surface and cause the two units to twist out of alignment. Motor and pump must be mounted on a plate at least 3/8" thick, thicker when necessary for larger motors. After mounting and aligning the two units, the plate is welded to the reservoir top.

To facilitate alignment on the plate, the motor can first be bolted *and* pinned to a pair of cleats so

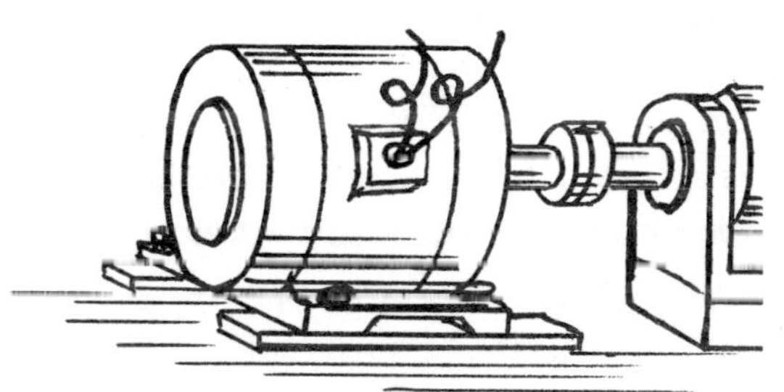

Mount the Motor on Cleats to Facilitate Alignment, Then Weld Cleats to Tank.

it can be moved around during alignment. Thin shims may also be required for perfect alignment. When alignment is completed, the cleats can be tack welded to the mounting plate, and the mounting plate can be welded to the tank top.

Catalog shaft heights on electric motors may vary from nominal dimension to –1/32" on larger frames, and it may be necessary to shim equally under each of the four legs.

TIPS ON INSTALLATION OF RESERVOIRS

Direct Sunlight on the Reservoir Can Cause Overheating in the Hydraulic Oil.

Cool Location. Install the reservoir (or the pumping unit) where it will be shielded against radiated heat from furnaces, steam pipes, and from direct sunlight, but where there will be free air circulation around it. Do not install it under benches, tables, or in cabinets where vertical convection air currents would be cut off. If necessary to install it in a cabinet, a forced air circulation cooling system should be provided. If the tank does not have legs about 6" high, mount it on blocks to allow air circulation under it.

Proper attention to the above points will make a big difference in whether the system overheats or whether a heat exchanger must be added.

Protect it also against adverse weather such as rain, snow, flooding, extreme cold, etc.

Drilling Holes. When installing a hydraulic reservoir, avoid drilling holes which would break through into the oil compartment. Mounting brackets can be welded to the side of the tank but only before it is filled with oil. Avoid any metal cutting operation which would cause metal dust to fall inside the tank.

Drain Valve. Every hydraulic reservoir should have been constructed with one or two drain connections. A globe valve should be installed in each of these drain ports. These drain valves often prove invaluable at some future time, either for draining the tank, for sampling the fluid for the presence of water, or for connection to an external filtering system.

Add-On Reservoir. Sometimes it may be necessary to increase the reservoir size. This may be necessary in order to increase the square foot tank surface for heat radiation, to increase the length of time return oil will stay in the reservoir for dirt and air purging before being picked up again, to reduce turbulence and emulsification of the oil, or because the reservoir is too small for the size cylinder to be operated. One solution to these problems is to add a second reservoir of any size to increase total reservoir capacity. When adding a reservoir, keep these points in mind:

(1). The add-on reservoir may be of a different gallonage capacity but should be approximately the same depth as the original reservoir, or in any event the high level marks of both reservoirs must be at the same elevation.

(2). The two reservoirs must be joined together at the lowest points in the tanks. The joining pipe must be very large — large enough to keep flow velocity to 1 foot per second. A rule-of-thumb is to provide one square inch of flow area for each 3 GPM of oil flow. Or, at least one pipe size larger than

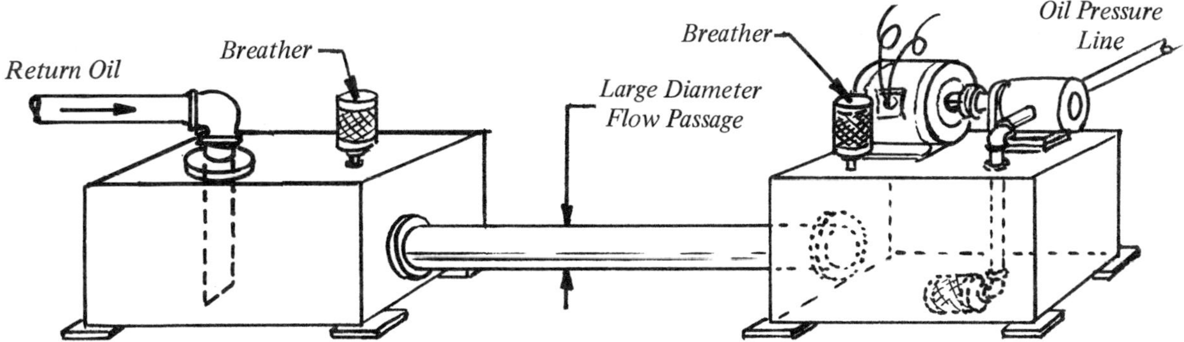

Reservoir Capacity Can be Increased With an Add-on Tank.

the pump inlet port. Often a drain port on the original tank is not large enough to use as a flow-through connection, and a new and larger opening must be made.

(3). The plumbing on the machine must be modified to discharge the main tank return flow into the add-on tank, with the pump pick-up point still in the original tank.

(4). Each tank must be independently vented to atmosphere through its own breather.

Reducing Reservoir Turbulence. First, the reservoir must be large enough to allow the return oil to remain in the reservoir for a few minutes before being circulated again. The rule-of-thumb for reservoir size on industrial system is a gallonage capacity 3 times the GPM of the pump. Mobile systems usually violate this rule because it is impractical from the standpoints of size and weight, to carry such a large tank.

Then, the return oil must be discharged near the bottom of the tank. An elbow at the bottom of the discharge line can be pointed to direct the flow parallel to one of the tank sides.

The tank return line can be enlarged to one or two pipe sizes larger for a distance of a few feet before the oil enters the reservoir. The drop line into the tank should also be of this enlarged size. This will reduce velocity and turbulence when the oil enters the tank.

Welding to Reservoir. Welding on the reservoir of an operating system is very bad practice and should only be done for emergency repair. Welding close to gasketed connections may ruin the gaskets. Welding on the outside of a tank filled with oil will burn the oil in contact with the metal, producing contamination in the oil. Some small reservoirs may have soft soldered or silver soldered seams which will come apart from welding heat. If reservoir is painted inside, the paint around the weld spot will be destroyed and a small area of metal surface will be exposed for rusting.

Captive Filler Cap. All filler caps should be chained to the tank. Caps which are not chained may be dropped and roll under the tank. Too often a shop man will plug the filler opening with a shop towel instead of retrieving the lost cap. A chained filler cap is a J.I.C. standards requirement.

Elevated Reservoir. It is always beneficial to have the reservoir mounted at a higher elevation than the pump inlet. This reduces the inlet vacuum at the pump. However, this will not eliminate pump cavitation caused by too small or too long a line, or too many bends.

Obviously, when using an overhead reservoir, a shut-off valve must be installed in the feed line to the pump, so it can be closed when the pump is to be repaired or replaced. A full-flow rotary ball-type plug valve can be used, one which is leaktight under vacuum conditions.

It is very important to be able to lock this valve in the open position to prevent someone from closing it while the pump is running. This would severely cavitate and destroy the pump in a very short time. Some plug valves are designed for padlocking; some others can be used by welding a pair of locking ears to the valve body and handle.

TIPS ON MAINTENANCE OF HYDRAULIC RESERVOIRS

Cleanliness in the Area Helps Keep Dirt Out of the Hydraulic System.

Cleanliness. Many years ago when most oil hydraulic systems were run with gear pumps, cleanliness was of less importance. The gear pumps were built with loose tolerances to operate at low pressures. They were capable of satisfactory performance over a reasonable life span because they were relatively insensitive to small particles of dirt in the oil.

Cleanliness has become vitally important in today's hydraulic systems because pumps and other components are designed with close clearances to operate in the 2000 to 5000 PSI range instead of at 250 PSI, and they have become much more sensitive to dirt.

Today's higher pressures enable more HP to be produced and transmitted with the same size components. At a given flow rate, HP is proportional to pressure. With the same size components we can today handle 10 to 15 times the HP we could many years ago with low pressure components.

Good housekeeping around the hydraulic reservoir should include but should not be limited to the following: keep surfaces, especially the top of the reservoir cleaned of oil and dirt. This will increase heat radiation and will reduce the chances of getting dirt into the filler opening. Keep oil level sight windows clean so oil level can be easily seen. Keep the floor around the reservoir cleaned of oil to prevent accidents caused by people slipping on the oil. Use oil absorbing granules, then clean them up after the oil has been absorbed. Check for leaks around fittings and make repairs when necessary.

It Is Difficult to Over-emphasize the Importance of Cleanliness in Handling the Hydraulic Oil.

Be sure the oil filler cap is replaced immediately after filling reservoir.

The oil must be kept meticulously clean when using certain piston pumps which have piston shoes riding against a cam surface. Pumps and motors of this kind are used on many high pressure applications, especially hydrostatic transmissions. They are especially sensitive to dirt and would not last long on the dirty oil that can be tolerated by mobile-type gear pumps operating at low pressure. However, cleanliness does increase the life of any pump, even those mobile-type gear pumps, and great care should be taken by using micronic filtration and by regular maintenance to keep the oil as clean as possible.

Filler Openings. Standard filler cap assemblies have an air filter built into the cap, and have a wire mesh screen to strain oil poured into the tank. They also have the cap chained to the assembly. These assemblies are good for most applications, but for hydraulic systems operating in locations where there are particles in the air, such as in steel mills, feed mills, cotton gins, and foundries, a non-vented filler cap will give better protection. The reservoir can either be operated with a slight internal air pressure, 1 PSI, as described earlier in this chapter, and/or a separate breather opening can be provided, protected by a $10\mu m$ filter. A separate breather opening has a further advantage: it provides an escape for trapped air in the reservoir while oil is being poured into the regular filler opening.

Air Breather. In all systems using a cylinder with rod coming out one end only (single-end-rod cylinder), the reservoir oil level falls and rises as cylinder(s) extend and retract. On ram-type cylinders (with no piston rod) and on 2:1 ratio (large rod) cylinders there may be quite a large rise and fall on each cycle. During a day there will be a very large volume of air passing in and out of the air space in the reservoir. Cheap air breathers, those with a coarse filter element of "steel wool" or wire mesh are not good enough. The air breather filter rating should be $10\mu m$ or finer.

Good Quality 10-micron Breathers Should be Used.

The rate of rise and fall of the oil level will depend on the rod displacement of the cylinder and on the cycle rate. A breather port size of 1/2″ NPT is sufficiently large to handle most systems except those pumping over 100 GPM, in which case a 3/4″ NPT opening size may be preferred.

Except in locations where the atmosphere is extremely dirty, the air breather need not be replaced often, perhaps once a year depending on atmospheric conditions. In exceptionally dirty environments it may need to be replaced more frequently. In dirty locations, an extension pipe can be connected to the breather port and run outside the immediate area to a point where the air is cleaner. Use an extension pipe diameter about the same size as the pump pressure port.

Suction Strainer. A rule-of-thumb is to clean the suction strainer about every 1000 hours of pump operation. On an 8-hour day, 5-day week of continuous running, this would be about once a month. In clean locations this period could be extended, but in exceptionally dirty locations more frequent cleanings may be required.

Remove suction strainers from the reservoir for cleaning. Use an air hose, blowing from the inside out. If there is a coating of brown varnish, the strainer will have to be cleaned with a solvent or replaced. The varnish means that the reservoir oil has been operated at too high a temperature.

Handling and Storage of Oil. To reduce contamination in the hydraulic system, these simple rules should be observed:

(1). Keep new oil stored in original containers and tightly sealed. Never store oil in an open container where it can collect dust.

*Keep Oil Tightly Sealed and Protected
from the Weather.*

(2). Protect storage vessels against exposure to rain, snow, standing water, excessive heat, etc.

(3). When transferring oil to a hydraulic power unit be sure all containers and hose in contact with the oil are clean and dry. A funnel used for pouring should have its spout cleaned.

(4). Even though oil is taken from a factory-sealed container, there is always the possibility of solid contamination being present. Pour oil through a strainer if there is not one in the throat of the filler assembly.

(5). Do not (except in an emergency) put used oil into a hydraulic power unit. Discard old oil unless there is a sufficient volume to justify the expense of filtering or centrifuging it. But remember that these proceses do not remove chemical contaminants dissolved in the oil.

(6). Do not try to salvage oil spilled on the floor unless it can be micronically filtered or centrifuged.

(7). Different brands of oil should not be mixed. While all petroleum base hydraulic oil is compatible, occasionally the additives put in the oils may not be compatible.

Testing Hydraulic Oil. The chemical condition of the oil can only be determined by laboratory tests. Your filter supplier or your oil supplier can have these tests performed by a laboratory. Contact them for directions and equipment necessary for drawing an oil sample to be tested. Samples must be taken in a clean glass bottle under specified conditions that will not permit external contamination to enter the sample to give erroneous test results. The user can make a few simple tests to judge whether the oil should be laboratory tested or changed:

(1). Feel the oil between thumb and forefinger. If it is badly contaminated with solids, the grit can be felt. Or, pour a few drops on a white painted surface and inspect with a powerful magnifying glass or a microscope.

Grit, in itself, is not a serious contaminant because it can easily be filtered out. But the entry point of grit should be determined and blocked. Check pump suction strainer; clean or replace. Replace elements in micronic filters. Add a micronic breather to the reservoir. Refer to Volume 1 "Industrial Fluid Power" on how to add an external filtering system to an existing machine.

*Simple Tests
May be Performed on Hydraulic Oil.*

(2). If the oil has a milky appearance this may indicate excessive turbulence, either in the reservoir from improper design, from one too small for the system, or from undersize plumbing somewhere in the system. Cloudy oil may also indicate emulsification with air or water. Draw an oil sample after the system has been running several hours. Draw the sample in a clear glass container and let it stand a few hours to see if dirt or water settles. If no water settles to the bottom, the problem may be an air leak into the system. Refer to Page 77 for possible entry points of air. If water settles, it may come from condensation in the reservoir due to ambient temperature changes. Purge the water daily with the reservoir drain valve. Or, the water may be entering from a leak in a water cooled heat exchanger.

Compare the Oil Color With That of a New Sample.

Let the sample stand a few days to see if there is any further precipitation of dirt or water. A milky appearance may also be due to the use of the wrong oil.

(3). Draw an oil sample in a clear glass container and compare its color with that of a new sample of the same oil. A darker color indicates that oxidation (chemical combination with oxygen) has taken place. Used oil is always darker than the same new oil, and a slight darkening is of no importance. But if the old oil is radically darker and the darkening is not due to entrained dirt, you may want to have the oil analyzed to see if it is unfit for further use. Smell the two samples. If the old oil is chemically contaminated it may have a strange odor.

Oxidation of the oil is caused primarily by two conditions which should be corrected if possible: (a). Operation of the oil at too high a temperature. Measure oil temperature in the reservoir after the system has been in full operation a few hours. The preferred temperature is 130° F or less for an industrial system. If temperature is higher than 150 or 160° F, corrective measures should be taken such as adding a heat exchanger. On mobile systems there may be no practical way to keep the oil temperature within the above limits, and frequent oil changes may be necessary. (b). Excessive turbulence in the oil. This produces a greater exposure area between oil and air and increases the rate of oxidation.

Chemically, there are many reactions that can take place under various conditions of temperature, pressure, and the presence of water and other contaminants. The simple tests described above may give an indication of whether the oil should be changed, whether corrective measures should be instituted, or whether laboratory tests should be made.

TIPS ON REDUCING OIL TEMPERATURE

Refer to Chapter 6 for installation and maintenance of heat exchangers. Information in this section is limited to cooling methods other than heat exchangers.

Protect Against Heat Absorption. Direct sunlight can add a substantial amount of heat to a hydraulic system. Shield hydraulic reservoirs and other components from the sun with a sun shade which will not interfere with vertical convection air currents nor with free air circulation around and under the reservoir.

Long Pipe Runs May be Buried to Shield Them from Sun Radiation.

Hydraulic systems installed near heat sources such as blast furnaces, ovens, steam engines, etc., should be protected with a baffle to intercept heat radiated from those sources.

Long runs of outdoor piping carrying hydraulic oil should either be shielded from the sun or should be buried under the ground.

Painting the Reservoir Black Helps it to Radiate Heat.

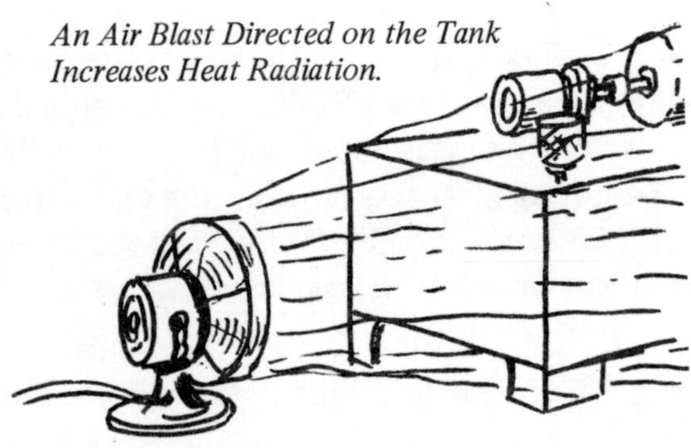

An Air Blast Directed on the Tank Increases Heat Radiation.

Cooling Fins Increase the Effective Radiating Surface of the Tank.

Painting the Reservoir. According to the laws of physics, black bodies absorb and radiate heat better than white ones, but white bodies reflect heat better than black ones.

To take advantage of these facts, painting a hydraulic reservoir black will help cool the oil if the surrounding air temperature is lower than tank temperature, but only if the reservoir is shielded from any source of heat radiation such as sun, ovens, etc. If the reservoir must be installed in direct sunlight, painting it white will reflect sunlight and reduce the amount of heat pick-up.

Black paint used must be very thin. A heavy coating of paint will tend to insulate the reservoir and it may lose some of its heat transfer capability, oil to air.

Provide Ample Air Circulation. Reservoirs serve to radiate heat from the oil, so they should not be installed in closed compartments such as consoles or cabinets or under shelving. They should be installed in the open, so rising convection air currents can carry away the heat. They can be installed in cabinets if there is a blower-type circulating air system to carry the heat to the outside. In cabinets, ample air space should be provided underneath the tank as well as around all sides.

Sometimes, on cabinets installed in the open, a little additional cooling may be required. An industrial electric fan can be installed to direct the air blast along the sides and underneath the tank.

Adding Cooling Fins. To increase heat radiation from the reservoir, ribs or fins can be welded vertically to the sides of the reservoir. This gives more heat radiating surface. They should be spaced apart about two times their width. They can be made from 1/4" steel strips and should be attached to the tank with a good heavy continuous weld seal on both sides for good heat conduction. Ribs should be open at top and bottom to encourage vertical convection air currents which will add further to heat transfer.

Water Coils Inside the Reservoir. For additional cooling, cold water can be circulated through copper tubing immersed in the reservoir below oil level. Heat transfer is increased if finned copper tubing is used.

Caution: Tubing should enter and leave below minimum oil level. This means water connections must be welded to the reservoir at the time it is constructed. The tubing must never extend above the oil at any time, and the oil level must never be permitted to drop below the minimum level. If any of the cold tubing becomes exposed to air, water will condense on it, enter the oil, and collect in a layer under the oil. If allowed to build up it will be drawn into the pump.

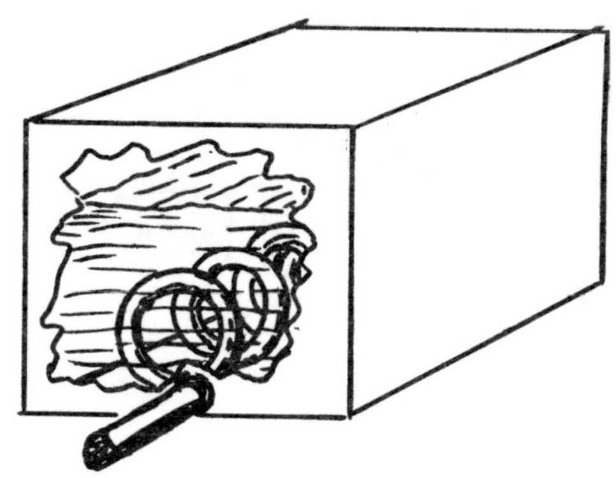

Connections to Cooling Coils Must Always be Below the Oil Level.

KEEPING THE OIL WARM IN COLD WEATHER

Cold Start-Ups. If the hydraulic system is exposed to extremely low overnight temperatures, the oil may become so thick that when the pump is started in the morning it may be damaged by cavitation. In a cavitated condition pump wear is accelerated and the machine cannot produce full tonnage and speed until the oil has had time to warm up and thin to its normal viscosity. If possible, the oil should be prevented from becoming too cold overnight.

On mobile equipment left outdoors it may be impossible to keep the oil warm overnight. To prevent damage to the pump, the engine and pump should be started and the pump should be run at engine idle speed and with little or no pressure

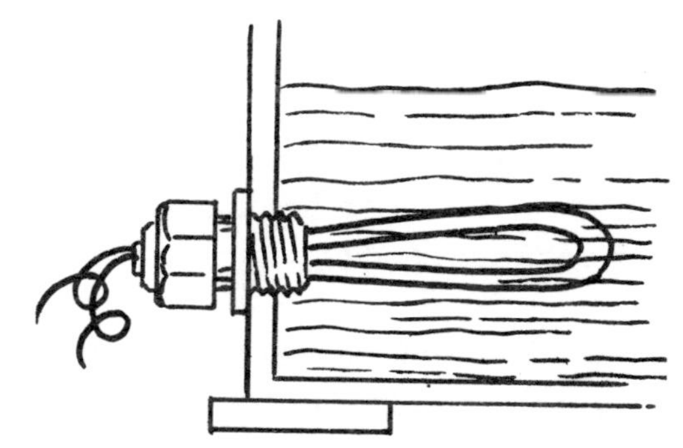

Heating Elements Should Be Limited to 10 Watts per Square Inch on Non-circulating Applications.

load on it for an hour or so depending on time required to bring the oil temperature high enough so the pump will not cavitate. Great damage will be done to the pump if it is run across the system relief valve to warm the oil. This will indeed warm the oil but it will also ruin the pump in a short time.

For mobile equipment the oil should be chosen with care. Its V. I. (viscosity index) should be as high as available. An oil with a higher V. I. will not get as thick when cold as an oil with a lower V. I. Its SSU viscosity at normal operating temperature should be no higher than required by the pump. Please refer to Volume 1 "Industrial Fluid Power" for explanation of viscosity and viscosity index.

Industrial hydraulic equipment is usually located in an area which does not get as cold at night, and the pump can usually handle the oil immediately on start-up. However, the oil can be kept warm overnight by installing one or more immersion electric heater elements below minimum oil level. Many of these elements have a mounting thread size of 2" NPT. They can be installed and removed from outside the tank through a 2" NPT half coupling welded into the side of the tank. If a high wattage element is used, it is a good idea to have a thermostat in the tank to cut off the heater if oil temperature should become too high.

Oil May be Circulated With a Small Immersion Pump.

Caution! Heating elements designed for oil (rather than water) should be used. They have a surface power of 10 watts per square inch. Water heating elements have a surface power of about 40 watts per square inch. This is too high for oil. Since oil is a poor conductor of heat, the oil in direct contact with the element would be burnt (carbonized) unless it is kept circulating with an auxiliary impeller pump installed below the oil level.

Water heating elements can be used without a circulating pump if they are operated at one-half their rated voltage, that is, a 440-volt element operated on 220 volts. This cuts the wattage to one-fourth nameplate rating. An element rated for 5 kW on 440 volts would produce 1.25 kW when operated on 220 volts. (When the voltage is cut in half, the current is also cut in half. This reduces the wattage to one-fourth).

A small impeller-type immersion pump can be used to circulate the oil if a water heating element must be used at full wattage. The output stream should be directed against the heater element. The electric circuit should be arranged so the circulating pump is turned off automatically when the main pump is started. Wiring to the circulating pump should come out the top of the reservoir through a small bushed hole. Plumbers putty can be used to seal air-tight around the wires.

Wattage for Maintaining Oil Temperature. The chart on the next page shows the approximate wattage required on immersion heating elements to maintain reservoir oil temperature at desired temperatures above the surrounding atmosphere. This is not the power needed to heat the oil; it is the power to just make up for radiation losses from the sides of the steel reservoir.

To use the chart, the reservoir will have to be measured to find the total number of square feet of radiating surface. Measure area on all four sides plus top, and plus bottom if reservoir is elevated above floor level. This may give a little more area than is effective for the reservoir itself, but the cylinders and other components will add a small additional radiation area. So, the bottom of the reservoir is included because, although it may be a less effective radiator than the tank sides, the radiating surface of components should make up for it.

Having measured and calculated the total surface area in square feet, use this area to enter the chart in the left column.

Example of use of Chart: If the reservoir has 70 square feet of surface, and if the oil is to be kept at 90° F when the room temperature drops to 60° F, find the wattage required.

The temperature difference to be maintained between oil and surrounding air is 90 – 60 = 30° F. Use the 30° column in the chart and follow down to the 70 square feet line. This shows 1575 watts of electrical power will be required to make up for heat lost through radiation from tank walls. To maintain a 100° difference in temperature on this same tank would require 5250 watts.

Values in the chart are approximate because the cleanliness and condition of the steel tank surfaces, the amount of air circulation around the tank, and the proximity of nearby heat or cold objects will influence the radiation. The chart has been calculated from this formula:

HP heat radiation = 0.001 x A (tank surface area, sq. ft.) x TD (temperature difference)

or, W (watts heat radiation) = 0.001 x A x TD x 746

ELECTRICAL POWER TO MAINTAIN ELEVATED OIL TEMPERATURE

Figures in the chart are watts required on heating elements

Sq.Feet Tank Surface	Air/Oil Temperature Difference, Degrees F									
	10°	20°	30°	40°	50°	60°	70°	80°	90°	100°
10	75	150	225	300	375	450	525	600	675	746
20	150	300	450	600	746	900	1050	1200	1350	1500
30	225	450	675	900	1120	1350	1575	1800	2025	2250
40	300	600	900	1200	1500	1800	2100	2400	2700	3000
50	375	746	1120	1500	1875	2250	2625	3000	3375	3750
60	450	900	1350	1800	2250	2700	3150	3600	4050	4500
70	525	1050	1575	2100	2625	3150	3675	4200	4725	5250
80	600	1200	1800	2400	3000	3600	4200	4800	5400	6000
90	675	1350	2025	2700	3375	4050	4725	5400	6075	6750
100	746	1500	2240	3000	3750	4500	5250	6000	6750	7500
125	935	1865	2800	3750	4675	5600	6550	7475	8400	9350
150	1125	2235	3360	4500	5600	6725	7850	8950	10,100	11,250

Some operators prefer to allow the oil to cool down at night or over week ends, then start the electric heaters several hours ahead of machine start-up. Thermostatic switches can be installed to prevent the tank from cooling below a certain minimum temperature and to prevent it from heating above a certain temperature.

Wattage for Heating Hydraulic Oil. To heat the oil preparatory to starting the pump, sufficient heat must be provided (1), to add heat energy to the oil to raise its temperature, and (2), sufficient additional heat to make up for radiation losses while the oil is being heated.

(1). The energy, in watts, can be found with the rule-of-thumb in the box. This is a close approximation, close enough for practical purposes. The formula has been mathematically derived for oil having a specific heat of 0.4.

(2). While the oil is being heated to raise its temperature, energy is being lost by radiation and will have to be supplied by the heating element. The maximum heat radiation (and the maximum wattage needed) will reach a maximum just as the oil reaches its elevated temperature. Find the wattage from

RULE-OF-THUMB

Wattage for Heating Hydraulic Oil

0.88 watts will heat 1 gallon oil 1°F in 1 hour

This rule has been mathematically derived for a petroleum oil of 0.4 specific heat and 0.9 specific gravity, using equivalents: 1 BTU/hr = 0.2931 watts; 1 gallon oil weighs 7½ lbs.

the chart on the preceding page, using as the oil temperature difference the elevated temperature minus the room temperature.

Example of Wattage Calculation. Assume a 100-gallon volume of oil in a tank which has 70 square feet of external surface area. Also assume the room temperature to be 50° F as the starting temperature, and the oil is to be heated to 90° F in a time of 3 hours.

First, use the rule-of-thumb to calculate wattage required to raise the oil from 50° F to 90° F: *100 gallons x 40° ÷ 3 hours = 1333 watts.*

Next, use chart on preceding page to estimate additional wattage to make up for heat lost due to radiation during the heating period: Use the 40° column (90 –50) and the 70 sq. ft. line. A power of 2100 watts will be required to supply radiation losses as the oil reaches a temperature of 90° F with the room temperature still at 50° F.

Therefore, the heating element will have to be capable of supplying 1333 + 2100 = 3433 watts or 3.433 kW (1 kW = 1000 watts).

Chapter 5

Hydraulic Accumulators

An accumulator is a storage vessel into which oil can be pumped under pressure and stored for later use. It is used on some hydraulic systems either to add to the pump flow for a short period, or to help dampen pressure shock which is produced by the pump, by a shifting valve, or by the release of a press. It is also used to supply oil to make up for leakage losses during the holding period of a press.

The material in this section is primarily for the installation and servicing of accumulators. Please refer to a more complete description of accumulator action, size selection, and circuits in Volume 1 "Industrial Fluid Power".

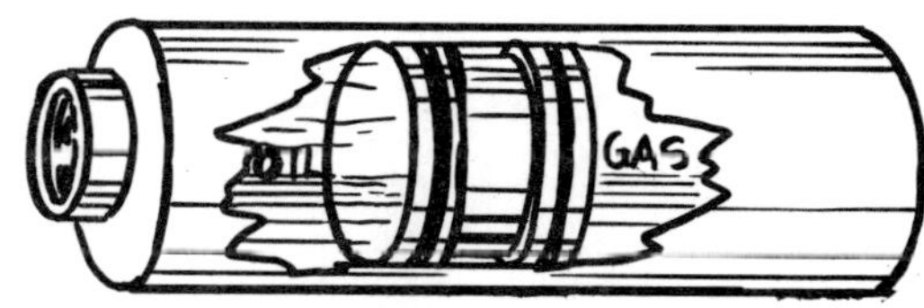

Piston-Type Accumulator

On applications where the accumulator is supplementing the pump flow, the system pump stores oil in it during periods when it normally would be unloaded or idling. The stored oil is then available at a later time in the cycle, and by adding to the pump flow, a cylinder can be moved faster than by pump oil alone.

On applications where an accumulator is used to reduce hydraulic shock, it is floated across the line and acts as a hydraulic cushion or spring. Details for use as a shock absorber are given later in this section.

Before oil is pumped into an accumulator it must have been previously pre-charged with dry nitrogen up to a pre-charge pressure level. Details of pre-charging are given later in this chapter.

Most hydraulic circuits which include accumulators are designed to work at 3000 PSI pressure to utilize the maximum capacity of 3000 PSI models. The nitrogen pre-charge pressure is not critical for most applications, and a level of 1500 PSI works out about right. When this pressure drops to 1000 PSI it is time to re-charge it back to 1500 PSI.

Energy is stored in an accumulator in the nitrogen gas. Oil pumped into the oil port compresses the nitrogen. Starting with a pre-charge of 1500 PSI with oil discharged, the gas (and oil) pressure increases to 3000 PSI as oil is pumped in. When fully charged to 3000 PSI the accumulator will be half full of oil and half full of compressed nitrogen. As oil is discharged from the oil port, its pressure decreases to the pre-charge (1500 PSI) level when discharge is complete.

Accumulators are not for every system. They add cost and they do add compressibility, so they should not be used except where they offer an important advantage over conventional circuits.

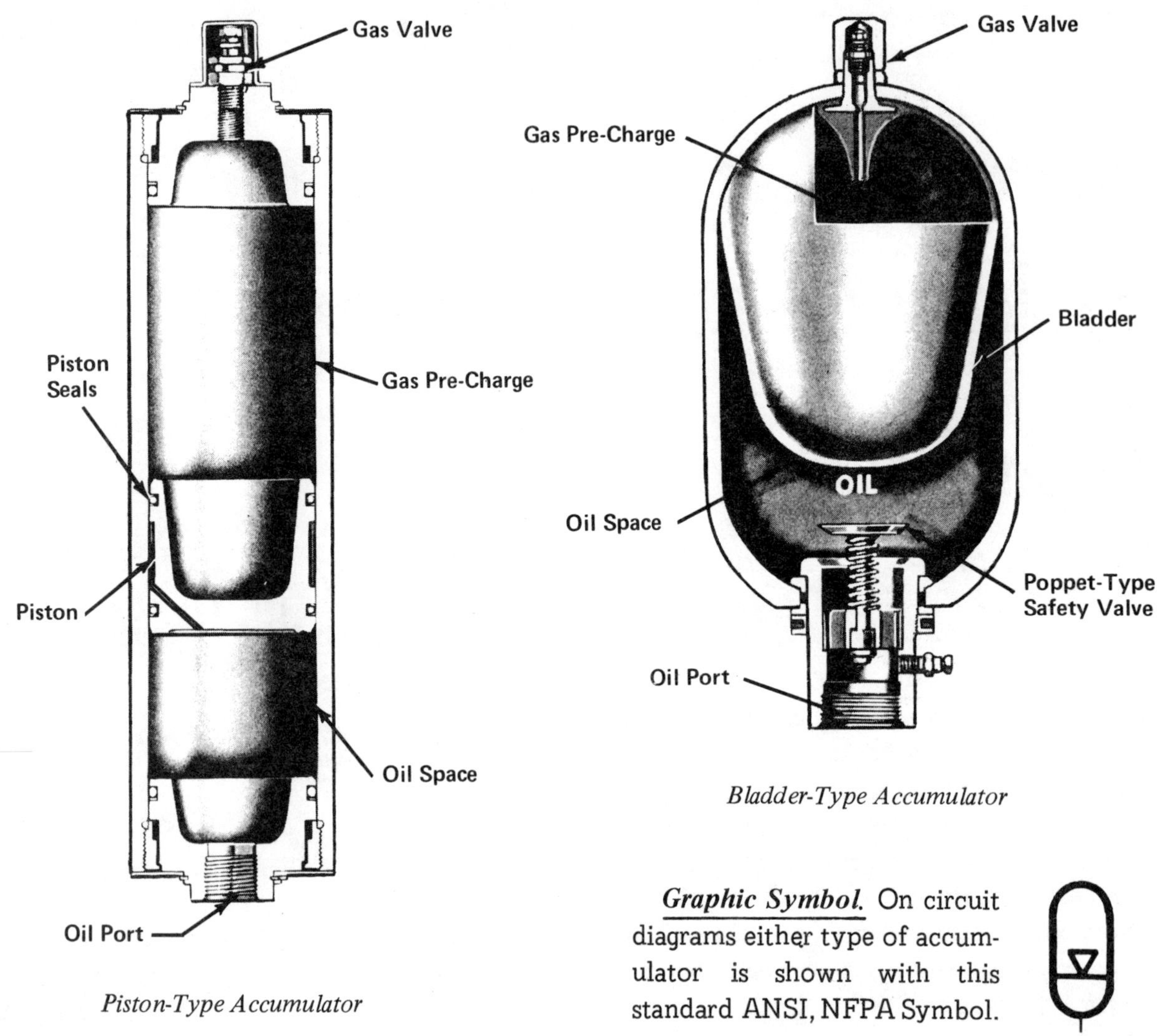

Piston-Type Accumulator

Bladder-Type Accumulator

Graphic Symbol. On circuit diagrams either type of accumulator is shown with this standard ANSI, NFPA Symbol.

Types of Accumulators. Two types of accumulators are currently popular for high pressure hydraulic applications. These are the piston type and the bag or bladder type. Each has advantages for various kinds of applications. Standard models of both types are designed to work at a maximum pressure of 3000 PSI, and this is considered the optimum design pressure for an accumulator system. A few sizes may be available for up to 5000 PSI.

Accumulator sizes start at a fraction of a gallon and go up to 10 gallons. Piston accumulators may be available up to 20 gallons. Capacity rating of an accumulator is based on the internal volume of the gas side when all oil is discharged.

All accumulators must have an internal barrier between the oil and the nitrogen pre-charge. At high pressure, without a barrier, the nitrogen would quickly dissolve in the oil and be carried away. A piston accumulator uses an iron piston, like a cylinder piston, sealed with O-rings. A bladder accumulator uses a synthetic rubber bag to separate nitrogen and oil.

Piston Accumulators. Built like a hydraulic cylinder but without a piston rod. The piston is sealed with two O-rings, with the space between the O-rings vented to the oil side. More than two O-rings do not seem to improve tightness of the seal and actually add to the friction drag of the piston.

The piston simply floats with very little pressure difference between opposite sides. A large oil port is on one end and a gas charging port is on the opposite end.

Advantages of a piston type over a bladder type are (1), they are capable of holding a greater volume of working oil between charge and discharge conditions (for the same rated size) because there is no bag to stretch on a full discharge nor to crumple on a full charge. Stretching and crumpling of the bag on each cycle reduces its life expectancy. A bladder accumulator should be operated over a more limited range than a comparable piston accumulator.

(2). A piston accumulator may not require re-charging as often. The length of time any accumulator will hold its charge is related to the area of the rubber seal or barrier. Nitrogen molecules pass between the rubber molecules by "osmosis". A piston accumulator has a small area of rubber on the O-ring seals compared to a relatively larger surface area on the bag of a bladder accumulator. Of course there may be slight leakage past the O-rings on a piston accumulator.

Bladder Accumulator. One of the important advantages of a bladder accumulator is the lower mass of its bag as compared to the heavier piston in a piston accumulator. It will usually give better results on a shock absorbing application because it can respond faster to pressure spikes which may pass through the system. A poppet-type safety valve in the mouth of the oil port is pushed closed by the bag when all oil is discharged. This prevents extrusion of the bag out into the line.

INSTALLATION OF ACCUMULATORS

On those applications where the purpose of the accumulator is to add to the pump flow during parts of the cycle requiring very high speed from the cylinder, the plumbing layout will be similar to this illustration. A schematic diagram of such an application is shown on the next page.

Plumbing connections between the accumulator and the inlet port of the 4-way control valve, Points A, should be oversize, larger than used in the line from the pump. Flow restriction in the accumulator line must be very low to obtain maximum cylinder speed. Eliminate as many bends and fittings in this line as possible.

The full flow valve, B, is optional and if used, it should be one with virtually zero flow resistance such as a rising stem gate valve.

The accumulator, C, should be mounted vertically with oil port down to prevent silt from settling in the accumulator.

A check valve, D, in the pump line prevents backflow from the accumulator into the pump while the pump is stopped.

A micronic filter, E, of 10 to

*Plumbing of Accumulator to
Supplement Flow From Pump.*

40μm, should be used in the pump line on circuits for piston accumulators. It may be installed either ahead of or following the check valve. Silt in a piston accumulator will score the barrel and cause the pre-charge to leak away.

Every installation should include a bleed-down valve, F. Oil must be discharged completely from the accumulator when the nitrogen pre-charge is checked or replenished, and before any plumbing line in the system is opened. For safety we recommend a 2-way, N. O. (normally open) solenoid valve of 1/4" size wired to the electric motor circuit. When the motor is shut down, the bleed-down valve will open and vent high pressure oil to the tank. This can prevent a serious accident to personnel who may not be aware that there is an accumulator in the system.

Further details of the hydraulic circuit are shown in the schematic diagram below.

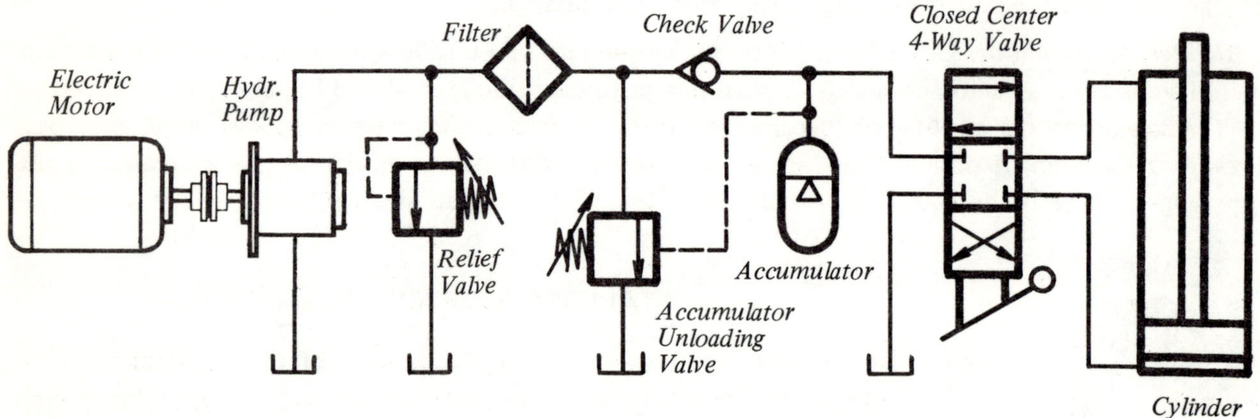

Accumulator flow adds to pump flow to produce a high cylinder speed.

Schematic Diagram. In addition to the components shown in the preceding illustration, the circuit should include a relief valve. For a 3000 PSI system this valve should be set for about 3500 PSI. It is a safety valve and normally should remain closed at all times. An accumulator unloading valve must be added, and adjusted to an unloading pressure of 3000 PSI. This is not a relief or by-pass valve; it is a special valve made specifically for this purpose. It is piloted to its open position by a pressure signal from the accumulator when pressure reaches 3000 PSI. This will allow the pump flow to unload to tank. When a volume of oil has been used from the accumulator sufficient to drop the pressure about 20%, or to 2400 PSI, the valve will close and will load the pump to re-charge the accumulator.

On some systems, a 20% differential may be too much for satisfactory operation. The unloading valve can be replaced with a pressure switch operating a 2-way solenoid valve to unload the pump. The pressure switch can be adjusted to load the pump after only a 5% drop in accumulator pressure.

Shock Suppression Applications. An accumulator, properly placed in the circuit, can usually reduce the intensity of hydraulic shocks passing through the system, but cannot entirely eliminate them. For most effective results, the accumulator must be *physically* plumbed as close as possible to the point where the shock is generated.

Refer to the diagram on the next page. If the shocks originate in the pump, the accumulator should be teed into the pump line as close as practical to the pump outlet port. If the shocks originate from shifting the 4-way valve, the accumulator should be teed into the line as close as practical to the valve inlet port. If the shocks are produced by a press on its extension stroke as it contacts the work, or as the tool breaks through metal, tee in the accumulator close to the blind end port. If shock is generated

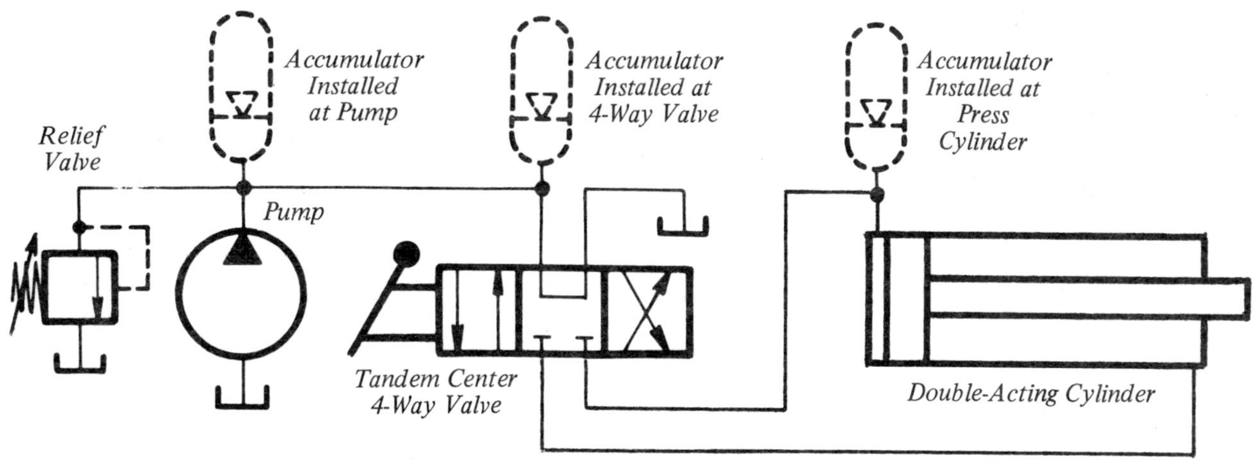

Hydraulic shock is usually produced at one (or more) of these three locations. By physically placing the accumulator near the point of shock generation, shock suppression is more effective.

on the retraction stroke, tee the accumulator into the rod port line. However, shocks from the press are often produced as the frame members spring back into normal position from a stressed position. An accumulator will not minimize these shocks. The fluid circuit will have to be modified to release the fluid more slowly to allow the mechanical members to resume their normal positions more slowly. Additional information will be found in the Womack book "Electrical Control of Fluid Power".

The best size accumulator for shock suppression cannot be easily calculated, but the rule-of-thumb shown here may be used as a starting point for further experimentation. Remember that oil working capacity is about one-half the rated size, the other half being filled with Nitrogen. Start with pre-charge Nitrogen pressure equal to about one-half the maximum hydraulic pressure. With the system running, bleed down the pre-charge and note any improvement in shock reduction.

We recommend bladder-type accumulators for shock suppression because they can respond faster to the pressure spikes. But before adding accumulators to any hydraulic circuit, be aware that they add compressibility to the oil circuit which may not be acceptable on some applications.

RULE-OF-THUMB

Accumulator Size for Shock Suppression

If practical, provide an accumulator with oil working capacity up to about 1/10th the GPM flow at the point where the shock is generated. See text for adjustment of pre-charge pressure.

Mounting Position. The preferred mounting position for all accumulators is vertical with oil port down. This tends to keep sediment from collecting in the accumulator. Functionally, an accumulator will work in any mounting position, and on systems using micronic filters, horizontal mounting is

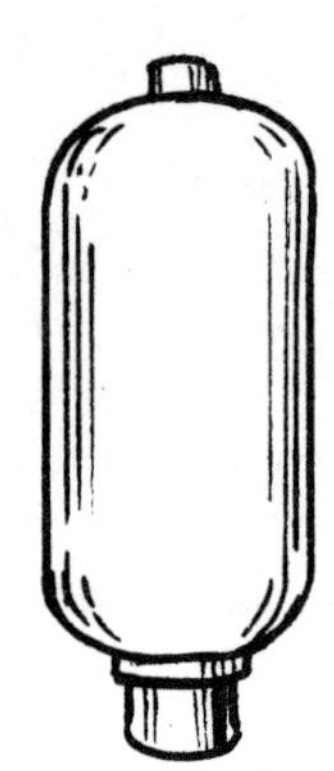

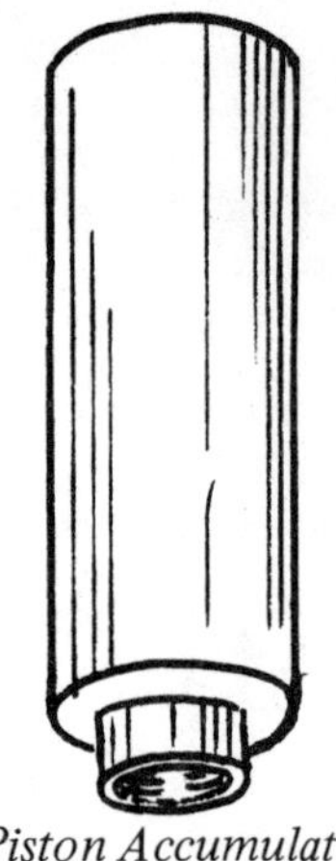

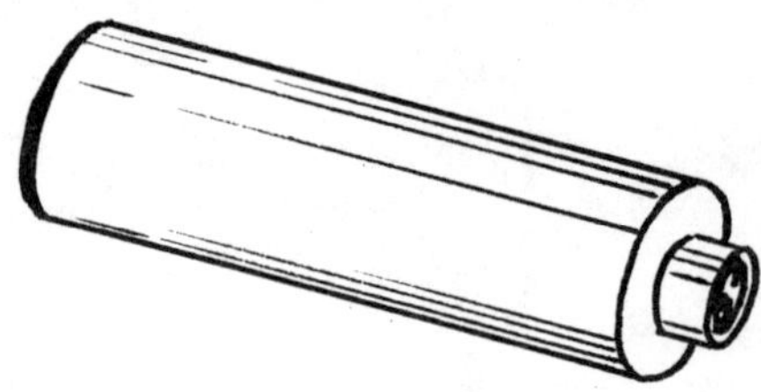

Preferred mounting position for all accumulators is vertical with oil port down. Horizontal mounting is acceptable only if micronic oil filters are used in the system.

Bladder Accumulator   Piston Accumulator

acceptable if there is a special reason for this mounting position. Piston accumulators can be damaged if solid particles lodge in the O-ring seals, producing longitudinal scratches in the barrel.

Protect Against Heat of Sunlight.

Use Nitrogen for Pre-charging Accumulators.

Mounting Location. For applications where accumulator will be required to discharge a high flow for a brief period, locate it as close as practical to the pressure port of the 4-way valve to hold line flow losses to a minimum. For shock suppression, locate the accumulator close to the point where the shock is generated.

Observe the same precautions as for cylinders in protecting them from physical damage. Do not expose them to direct sunlight. Install a sunshade. Protect them with a baffle from other radiant heat sources such as die casting machines, furnaces, steam boilers, etc.

Caution! Do not operate an accumulator until it is securely anchored to a strong structure. Unsecured accumulators will take off like a rocket if a connecting hose or line should break.

CHARGING ACCUMULATORS

Energy is stored in an accumulator by compression of an inert gas. Any inert gas could be used but nitrogen is nearly always used because it is less expensive and is readily available in high pressure bottles at approximately 2200 PSI.

Use oil pumped "dry" nitrogen. Do not use water pumped nitrogen because it contains water vapor which could damage a piston accumulator. Do not use air because it contains oxygen. The oxygen will deteriorate rubber seals, and in some applications might even present a fire or explosion hazard.

Pre-Charge Level. The pressure level of the pre-charge does not have to be exact. A fairly wide range of pressures will give satisfactory operation on most systems. We have found that a good value for pre-charge pressure is one-half the maximum hydraulic pressure of the system. Accumulator systems are usually designed to operate at 3000 PSI, so a pre-charge pressure of 1500 PSI should give good results. This pressure should be checked from time to time and when it has dropped to one-third the hydraulic pressure it is time to replenish it. Be sure to have all oil discharged when pre-charging or replenishing pre-charge pressure.

Low Pre-Charge Pressure. If the pre-charge pressure should drop to an unsatisfactory level, the hydraulic pressure may drop too low before the machine cylinder completes its stroke, perhaps so low that it stalls against the load. Raising the pre-charge pressure should restore the machine to normal operation. A pressure gauge installed at the cylinder port will provide a quick and easy way to estimate whether low pre-charge pressure may be the cause of loss of power.

SELECTION OF ACCUMULATOR SIZE

Starting with a full charge (usually at 3000 PSI), the pressure delivered from the accumulator falls as oil is discharged. When the entire volume of stored oil has been delivered, the last oil will have dropped to pre-charge pressure (usually 1500 PSI). Assuming a hydraulic cylinder is being extended, an accumulator must work between maximum fully charged pressure at the start of the cylinder stroke and a lower pressure by the time the cylinder reaches full extension. The larger the accumulator size, the less the pressure will fall by the end of the stroke. Therefore, the problem of accumulator size selection is to have one large enough so the pressure will still be high enough at the end of the stroke to finish the job.

The job of calculating accumulator size involves three steps: (1), to select a cylinder bore which will deliver sufficient force to finish the job after the initial 3000 PSI has fallen to, say, 2500 PSI; (2), to calculate the cubic inches of oil needed to extend the chosen cylinder the length of stroke needed; (3), and to select an accumulator which will deliver this amount of oil without its pressure falling below 2500 PSI. Usually, the oil delivered by the pump during the cylinder stroke can be neglected. It will be small compared with volume from the accumulator, and can be considered as a "bonus" or "safety factor".

This chart shows the number of cubic inches of oil which will be delivered by a *1 gallon* size accumulator as its pressure falls from 3000 PSI to various lower pressures. A 3-gallon accumulator will deliver 3 times this amount, a 10-gallon accumulator will deliver 10 times, etc.

This is the minimum size required. A larger size will give even better performance. The above is but a brief summary of accumu-

Minimum Acceptable System PSI	Cubic Inch Discharge
2700	12
2600	17
2500	22
2400	27
2300	33
2200	40
2100	46
2000	53
1900	63
1800	73
1700	84
1600	96
1500	109

lator size selection. A more detailed description with a calculation example is given in Appendix A of "Industrial Fluid Power — Volume 1".

CHARGING AND GAUGING

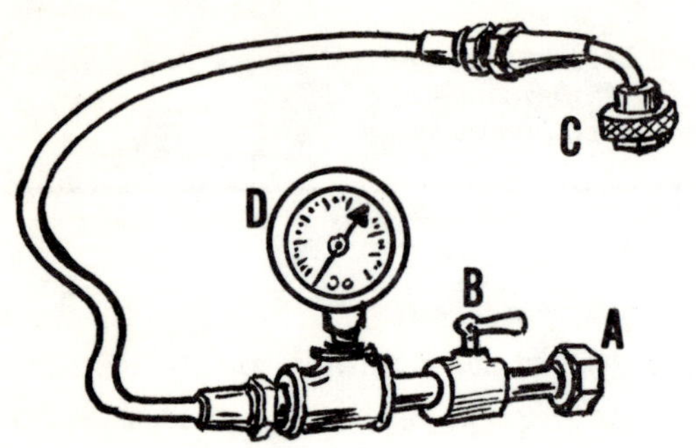

Pre-charging and Gauging Assembly.

Pre-Charging Equipment. A charging and gauging hose assembly should be purchased from the accumulator manufacturer, and should be kept on hand as a service tool. Because of variations in construction of accumulator air valves and of bottle connections, a charging assembly for one brand of accumulator may not fit another brand. This diagram shows one popular charging assembly.

Coupling "A" is a swivel fitting with left hand threads for attaching to the nitrogen bottle. It has straight threads of 0.908 - 14 to Specifications NGO-LH-(CGMA-550). Most nitrogen bottles have left hand threads so oxygen bottles cannot accidentally be connected to a nitrogen device. If supplied with right hand threads, the bottle supplier should be requested to supply an adaptor for left hand threads.

Valve "B" is a safety shut-off valve, plug-type, rated for high pressure. It has good throttling characteristics so the nitrogen can be slowly metered into the accumulator. When it is closed, the gauge will show the pressure inside the accumulator.

Fitting "C" is for connection to the accumulator. This particular assembly has a 0.312–32 UNEF right hand thread which fits the standard thread on Schrader automobile tire stems.

Pressure gauge "D" is any air gauge with a range sufficient for the level of pre-charge pressure to be used. A 5000 PSI range is suggested.

Checking the Pre-Charge Pressure. This same charging hose assembly is used for checking the pre-charge pressure from time to time. Valve "B" should first be closed. Then fitting "C" should be attached to the accumulator gas valve.

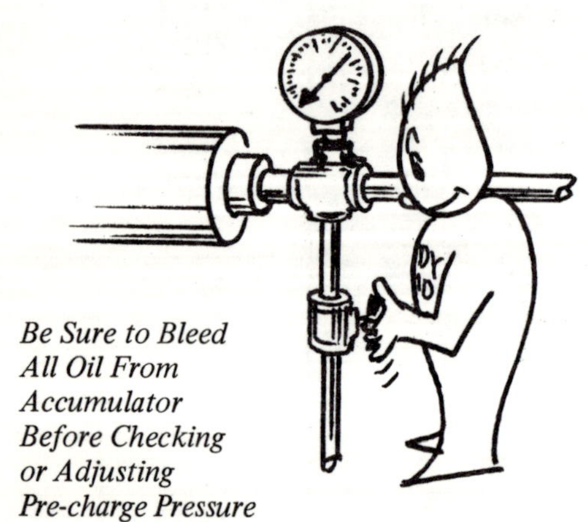

Be Sure to Bleed All Oil From Accumulator Before Checking or Adjusting Pre-charge Pressure

Procedure for Pre-Charging. The accumulator can be pre-charged either before or after it is connected into the system.

First, make sure all oil has been discharged. If it is connected into a hydraulic system, open the bleed-down valve to discharge all oil into the reservoir.

Next, close Valve "B" on the charging hose and connect the hose to the accumulator gas valve. Note the reading of existing gas pressure, if any.

Connect the other end of the hose to the gas bottle. A pressure regulator can be used if available but its use is not necessary. Flow can be metered either with Valve "B" or with the valve on the bottle.

Very slowly open the valving and allow gas to flow into the accumulator. The pre-charge level cannot be read accurately while gas is flowing. Shut off Valve "B" frequently to observe the gas pressure level in the accumulator.

After bringing the charge pressure to the desired level, close all valves and remove the charging hose. For safety, replace covers on both the gas bottle and on the accumulator gas valve.

Caution! Some accumulators have gas shut-off valves built into the gas port. Be sure that this valve is open during charging operations. If it is not, a false indication of pre-charge may be obtained, since only the hose will have been pressurized. Some accumulators have a cut off valve which must be opened with a hex wrench; others must be serviced with a charging hose which has a T-handle for depressing the Schrader valve stem after the fitting has been connected.

Make a simple test to see whether the gas has actually gone into the accumulator. With Valve "B" closed, slightly crack a fitting between the valve and the accumulator. If the pre-charge is just in the hose, the gauge will immediately fall off sharply and the reading will not be restored when the fitting is tightened.

Slightly Crack a Fitting to Test for False Pre-charge.

Frequency of Servicing. On any accumulator the nitrogen pre-charge will gradually leak away. The gas molecules transfer through the pores of the rubber seals, O-rings or bladder, by osmosis.

Piston accumulators will hold their pre-charge often for many years while on the shelf. When in operation the frequency of servicing depends on several factors including the total number of cycles, cleanliness of the oil, metal finish on the inside of the bore, etc. We recommend checking the pre-charge when an estimated 100,000 cycles have been completed. If pressure has not been lost, the next interval can be longer.

Bladder accumulators may require more frequent checking of the pre-charge. The large area of rubber exposure may allow the nitrogen to leak away at a faster rate. They do not hold the pre-

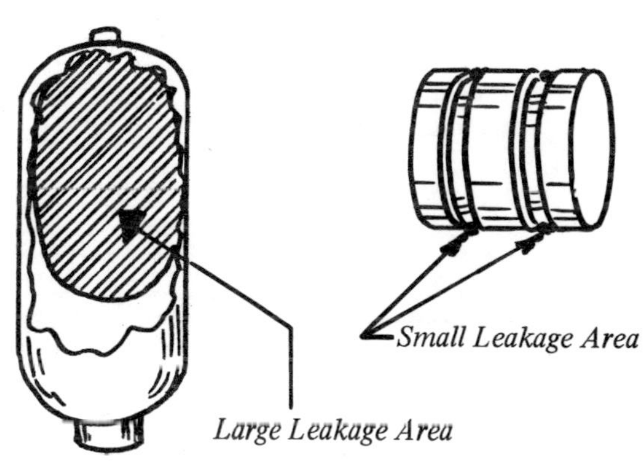

Leakage Paths Through Rubber in Piston-type and Bag-type Accumulators.

charge as long while on the shelf. They are usually furnished to the user in an uncharged condition.

With piston accumulators the piston O-rings should be replaced as a maintenance routine about every 1 million cycles because of possible mechanical wear. Some accumulators may use O-rings of a special cross section or special rubber compound. It may or may not be possible to use standard stock O-rings. In any event it is better to obtain exact replacements from the accumulator manufacturer.

ACCUMULATOR SAFETY PRECAUTIONS

Since railroads and motor freight lines permit accumulators to be transported while charged with high pressure nitrogen, they are not considered unduly hazardous. But when installing, pre-charging, or servicing them, these safety rules should be followed:

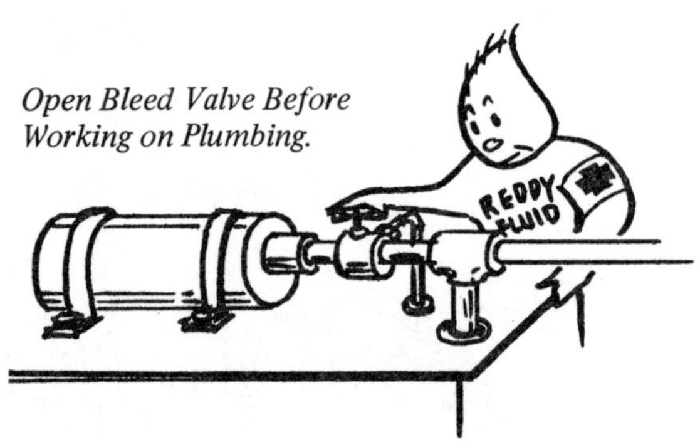

Open Bleed Valve Before Working on Plumbing.

Securely Anchor All Accumulators.

Bleed-Down. Before disconnecting any plumbing on a hydraulic system containing accumulators, find and open the bleed-down valve to discharge all high pressure oil to the tank. This will usually be a small needle or plug valve teed directly into the pressure line near the accumulator. Also bleed off the accumulator before unfastening it from its mounting. Some systems may have a solenoid bleed-down valve to automatically discharge oil when the electric motor is shut off.

Important! Before the initial start-up of an accumulator hydraulic system, make sure there is a bleed-down valve. If not, there may be no way to get oil out of the accumulator safely, and it may be difficult to check or replenish the nitrogen pre-charge pressure.

Anchoring. Never operate a hydraulic system without having the accumulator securely tied down. Do not depend on the plumbing connections to keep it tied down. An unanchored accumulator will take off like a rocket if a hose or fitting in the accumulator line should break or should accidentally be disconnected. This is a serious hazard to nearby personnel.

Gas for Charging. Do not ever use oxygen or other flammable gas for pre-charging. Oxygen causes rapid deterioration of the rubber bag or seals, and is also a fire hazard if the gas valve should blow out or if the gas should leak out in the presence of combustible material and become ignited. Do not use compressed air. Oxygen in the air will deteriorate the seals and be a fire hazard.

Gas Valve. Keep protective covers on the accumulator gas valve and on the nitrogen bottle except when checking or adding gas to the accumulator. Bumping into an exposed gas valve with a fork lift may cause a serious accident. Always be sure the gas valve cover is in place when transporting the accumulator if it is gas charged.

Automatic Shut-Down. For protection of personnel, we recommend that the bleed-down valve, mentioned above, be a 2-way, normally open solenoid valve. It should be wired to the electric motor on the system so it will automatically open when the motor is shut off. Wiring of a solenoid bleed-down valve is described in the Womack book "Electrical Control of Fluid Power" in the chapter on safety circuits.

USING ACCUMULATORS ON SPECIAL FLUIDS

These are general observations. Be sure to read accumulator specifications for information which may apply to particular models or to particular situations.

Water. An accumulator built for service on hydraulic oil should not be used on water or high water base fluids because of rusting. The manufacturer can furnish special models with internal parts nickel plated or otherwise protected against rusting. Standard seals or bags are usually compatible on these fluids.

Fire-Resistant Fluids. Synthetic fluids such as phosphate esters may be compatible with metal parts but are completely incompatible with buna-N rubber O-rings or bags normally used for hydraulic oil service. Special models must be used with butyl or Viton rubber seals or bags. Buna-N rubber internal parts are usually compatible with water base fluids but internal metal parts should be rust proofed.

High Temperature Oil. Ordinary synthetic rubber seals like buna-N should not be used at temperatures above 200° F. Viton or other high temperature rubber must be used for O-rings and bags.

MISCELLANEOUS ACCUMULATOR APPLICATIONS

Hand Pump Oil Charging. Where there is no source of electric power to run a hydraulic pump, a hand pump can be used to store oil at 3000 PSI in an accumulator. This is useful on applications where a short burst of high pressure oil is needed once in a while for a brief period. Even where electric power is available, a hand pump is often installed as an emergency charging source in case of electrical failure.

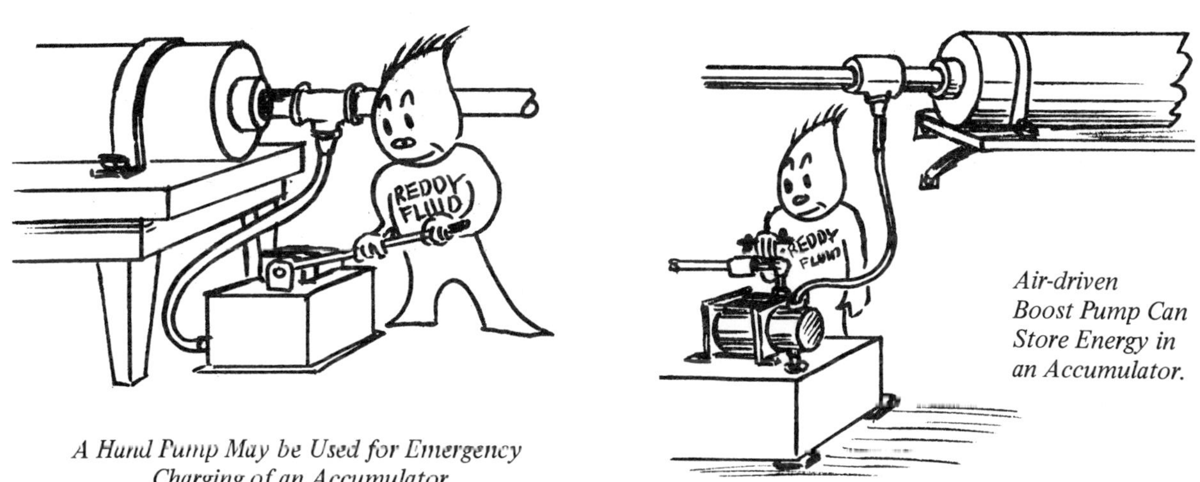

Air-driven Boost Pump Can Store Energy in an Accumulator.

A Hand Pump May be Used for Emergency Charging of an Accumulator.

Air-Driven Charging Pumps. Hydraulic pressure intensifiers driven from a shop air line can also be used in an emergency for charging an accumulator to 3000 PSI oil pressure. Charging rate is, of course, slower than with a power driven rotary pump. These air-driven pumps are known as "booster" pumps or "pressure intensifiers". Typical pumping rate is about 1/2 GPM in a 3000 PSI model. On applications where a hydraulic discharge is needed occasionally, the intensifier can be connected to "float" across the line. It will charge the accumulator to the desired PSI level, then will stall and hold this pressure indefinitely. If the oil pressure should fall either because of circuit leakage or because oil is discharged, the intensifier will automatically start pumping until full oil pressure is restored.

Load Simulator. An accumulator or a bank of accumulators connected in parallel can be used as an artificial load for testing hydraulic power units built for use on presses used to compress material such as cotton, paper, etc. The slow build-up of pressure in the accumulator resembles the slow build-up of back pressure in a baling operation.

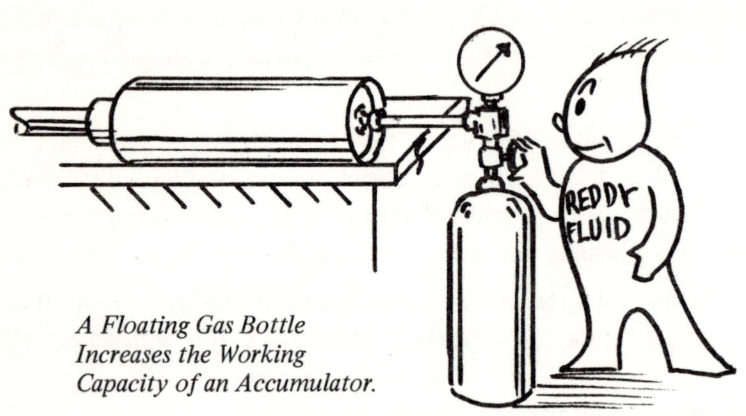

A Floating Gas Bottle Increases the Working Capacity of an Accumulator.

Improving Accumulator Performance. On piston accumulators only, the addition of an external gas storage vessel connected to the gas charging port has the effect of increasing the amount of oil which can be stored in the accumulator. A 5-gallon accumulator connected to a 5-gallon air storage vessel will give the same circuit performance as a 10-gallon accumulator. Normally, only about half the accumulator volume is available for oil storage, the other half containing nitrogen pre-charge which has been compressed. With an external gas storage vessel, the entire volume of the accumulator is available for oil storage.

A nitrogen bottle in which the pressure has dropped below the usable level for pre-charging is suitable as an add-on vessel. It should be permanently connected to the gas port with the gas valve left open or removed. A pressure gauge installed at this junction will show the pre-charge pressure when oil is bled from the accumulator.

The popular size gas bottle is 9¼'' diameter and 48 inches high. It has a volume of about 10 gallons. The smaller bottle 8½'' diameter and 30 inches high has a capacity of about 5 gallons.

Automatic Replenishment of Pre-Charge. On most applications, once the accumulator has been pre-charged with nitrogen up to desired pre-charge level, the gas bottle and charging hose are disconnected

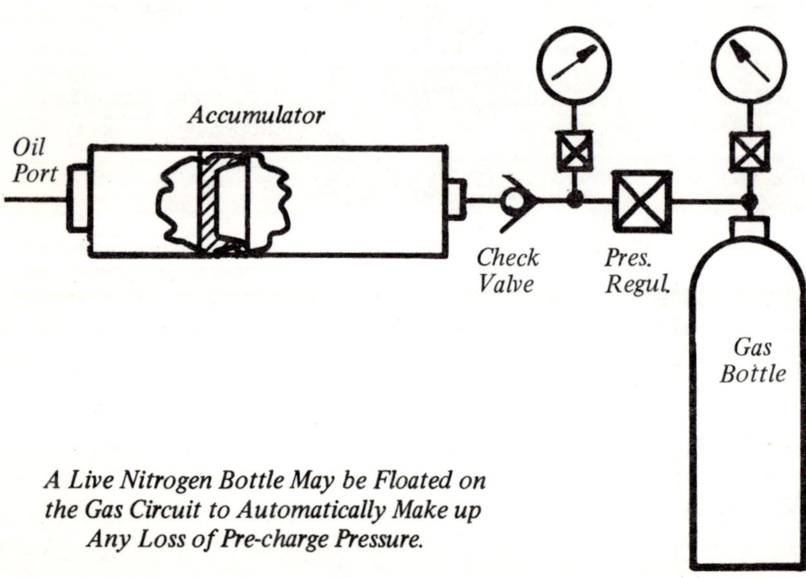

A Live Nitrogen Bottle May be Floated on the Gas Circuit to Automatically Make up Any Loss of Pre-charge Pressure.

and set aside for future use. Some users prefer to permanently connect a pressure bottle to the gas port through a pressure regulator. The regulator is set for the desired pre-charge pressure, 1500 PSI or whatever, and should be set when all oil has been discharged. Bottle pressure is fed through a check valve. For this application do not remove the stem of the gas valve in the accumulator.

Please note this application is different from the preceding one using an external bottle as part of the working system. Here, the pre-charge in the accumulator cannot back up past the check valve so operation is limited to the accumulator alone, and circuit operation is exactly the same as if the charging bottle were removed after pre-charging. The advantage of the floating bottle is to automatically keep the pre-charge from falling below a certain level. Any gas leakage past the accumulator piston will be made up from the bottle.

Before start-up each day, assuming the oil has been discharged, the two pressure gauges will show the state of the accumulator, the pre-charge on the left gauge and the pressure level of the gas remaining in the bottle on the right gauge.

SUMMARY OF SAFETY PRECAUTIONS!

High pressure gas is very dangerous. Be sure all hose and fittings are rated for at least 5000 PSI. Install a cut-off valve ahead of all pressure gauges and keep them closed except when a reading is being taken. If the gauge should be broken off by an accident this could cause a serious injury to personnel. Do not use any gas bottle that is not ASME approved, and do not use standard bottles at pressures over 3000 PSI. Securely anchor all accumulators and add-on gas bottles before putting pressure in them.

Chapter 6

Heat Exchangers for Oil Cooling

TYPES IN GENERAL USE

Air Blast Type. This is a radiator core mounted in a supporting frame. Oil is passed through the tubes and is cooled by the flow of air through the core. Industrial versions use an electric motor driven fan blade to produce an air flow. Mobile versions usually have the core mounted in front of the engine cooling radiator and use the same fan blast. The radiator core has an appearance similar to an automobile radiator, but is constructed to handle higher internal pressure. It uses a steel core rated for 100 to 150 PSI and temperatures up to 500° F.

An air blast type works well when air temperatures are moderate but it may be unable to keep oil sufficiently cool when ambient (surrounding) air temperature exceeds 110° F.

Since a water supply is not required, this type may be the only practical means of cooling hydraulic oil on a piece of moving machinery.

The cost and weight of an air blast heat exchanger is quite high compared with a water cooled model of the same capacity, so they are not as popular for in-plant industrial equipment.

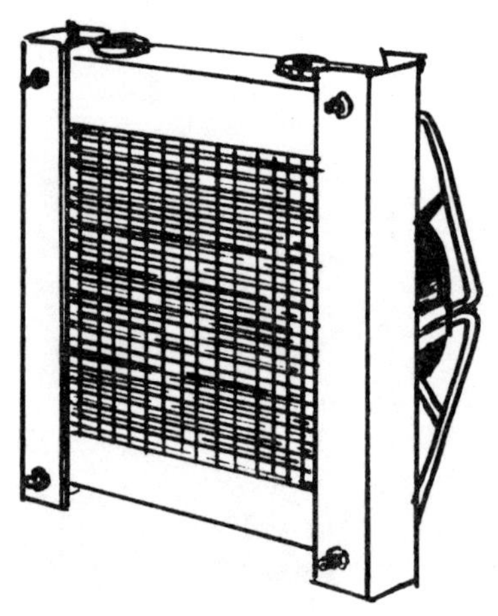

Air/oil Heat Exchanger With Motor-driven Fan.

Water-Cooled Type. These are called "shell and tube" heat exchangers. The cooling water passes through the inside of tubes in a tube bundle. Oil flowing inside a shell which surrounds the tube bundle is cooled by transfer of heat to the flowing water. Heat transfer is more efficient than for air blast models and does not depend on temperature of surrounding air, only on cooling water temperature, and volume of water and oil flow.

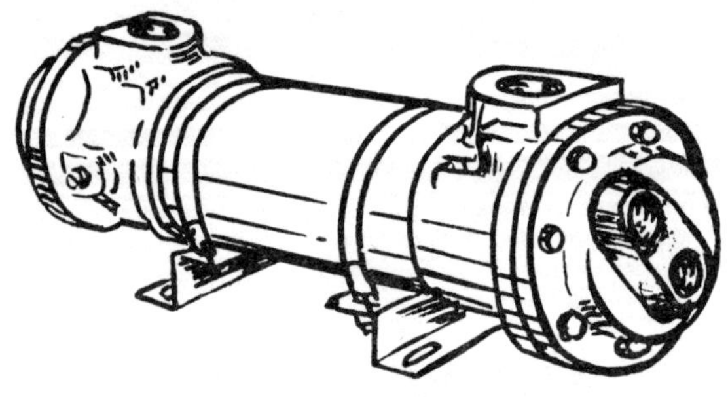

Shell and Tube Heat Exchanger.

Shell and tube heat exchangers are designed to work on fluid pressures of 100 to 250 PSI and on temperatures up to 350° F depending on brand.

__Automobile Radiators__. These are not recommended as hydraulic oil coolers. They have brass and copper cores designed for maximum 15 PSI, and would be in danger of blowing out on the pressure spikes which often pass through the return line of a hydraulic system.

__Convection Cooling__. Where there is no power to drive a fan blade and no water supply for a shell and tube heat exchanger, a limited amount of cooling can be obtained by taking advantage of rising convection air currents.

Remove the fan shroud and mount radiator in a horizontal position to allow rising convection air currents to flow through the core. Allow ample space above and below the radiator to avoid restrictions to normal air flow.

Shield the radiator from direct sunlight and paint it black to obtain maximum heat radiation.

Radiating capacity will be far less than the capacity of the same radiator core with fan blast but will be worthwhile if there is no other means of cooling.

A vertical mounting position may give more heat radiation when installed on a moving vehicle to take advantage of horizontal air flow.

__Mounting Position for Shell and Tube Heat Exchangers__. They may be mounted in any position but horizontal mounting is usually more convenient. However, if used on steam service they should be mounted vertically to allow condensate to drain off through the drain vents.

Plumbing to the oil side should be the same size as the oil connections on the heat exchanger.

__Direction of Plumbing__. Heat transfer is greatest on a shell and tube heat exchanger when piped "counterflow". True counterflow is when the water and oil flow in opposite directions. If piped to flow in the same direction the heat transfer is not quite as good.

A single-pass model is the only one which can be piped in a true counterflow direction, with water entering at one end of the tubes and oil entering the opposite end of the shell.

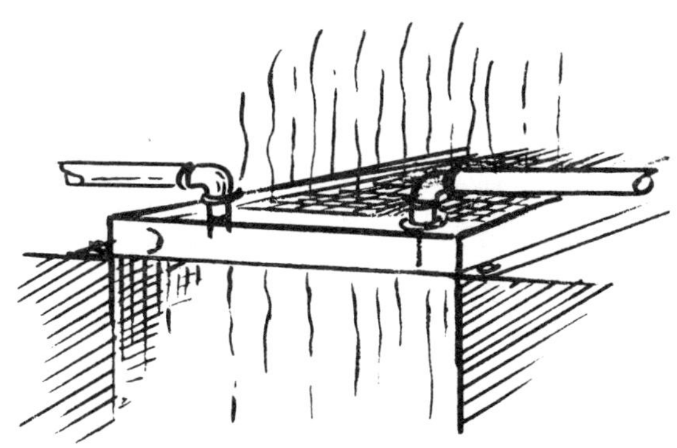

Mount Radiator Horizontally for Best Convection Cooling.

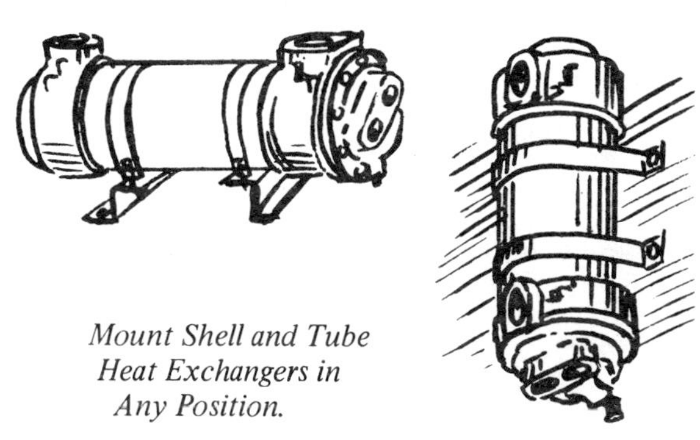

Mount Shell and Tube Heat Exchangers in Any Position.

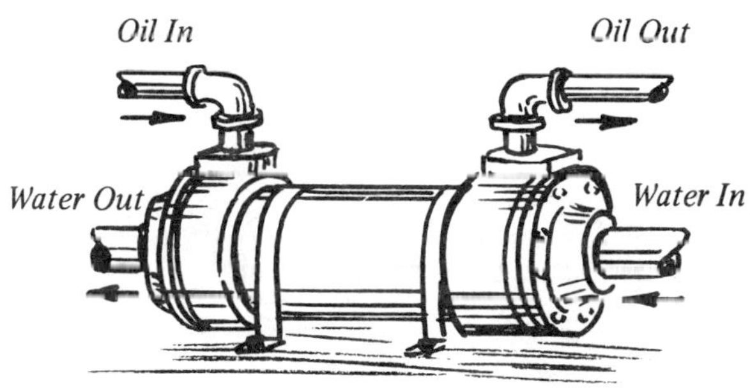

Plumb Shell and Tube Heat Exchangers in Counter-flow Direction.

This arrangement gives the greatest average temperature difference throughout the circuit, and maximum average temperature difference gives the best overall heat transfer.

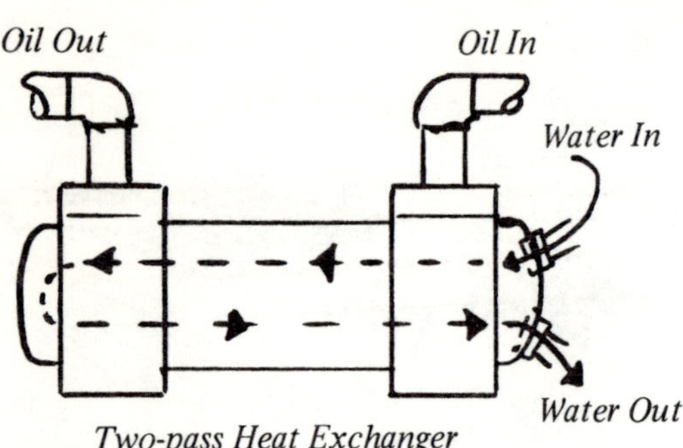

Two-pass Heat Exchanger

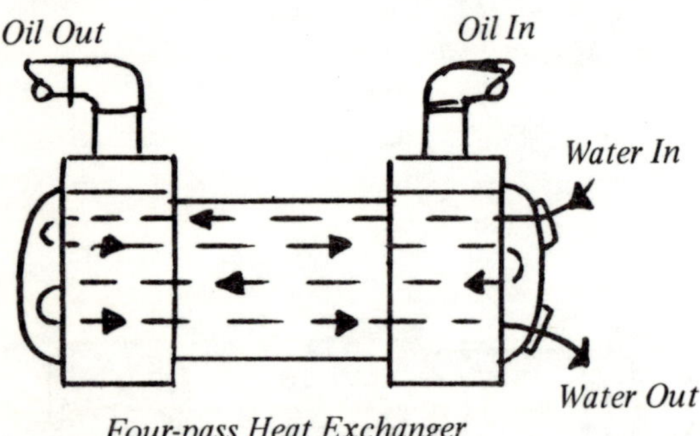

Four-pass Heat Exchanger

*Multi-pass Models Should Have Oil Inlet
on End with Water Connections.*

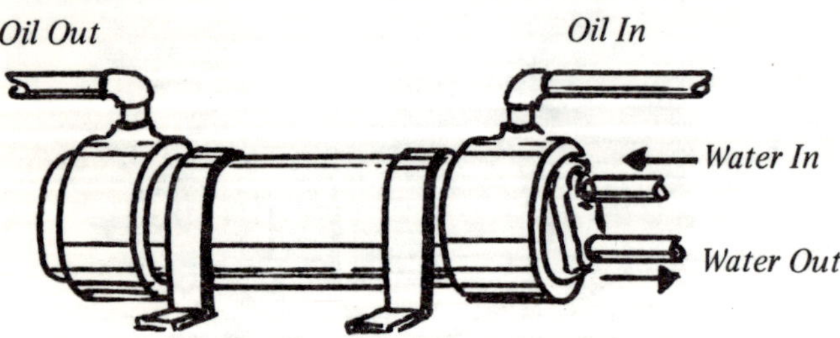

*Preferred Method of Plumbing a Multi-pass
Heat Exchanger.*

Plumbing Multi-Pass Heat Exchangers. Shell and tube heat exchangers are also made so the water can make two or four passes through the shell. On these models both water connections are on the same end bonnet, so true counterflow is not possible, but for these models the best heat transfer is with the oil entering on the same end as the water connections.

Either fluid can be passed through the inside of the tubes, but for hydraulic oil cooling, the water should flow through the inside of the tubes and the oil should flow through the shell to obtain the following advantages:

(1). The water is more likely to create fouling from corrosion or chemical deposits. Therefore it should be passed through the tubes. The tube bundle can be removed and cleaned but the inside of the shell cannot.

(2). The shell side can tolerate a higher pressure. Since the hydraulic oil is more likely to be at higher pressure due to surges in tank line flow and because it may have a high back pressure for a short period after start-up due to the oil being cold, it should flow through the shell.

(3). From the standpoint of expansion and contraction of the shell with respect to the tube bundle, it is better to pipe the hotter fluid into the shell.

(4). Shell and tube heat exchangers have a provision for installing zinc anodes in the end bonnets to prevent galvanic corrosion. There is no provision for zinc anodes in the shell. Since water is the fluid which produces galvanic corrosion, it should be piped into the end bonnets and through the tube bundle.

For these reasons, on hydraulic oil cooling, the water should flow through the tubes and the oil through the shell, and should be piped as shown here.

INSTALLATION OF HEAT EXCHANGERS

As with all hydraulic equipment, mount heat exchangers where they are shielded from sources of heat radiation, including direct sunlight. Protect water cooled heat exchangers from freezing. Air blast radiators should also be located where they will not be damaged by moving machinery such as swinging booms, lift trucks, etc.

Installation of Air Blast Types. If the blast comes from an electric-motor-driven fan, the motor rotation should be so air flow, as shown in this diagram, is forced under pressure through the radiator core rather than being pulled by a partial vacuum. This produces a larger volume of air for greater cooling and is a safety precaution against damage to the core if the fan blade should become loose on its shaft.

A 3-phase electric motor can be reversed by interchanging any two of its three line wires. When moving the heat exchanger to a different location, check the motor for correct rotation. A single-phase motor can be reversed by following wiring instructions on its nameplate or inside its junction box. It involves reversing the starting winding with respect to the running winding. Once the rotation is correct it will not change if the motor is plugged in at a different location.

Radiator Only. On moving vehicles driven with a water cooled engine, a radiator core (without fan) can be mounted in front of the engine radiator to take advantage of the air flow created by vehicle movement and by the engine fan. This should not seriously affect the efficiency of the cooling to the engine jacket water.

On some applications, engine jacket water can be circulated through a shell and tube heat exchanger. This requires a special and enlarged lower reservoir on the engine radiator core, with internal baffles so the water can be picked up and returned to this reservoir. However, the amount of cooling is limited because the jacket water is already hot.

Additional Cooling Capacity. By choosing a favorable location the cooling capacity can be increased if the discharge of air through the core is directed toward the hydraulic power unit, against the sides of the tank and across the pump and other components.

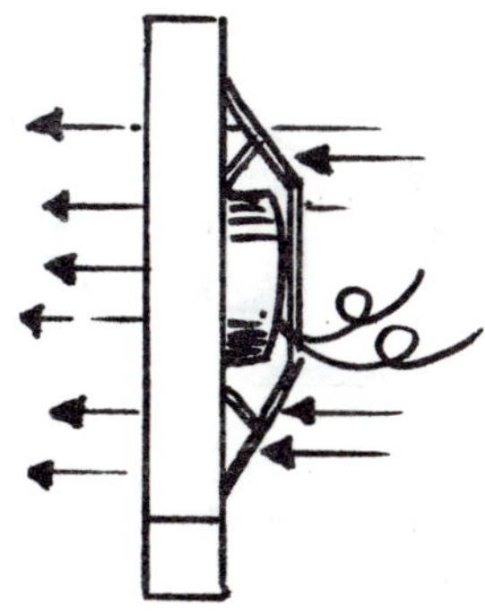

Air Flow Should be in the Direction Shown.

Mount Heat Exchanger in Front of Engine Radiator.

Extra Cooling is Obtained by Directing Fan Blast Against Reservoir.

PLACEMENT OF A HEAT EXCHANGER IN A HYDRAULIC SYSTEM

A heat exchanger must always be installed in a low pressure part of the system, less than 100 PSI. The lower the oil pressure the less likely to develop a leak. It must be where there is a continuous flow of a large volume of oil. The rate of heat removal is related to the volume of oil flow.

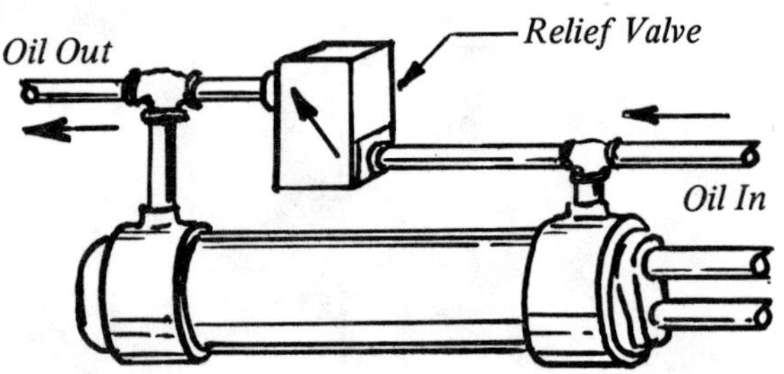

Protect Heat Exchanger with a Relief Valve.

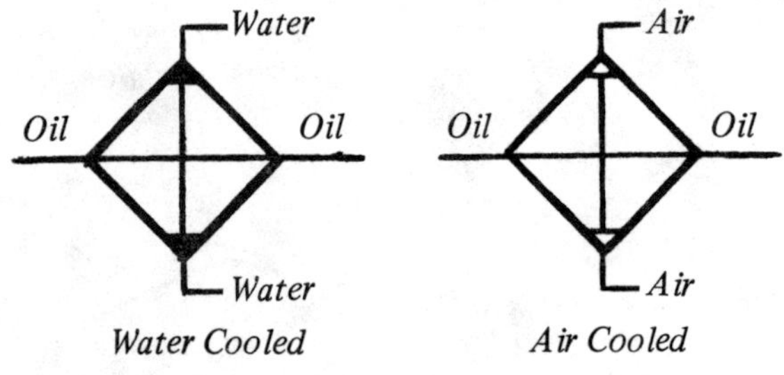

Graphic Symbols for Heat Exchangers

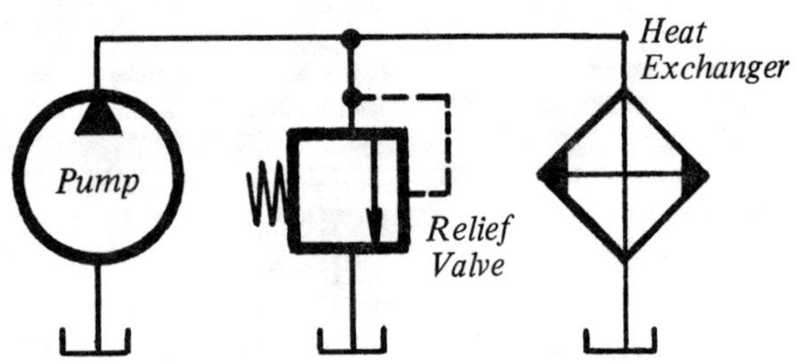

A Separate Pump May be Used for Cooling

Relief Valve Protection. Every heat exchanger should be protected from pressure surges with a pressure relief valve on the oil side. Since oil viscosity becomes high at low temperatures, the pressure drop through the oil loop can become quite high when the oil is cold. A relief valve is not necessary on the water supply provided its pressure is less than 100 PSI and is relatively constant.

The relief valve must be non-adjustable to prevent field tampering. There is no easy and convenient way, on location, of re-setting it to correct value. A hydraulic check valve makes an excellent non-adjustable relief valve with cracking pressure of about 20 PSI. When used as a low pressure relief valve, be sure its free-flow direction is the same as the direction of oil flow.

Heat Exchanger Symbols. On graphic symbols for circuit diagrams, a water cooled heat exchanger is shown by solid traingles and air cooled models by open triangles on the cooling medium line. In diagrams to follow, the triangles are omitted because in many cases either medium could be used.

Separate Cooling Circuit. On existing systems where a heat exchanger is to be added, it may be more convenient to add a separate cooling circuit using either an impeller pump to circulate oil, or a gear pump protected with a non-adjustable relief valve. A separate loop is also an excellent place to add both a heat exchanger and a micronic oil filter. The pump, driven with a fractional HP electric motor can be mounted alongside the main tank on a picture frame skid made by welding angle iron.

Oil can be picked up from one of the tank drain ports, or if there are no drain ports, by teeing into the main tank line before it goes into the reservoir. After passing through the heat exchanger (and filter) it can be discharged back to the reservoir through the other tank drain port, or by teeing into the main tank return or into the filler opening. A low pressure, non-adjustable relief valve with cracking pressure of 15 to 50 PSI should be installed across the heat exchanger.

The pump GPM in the cooling circuit is related to the heat exchanger selected. For most efficient cooling it should be the flow which will pass through the heat exchanger at a velocity of 3 feet per second.

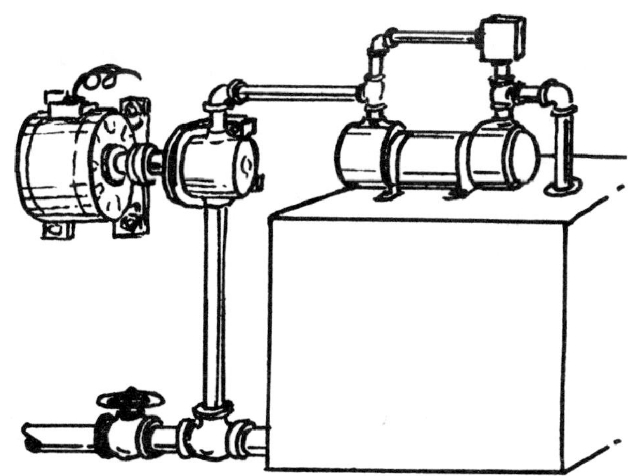

If separate oil pump is used, oil may be picked up from the tank drain to circulate through the heat exchanger.

Cooling in the Tank Return Line. To be most effective there should be a high volume continuous flow through the heat exchanger. On open loop hydraulic circuits the tank return line is by far the most effective place for installation. Return oil from 4-way valves, relief and by-pass valves should be combined into one return line and routed through the heat exchanger. Drain oil from sequence, by-pass, and pressure reducing valves should not be combined with the main return flow but should be run to a separate drain connection on the reservoir. This is a relatively small part of the return oil and should be kept separate from the main flow so it will not be subject to back pressure spikes which may appear in the main return line.

The non-adjustable relief valve across the heat exchanger, shown in the circuit below, is particularly important because during parts of the cycle, when large-rod cylinders are retracting, the return flow could be twice or more the flow when cylinders are extending. These high flow periods could produce a high back pressure to damage the heat exchanger. The relief valve will prevent these high pressure build-ups. The relief valve must be non-adjustable so it cannot be tampered with.

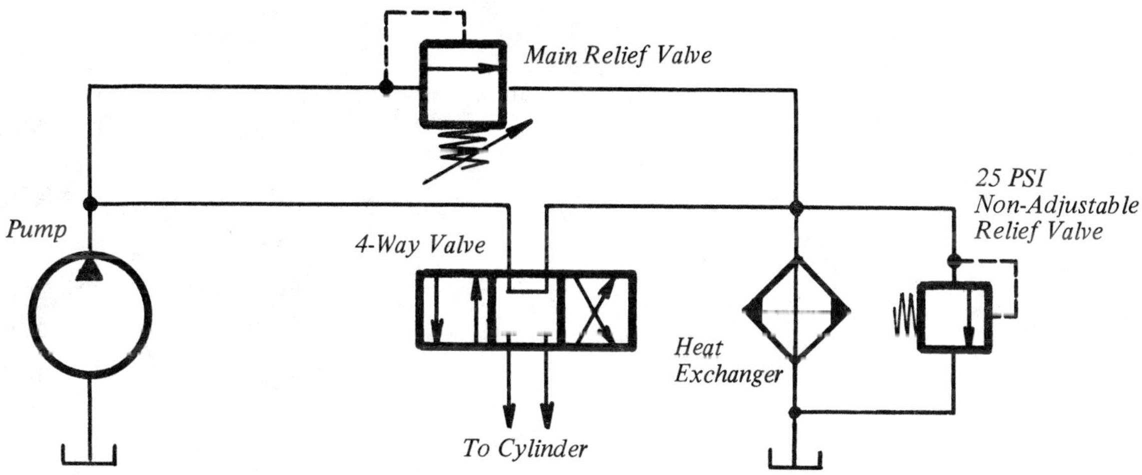

When Cooling in the Tank Return Line, be Sure to Size the Heat Exchanger to Take Care of the Peak Flow From the Blind End of a Large-rod Cylinder.

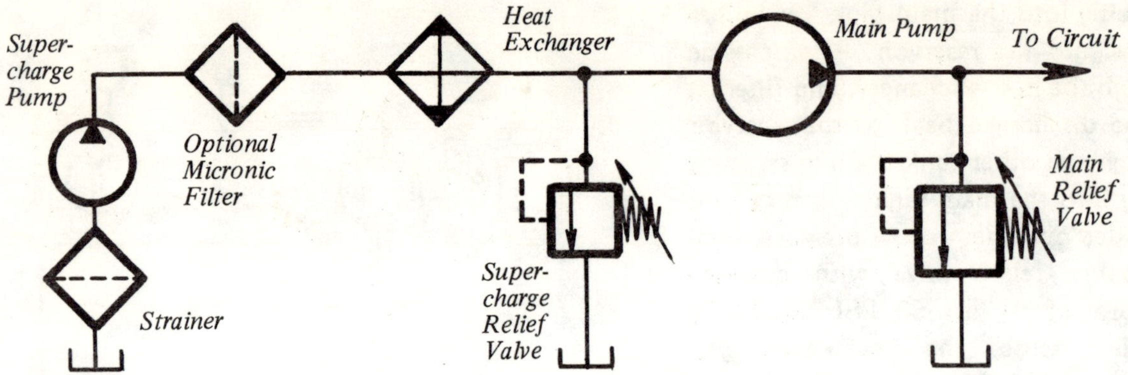

On circuits where a supercharge pump is used, the supercharge circuit is a good place to install either (or both) a micronic filter and a heat exchanger.

<u>*Cooling in the Supercharge Circuit*</u>. Some pumps are unable to develop sufficient suction to draw in a full charge of oil on each shaft revolution. This includes some kinds of high pressure piston pumps. It also includes other pumps which can run at low to moderate speeds with atmospheric pressure inlet but which must be supercharged if run above a certain specified speed. In these cases an auxiliary pump, called a supercharge pump, can be added to feed oil to the inlet of the main pump at low pressure, 5 to 10 PSI, no higher. The supercharge pump is usually a positive displacement gear pump which can produce a higher volume than can be accepted into the inlet of the main pump. A low pressure, non-adjustable relief valve is added across the main pump inlet to protect it from excessive pressure by discharging to tank all the charge pump oil which is not accepted by the main pump. Too high a pressure on the main pump inlet would, on some pumps, blow out its shaft seal. These components, the supercharge pump and its relief valve can be seen on this circuit.

A heat exchanger can be added in the supercharge pump pressure line. The low pressure relief valve also protects the heat exchanger from high pressure. This is also an excellent location to install a low pressure micronic filter.

<u>*Cooling in the Case Drain Lines.*</u> On reversible hydrostatic transmissions the only place where there is sufficient flow at low pressure for installation of a heat exchanger is in the case drain and low pres-

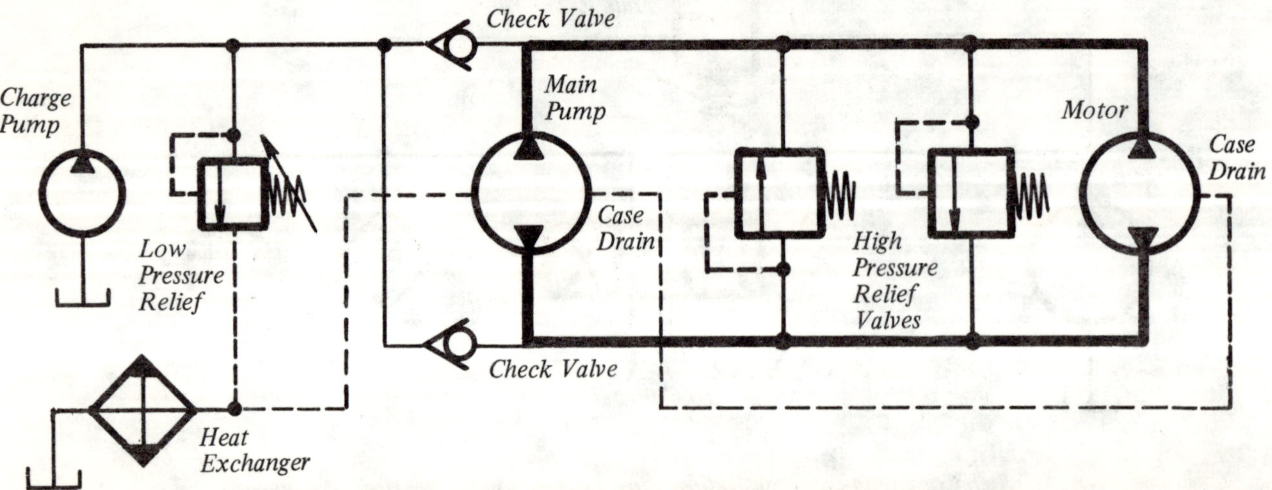

Using a heat exchanger in the case drain lines of a hydrostatic transmission.

sure relief valve return lines. In the preceding circuit all available low pressure flow is combined and routed through a heat exchanger to tank. For most effective cooling, start with drain flow from the motor, route this into one case drain port on the pump for additional pump cooling. The case drain flow from another drain port on the pump is then combined with discharge from the charge pump relief valve. The combined flow then goes through the heat exchanger. If this does not give sufficient flow volume for adequate cooling, the alternative is to add a separate cooling system external to the hydrostatic circuit, using a fractional HP electric motor and a low pressure gear pump, to take oil from the tank, run it through a heat exchanger and return it to tank.

Closed loop hydrostatic transmissions are described in the Womack book "Industrial Fluid Power — Volume 3". Please refer to that book for additional information.

Cooling in the Relief Valve Discharge. On selected applications described below, a heat exchanger can be installed in the tank return line from the main relief valve. As in all heat exchanger installations, we suggest a non-adjustable relief valve of 25 to 75 PSI cracking pressure be shunted around the heat exchanger to protect it from pressure spikes generated by sudden surges in flow.

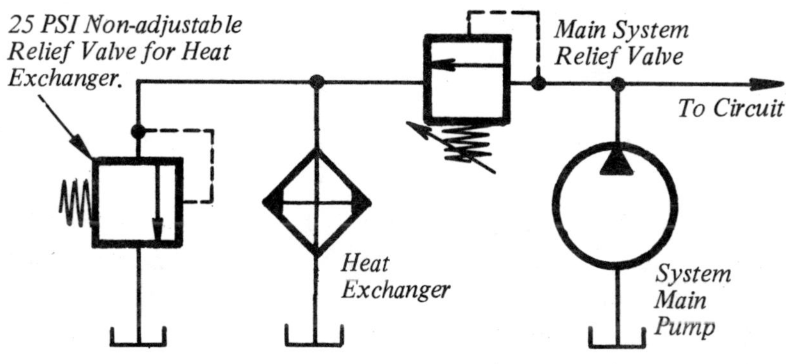

Cooling in the Relief Valve Discharge is Effective Only if There is Reasonably Continuous Flow in This Circuit.

This location is only effective on those applications where there is a continuous discharge of at least one-quarter of the pump flow across the relief valve. A typical application would be where a series connected flow control valve for cylinder or hydraulic motor speed control guaranteed that there would always be a partial flow of pump oil across the relief valve and through the heat exchanger.

THERMOSTATICALLY CONTROLLING A HEAT EXCHANGER

Air Blast Type. If a model is used which has an electric motor fan drive, the usual procedure is to switch off the motor when oil temperature has dropped to a desired lower level, then to switch it on again when the oil has reached a pre-set high temperature.

Oil temperature is sensed with a temperature switch having an adjustable differential between on and off. For example, it could be set to break the motor circuit at 140° F and to start the motor at 160° F.

A temperature switch is supplied with a capillary tube about 6 feet long. The bulb

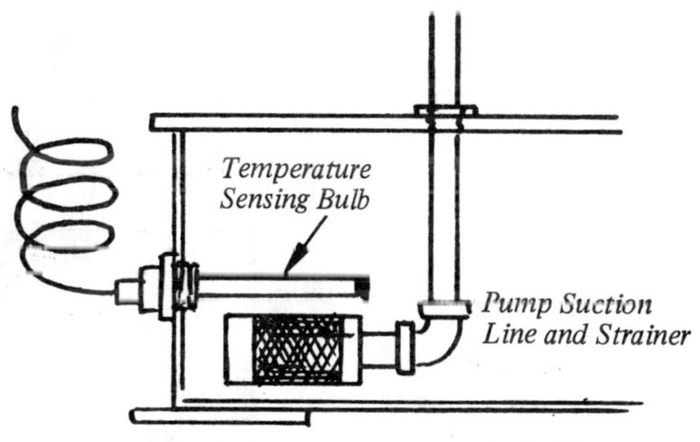

Sensing Bulb Should be Near Oil Pick-up.

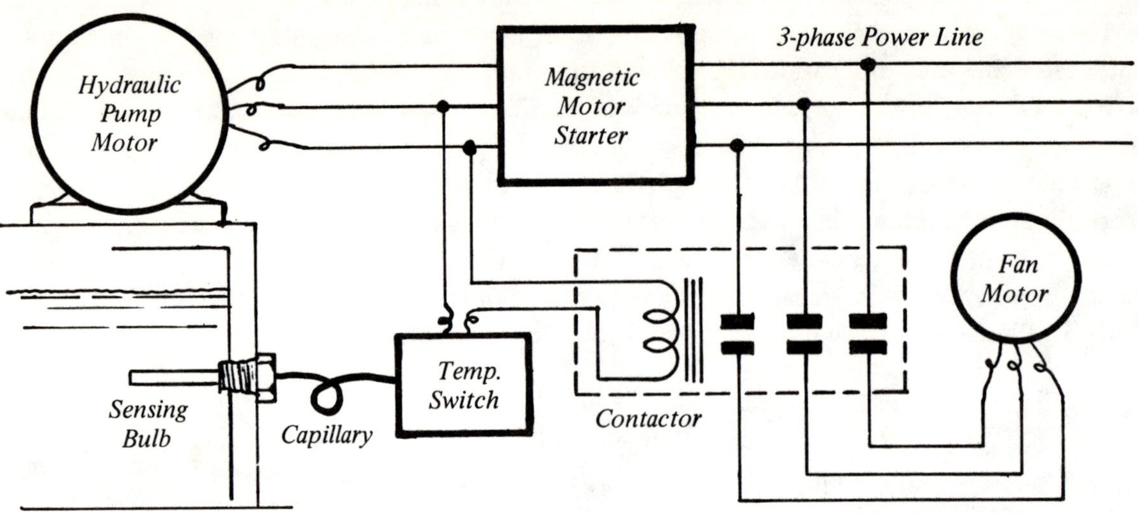

Fan Motor Should be Wired so it is Switched off Automatically When the Pump Motor is Switched Off.

at the end of the tube should be installed in the reservoir under the oil level. The preferred location for the sensing bulb is where the oil is circulating and near the point of oil pick-up.

The temperature switch can be wired as shown in the diagram above. It should be wired to the electric motor which drives the hydraulic pump so when the pump is shut off, the heat exchanger fan motor will also be shut off. The temperature switch will not directly control the fan motor but should be wired to a relay or contactor which, in turn, controls the fan motor.

On moving vehicles where the radiator core is mounted in front of the engine radiator, it may be necessary to provide a means for disconnecting the cooling system during cold weather or during start-up of the vehicle. This can be done by by-passing the oil around the radiator core when it is below a certain temperature.

In the diagram below, the modulating flow valve is the same kind used to control wate flow in a shell and tube heat exchanger except that it is a normally open type. At room temperature the valve orifice is wide open, allowing all the oil to go directly to the tank instead of through the radiator core. As temperature rises, the valve orifice starts to close, routing part of the oil through the radiator. On a further temperature rise the valve orifice completely closes, routing all the oil through the radiator. When purchasing a modulating flow valve be sure to specify the orifice must close on a temperature rise.

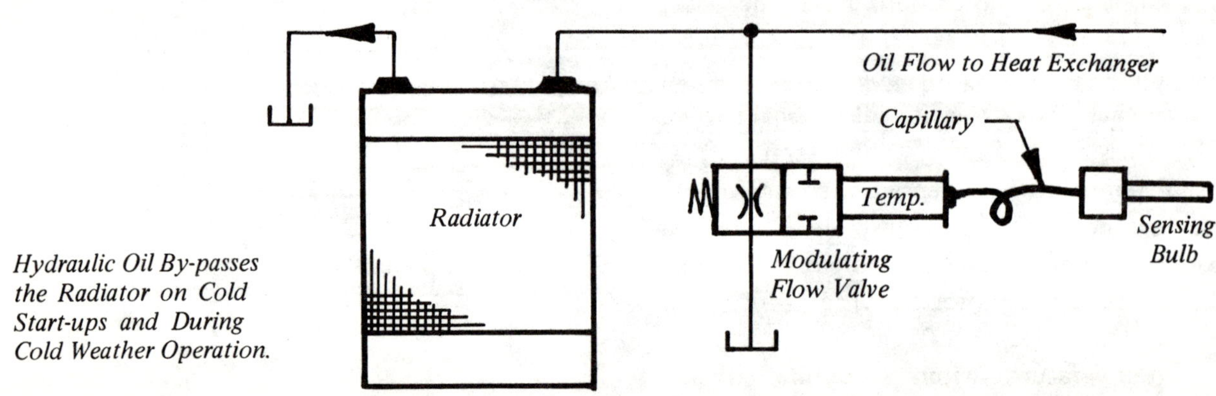

Hydraulic Oil By-passes the Radiator on Cold Start-ups and During Cold Weather Operation.

Thermostatic Control of Shell and Tube Heat Exchangers. For most efficient heat transfer, the *available* water flow should be approximately one-half of the oil flow through the heat exchanger. This volume of water will seldom be used but should be available in case it should ever be needed. Water flow is controlled by a temperature sensitive modulating flow valve which meters in only the amount of water actually needed. So no water is wasted. At start-up, no cooling water will be needed for perhaps the first hour of operation. The flow valve will keep the water cut off. If the heat exchanger has been adequately sized, only rarely will the maximum flow of water be needed. The modulating flow valve opens gradually, as oil temperature rises, to keep the hydraulic oil very close to the desired temperature. This valve has a 6-foot or 10-foot capillary tube which should be installed in the hydraulic reservoir in a bulb well near the point of oil pick-up. This valve has the opposite action from the one previously described. Its orifice must open on a temperature rise.

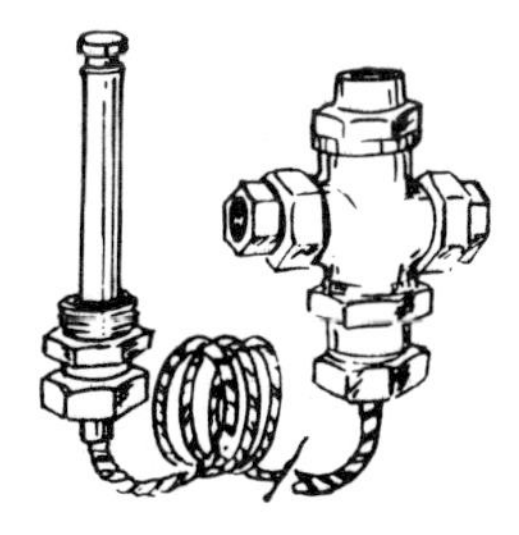

Modulating-type Water Valve.

The modulating flow valve is limited to a remote distance of 6 to 10 feet between sensing bulb and water valve. On longer runs the air temperature surrounding the capillary tube may influence the control and cause it to be less accurate. To avoid this, the capillary tube should be protected from heat or cold sources with insulation or a heat shield.

A Bulb Well May be Used.

The Use of a Bulb Well. A bulb well is a fluid tight receptacle which can be installed in the side of the tank. The capillary tube and sensing bulb can be slipped in and out of the bulb well without draining the oil. A bulb well can usually be purchased along with the flow modulating valve. It has mounting threads and can be screwed into a pipe half-coupling which has been welded into the side of the tank.

Water Plumbing. Suggested plumbing of the water supply to the heat exchanger is shown in this diagram. All valving should be installed upstream to keep water pressure off the heat exchanger when no water is flowing through it. First, a cut-off valve, globe or plug type, can be installed to shut off water when the system is shut down or when the heat exchanger or flow valve has to be installed or

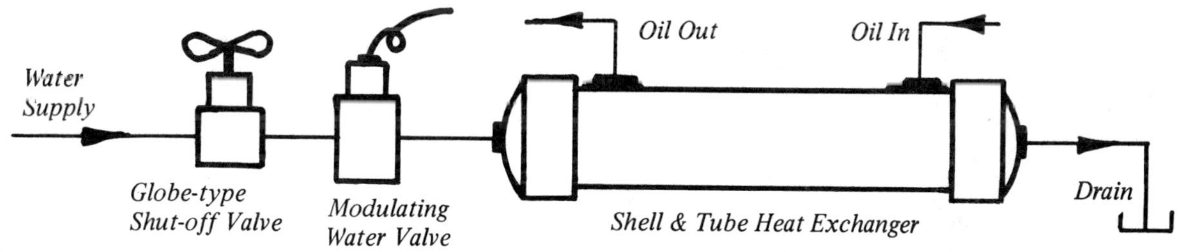

Install All Water Valving in the Upstream Side of the Heat Exchanger, to Reduce Internal Pressure.

removed. Take care that this valve is not accidentally left closed when the system is started. A rotary plug valve with ears to accept a padlock can be used.

Discharge water from the heat exchanger can either be discarded, it can be circulated to a cooling tower by means of a water pump, or it can be piped within the plant for other uses. Please note that in many cities it is illegal to dump heat exchanger water into the city sewer system.

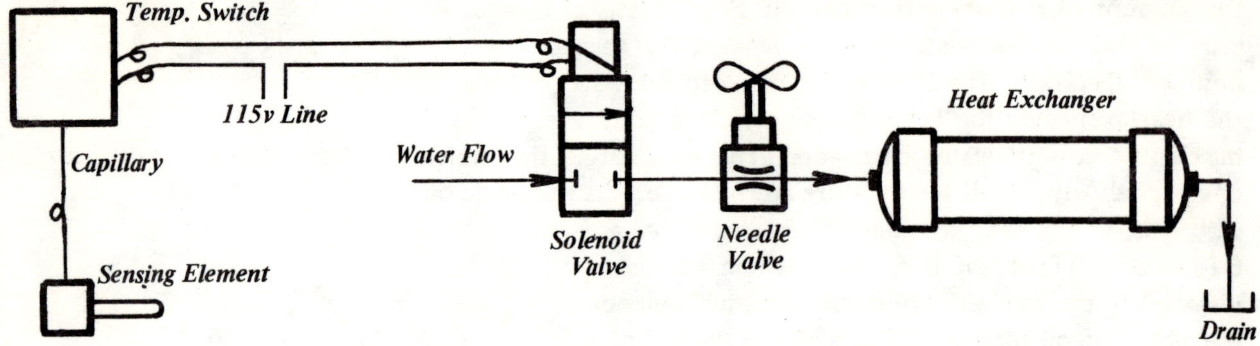

Solenoid valve water control can be used where sensing distance is greater than 10 feet.

Solenoid Valves for Water Control. On applications where there is a separation of more than 10 feet between the temperature sensing point and water control point, a flow modulating valve may not work well because of the sensitivity of its capillary tube to atmospheric temperature. Better control may be obtained with a 2-way solenoid valve to start and stop the water flow to a shell and tube heat exchanger. It can be actuated by a temperature switch with sensing element in the oil tank. The sensing element should be installed in a bulb well as described for the flow modulating valve. The switch should have an adjustable differential for setting high and low temperature points.

The solenoid valve must be suitable for use on water. We suggest using a needle valve following the solenoid valve to limit GPM water flow when the solenoid valve is open to approximately half the GPM oil flow.

The solenoid valve should be a 2-way, normally closed (N.C.) type and should be wired so it will be de-energized when the hydraulic pump motor is switched off. This will prevent accidentally leaving the water turned on. The temperature switch contacts should close on a rise in temperature.

For most applications a flow modulating valve should be used in preference to a temperature switch and solenoid valve. It tends to conserve the water and can hold the oil temperature to a closer tolerance. A solenoid valve, having an ON-OFF action, will allow oil temperature to fluctuate between high and low limits.

MAINTENANCE AND SERVICING OF HEAT EXCHANGERS

Replacement of Zinc Anodes. Shell and tube heat exchangers have portholes on one end bonnet for installation of zinc anodes. The anodes extend into the water which circulates through the tubes. Their purpose is to prevent galvanic corrosion from attacking the brass tubes. Galvanic corrosion is like a battery — two dissimilar metals in contact with an electrolytic solution. Molecules from one metal break away and attach to the other metal. The water is the electrolyte. The zinc anode is eaten away instead of the brass tubes.

Zinc anodes may not be necessary when fresh water is used, and the anode portholes may be plugged. But if the water has a mineral content, especially salt, which makes it electrically conductive, the anodes serve an important function in preserving the life of the brass tubes. They should be checked from time to time, the frequency of checking will depend on the mineral content of the water and the duty cycle of the heat exchanger. When about half of the zinc anode has been consumed it should be replaced. Anodes should be kept in stock as spare parts.

Cleaning Shell and Tube Models. If the cooling water contains minerals which foul the inside of the tubes, the ability of the heat exchanger to cool the oil will diminish, and the mineral deposits must be removed.

These heat exchangers are of two kinds: those with a fixed tube bundle and those with a removable tube bundle.

Fixed Tube Bundle. The tube bundle cannot be removed from the shell, so the inside of the tubes can be cleaned by removing both end bonnets, then using a suitable wire brush or cleaning rod and running it through each tube. Some mineral deposits can be removed by steam cleaning.

If hydraulic oil is handled in the shell, the inside of the shell should not have to be cleaned. If deemed necessary to clean inside the shell, flush with a solvent or clean with steam, whichever is more effective.

Suggestion: When making an original selection, choose a model with smallest diameter tubes if clean water is to be used. This will give the best heat transfer for a given size heat exchanger. If the water is apt to foul the tubes, use a model with larger diameter tubes so it will be easier to clean.

Zinc Anodes Should be Inspected Regularly.

Use a Wire Brush for Cleaning Inside of Tubes.

Removable Tube Bundle. The construction of these units is such that the tube bundle can be completely removed from the shell for easier cleaning, or replacement if necessary. They are more costly than fixed bundle models but are much easier to clean if both fluids may cause fouling or if, for some reason, water must be circulated inside the shell. If a leak should develop between oil and water sections, they are much easier to repair.

Cleaning Air Blast Models. If used for cooling oil, the inside of the tubes should never need cleaning. Dirt or trash which may lodge between the tubes can be removed with a brush, with compressed air, or by steam cleaning. These models should be inspected occasionally to see if there is a dirt build-up which may restrict the air flow. Inspect to be sure objects which would restrict full air flow are not allowed near the radiator core.

*Check Oil Sample
to see if Water
Settles Out.*

Check for Leaks. If a shell and tube heat exchanger should develop a leak between water and oil sections, oil will leak into the water if oil pressure is the higher. Water will leak into the oil if oil pressure is lower.

Water leaking into the oil will emulsify with the oil and the oil will take on a milky appearance. Some of the water will settle to the bottom of the reservoir. Drawing a sample from the reservoir drain valve should show whether there is water in the oil.

To test for water in the oil, draw a sample from the reservoir in a clear container after the system has been running an hour or so. Set the sample aside for a few hours to see if water settles. If the settling does not produce a clear oil sample, a further investigation should be made or a lab test should be conducted to find the nature of the contaminating material. It may turn out to be an air leak into the system. If the hydraulic oil is not clear, although it may be various shades of yellow or brown, this definitely indicates a problem which should be corrected as soon as possible.

If oil leaks into the water supply this will soon become apparent if the water re-circulates through a cooling tower. If the used water is piped to a drain the oil leak will not be so obvious, but can be verified by drawing a sample of the water to see if an oil film forms on top.

After a leak has been repaired it may be more economical on small systems to replace the oil. If a large quantity of oil is involved, allow it to stand in the reservoir a few days, if possible, then tap off the settled water from the tank drain. Do this for several times until the oil remains clear while the system is running. If it will not clear, check for air leaks or for additional water leaks.

HEAT EXCHANGER SIZE SELECTION FOR A NEW DESIGN

Heat is generated in all hydraulic systems because of mechanical losses in pumps and hydraulic motors, and flow losses through valves and plumbing runs. In small systems, this heat can be radiated from the metal surfaces of hydraulic reservoirs, cylinders, valves, and piping, without the oil temperature rising above an acceptable temperature. But on larger systems there is not sufficient surface area on tank and components to dispose of the generated heat without the system temperature rising to an unacceptable level. On these applications a heat exchanger must be added to supplement the natural cooling.

How can a designer determine whether a heat exchanger will be necessary on a new design, and if it is, how much capacity must it have? In this section we will try to give some common sense answers to these questions. Later in the chapter we will give a method of calculating the additional heat exchanger capacity to be added to an existing system to bring the oil temperature down from an unacceptable high temperature to an acceptable lower temperature.

Selecting the size of a heat exchanger is not an exact science because there are too many unknown factors. It should always be selected oversize because, in the case of a shell and tube type, the water modulating valve will meter just enough water to it for the amount of cooling needed and without wasting water. If not all of its capacity is used, there will be a reserve for use during unusual periods of operation or during periods of unusually high ambient temperature.

Heat Radiation From a Reservoir. In a hydraulic system without a heat exchanger, the reservoir oil will be at room temperature when the machine is started. As time passes the oil temperature will start rising and will continue to rise for an hour or so of continuous operation until a levelling-off temperature is reached, then it will rise no higher no matter how long the machine is operated. This levelling-off temperature is the balance point at which heat is radiated as fast as it is produced. If the levelling-off temperature is acceptable, then a heat exchanger is not needed; the system will take care of itself. But if the levelling-off temperature is too high, then a heat exchanger should be added to bring it down to an acceptable level.

Heat Radiating Capacity of a Reservoir. The radiating capacity in HP or BTU per hour, is directly related to the square foot external surface including top, bottom, and all four sides, and to the temperature difference between the oil and the surrounding air. This chart illustrates one set of conditions: a temperature difference of 80° F between room temperature and oil temperature. For example, the chart shows that a 90 square foot tank can radiate 9.2 HP of heat when room temperature is 80° F without allowing the oil temperature to exceed 160° F. A complete chart is given in the Appendix to Volume 1 "Industrial Fluid Power", or special values can be computed from this formula:

$$HP = 0.001 \times A \text{ (sq. ft.)} \times TD \text{ (temp. dif., °F)}$$

HP Heat Radiation From Reservoir at 80° F Temperature Difference Between Reservoir and Air.

Sq. Ft. Tank Surface	HP Heat Rad.
10	0.80
20	1.60
30	2.40
40	3.20
50	4.00
60	4.80
70	5.60
80	6.40
90	7.20
100	8.00
120	9.60
140	11.6

How to Estimate Heat Load. Use the input HP to the system as a guide (electric motor nameplate or engine rating). Common sense tells us that the maximum theoretical capacity for the heat exchanger would never exceed the input HP even if no output were delivered and the entire input was converted into losses. In an actual system we can count on about a 15% loss in each pump and hydraulic motor, most of which would end up in the oil. The additional valving and fluid flow friction losses should amount to another 15 to 20%. Cylinder friction loss should be about 5%. Therefore, in a system where one pump is operating one cylinder, system losses would be about 30% of the input HP. The reservoir would radiate a part of this heat. The remainder could be handled with a heat exchanger.

On typical hydraulic systems a heat exchanger capacity of 25 to 50% of the input HP should be sufficient. During the construction of hydraulic power units of 25 HP or more, space should be provided and connections installed so a heat exchanger could be added later if needed.

How to Select a Heat Exchanger. For most applications heat exchanger size is not so critical that it has to be calculated with mathematical precision. On new designs for cooling hydraulic oil it is not possible to accurately predict the load on the heat exchanger. Be generous in your size selection. If the machine is a prototype and if the heat exchanger is found to be larger than necessary, its size can be reduced on production machines. If the following rules-of-thumb are observed, heat exchanger selection will be easy and should be quite adequate.

Determine the GPM oil flow at the point where the heat exchanger is to be installed. Estimate heat load as 25 to 50% of system HP input. Calculate the surface area of your reservoir and from the chart

Estimate Heat Generation Before Selecting Heat Exchanger.

Cooling Capacity of Existing System Can Be Determined From Very Simple Calculations.

Use This Check List to Order Heat Exchanger.

or formula on the preceding page calculate how much of this heat load can be radiated without a heat exchanger. Subtract this from the heat load to estimate heat exchanger capacity.

Consult a heat exchanger catalog to find a model rated for your oil flow and which has equal or greater than your estimated HP load. Check catalog rated pressure drop through this model to be sure it does not exceed about 10 PSI. If it does, select the next larger model. On small sizes, a single-pass model is preferred. On larger sizes a 2-pass model will give good results. Be prepared to have a water supply *available* on shell and tube models equal to about half the GPM oil flow, although it is unlikely this volume will ever be required or used.

One important exception to the above rule is for the pressure drop on a model to be installed in case drain lines of piston pumps or motors such as used on hydrostatic transmissions. Pressure drop should not exceed 1 to 2 PSI. A model with 10 PSI pressure drop might blow out the shaft seals by putting this much pressure inside the case. When determining pressure drop from catalog information be sure to adjust for the viscosity of your oil if it is different from the viscosity for which the charts were prepared. Catalog instructions will tell how to do this.

Factory Selection. On special applications or those for cooling fluids other than hydraulic oil, you may want the manufacturer or his sales representative to select a suitable size. Information you will need to furnish is:

(1). You must estimate your heat load in HP or in BTU per hour.

(2). The fluid to be cooled. Give its viscosity, specific gravity, and the GPM flow. Specify whether viscosity is at 100°F or at operating temperature.

(3). How much cooling water is available and what is its temperature? For oil a minimum of 1/4th and a maximum of 1/2 the GPM of the oil is required. For fluids with higher specific heat, more water may be required.

(4). How much pressure drop is permissible in the cooled fluid?

Optional information might also include preferred tube diameter, whether zinc anodes should be supplied, and any other pertinent information.

ADDING A HEAT EXCHANGER TO AN EXISTING INSTALLATION

The HP capacity of a heat exchanger to be added to an existing hydraulic system to bring the levelling off temperature to a desired lower value can be estimated to a fair degree of accuracy because we can measure the present performance.

The chart below is a condensed version of a more complete chart in the Appendix of "Volume 1 — Industrial Fluid Power". It was calculated from the formula on Page 139, and shows the estimated HP heat which can be radiated from sides, top, and bottom of various size reservoirs at various differences in temperature between oil in the reservoir and surrounding air. By making air and oil temperature measurements, then using the chart, we can calculate the additional heat exchanger capacity needed.

First, after the system has been running at full load for several hours or long enough for the oil temperature to stabilize, measure temperature of the oil and surrounding air. Oil temperature can be measured with a thermometer suspended in the oil through a breather or filler opening. Take a reading from time to time until the temperature rises no further. If the system already has a heat exchanger, it should be running during these measurements. Take air temperature near the reservoir.

Next, measure your reservoir and calculate the number of square feet surface on top, bottom, and four sides. Then use the chart below or the more complete one in the Appendix to Volume 1 "Industrial Fluid Power", to determine the HP which is presently being radiated from reservoir surfaces at the air and oil temperature difference measured, and for the number of square feet surface area.

Now, to find how much additional HP capacity should be added to obtain your desired lower temperature difference, use the chart again in the column for the desired temperature difference and for the same tank surface area. This will show the maximum capability of the tank, by itself, for HP heat radiation not to exceed the desired temperature difference. Subtract this figure from the one determined above to find the heat exchanger capacity needed.

Calculation Example. A hydraulic system is driven with a 10 HP electric motor. The system is operating in a room where the air temperature near the tank is 90° F. The systems runs overheated. The oil temperature levels off at 190° F after a few hours operation. After measurement it has been found that the reservoir has 20 square feet of external surface. Find how much HP is presently being radiated

---

HP HEAT RADIATION FROM STEEL HYDRAULIC RESERVOIRS

Sq. Ft. Surface	Temperature Difference, in Degrees F, Between Oil in Tank and Surrounding Air									
	30	40	50	60	70	80	90	100	120	140
10	0.30	0.40	0.50	0.60	0.70	0.80	0.90	1.00	1.20	1.40
20	0.60	0.80	1.00	1.20	1.40	1.60	1.80	2.00	2.40	2.80
30	0.90	1.20	1.50	1.80	2.10	2.40	2.70	3.00	3.60	4.20
40	1.20	1.60	2.00	2.40	2.80	3.20	3.60	4.00	4.80	5.60
50	1.50	2.00	2.50	3.00	3.50	4.00	4.50	5.00	6.00	7.00
60	1.80	2.40	3.00	3.60	4.20	4.80	5.40	6.00	7.20	8.40
80	2.40	3.20	4.00	4.80	5.60	6.40	7.20	8.00	9.60	11.2
100	3.00	4.00	5.00	6.00	7.00	8.00	9.00	10.0	12.0	14.0
120	3.60	4.80	6.00	7.20	8.40	9.60	10.8	12.0	14.4	16.8

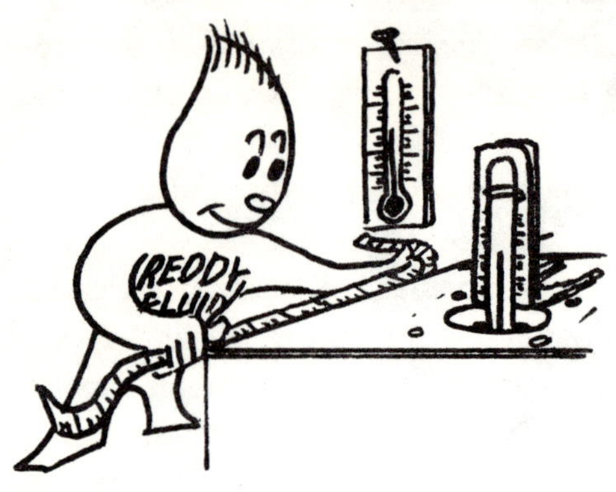

*Measure air/oil temperature
difference and tank area.*

*Use chart to
find heat radiation
of existing reservoir.*

from the reservoir and how much heat exchanger HP capacity would have to be added to bring its levelling off temperature down from 190° F to 160° F.

Solution. The present temperature difference between oil and air is 190 – 90 = 100° F. Enter the chart in the 100° F column. On the 20 square foot line a HP of 2.00 is shown. This means that the reservoir will radiate 2 HP but the oil temperature must rise to 190° F for it to do so. But we do not want the oil to rise to over 160° F, or to a temperature difference of more than 160 – 90 = 70° F. Enter the chart again in the 70° F column. A HP of 1.40 is shown. This means the tank will only radiate 1.40 HP without allowing the oil to exceed 160° F. A heat exchanger will have to make up the difference of 2.00 – 1.40 = 0.60 HP. Select a heat exchanger from the catalog in the same way described on Page 139.

<u>*Making Power Loss Calculations.*</u> In most cases the heat exchanger load derived from rules-of-thumb given previously will be about as realistic as an estimate made from more involved calculations, but for those who wish to calculate, calculate for each component separately and add them for a total. Be sure to average each one for the time the power loss occurs. Example: a power loss of 5 HP which occurs 1/3rd of the total time over, say, a 1-hour or 1-day period will amount to 5 ÷ 3 = 1.67 HP average for the hour or day.

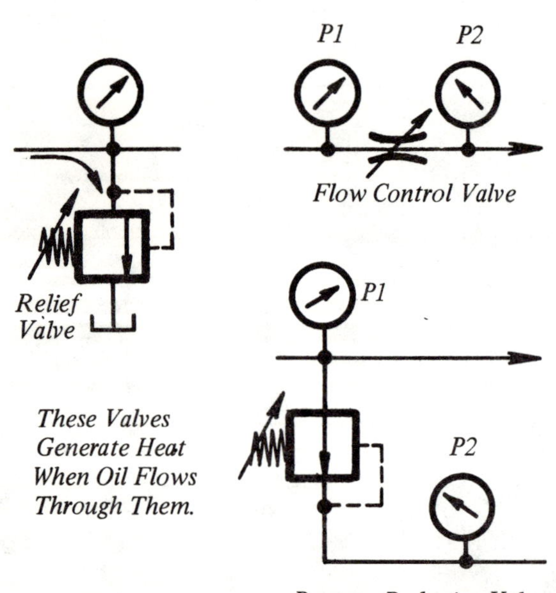

P1 *P2*

Flow Control Valve

*Relief
Valve*

*These Valves
Generate Heat
When Oil Flows
Through Them.*

P1

P2

Pressure Reducing Valve

Hydraulic Pumps and Motors. Allow about 15% loss of input power to heat for each pump or motor while it is running under load. High efficiency piston units may have only 10% loss and mobile gear pumps may have at least 20% loss. Average the loss over an hour or a day.

Hydraulic Cylinders. Allow 5% loss during the time the cylinder is moving under full load.

Pressure Relief Valves. Calculate loss only for the time discharge is taking place and average this for an hour or a day. Use the formula: *HP loss = PSI x GPM discharge ÷ 1714.*

Flow Control Valves. Take the upstream pressure, P1, minus the downstream pressure, P2, to use as PSI in the formula on the preceding page. Average this loss over an hour or a day.

Pressure Reducing Valves. Calculate loss only while flow is passing through the valve. Take inlet pressure, P1, minus outlet pressure, P2, to use for PSI in the preceding formula.

Plumbing Losses. This is difficult to calculate and must be estimated on the assumption the system plumbing sizes have been chosen wisely. The majority of these losses will occur while the cylinder is moving either loaded or unloaded. Usually they will amount to 10 to 15% of the input HP.

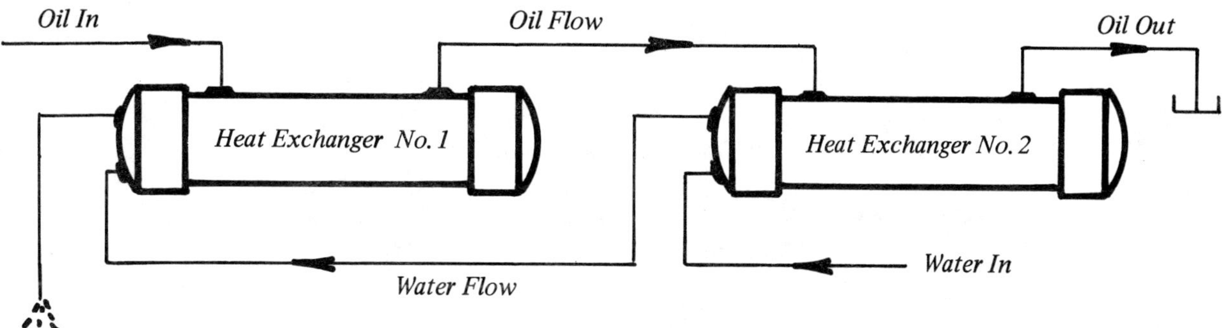

If the two heat exchangers are identical, water and oil circuits can be connected in parallel for lower flow resistance or in series for more economical use of water. See text for details.

<u>*Adding a Second Heat Exchanger*</u>. If a further increase in cooling capacity is needed on a system where a heat exchanger is already in use, the existing one can be replaced or a second one can be added. Oil and water circuits to two heat exchangers can be connected either in series or in parallel, but the greatest compatibility with the existing system is to have the add-on heat exchanger identical to the existing one.

Oil Circuit. Connecting oil circuits in series, as in the diagram, will result in the correct velocity for best heat transfer in both units, but the back pressure to the hydraulic system will be doubled. On ordinary open loop circuits, measure the back pressure on the existing unit at normal maximum system load. If it does not exceed 10 PSI, then a series connection to the new unit will probably work best. For installation in the case drain lines of piston pumps and motors, the two units must have oil circuits in parallel — a series connection would give too much back pressure for the pump or motor shaft seals.

Water Circuit. To make the best use of cooling water, if the two units are single-pass heat exchangers, water circuits should be in series. If they are 4-pass units, the water circuits should be in parallel. If they are 2-pass units, the water circuit can be connected either series or parallel.

Counterflow Plumbing. Notice the preferred method of plumbing water and oil circuits in the diagram above. For best heat transfer they should be connected counterflow or as nearly counterflow as possible. Inlet hydraulic oil should be connected to the oil port which is closer to the end with the water connections in the case of a 2-pass or 4-pass unit and to the end closer to the water outlet on a single-pass unit.

Different Size Heat Exchangers. If the add-on heat exchanger is different in size than one which is already on the machine, the water and oil circuits of both heat exchangers should be piped in parallel.

INEXPENSIVE COOLING WATER

In many industrial plants an ample supply of cooling water is available for a heat exchanger at little or no extra expense. Water flow to other functions such as toilets, batch mixing, lawn sprinkling, washing, and similar purposes can be intercepted and run through a heat exchanger then returned to its intended use at 10 to 20° F higher temperature, and with a loss of pressure of no more than 10 PSI.

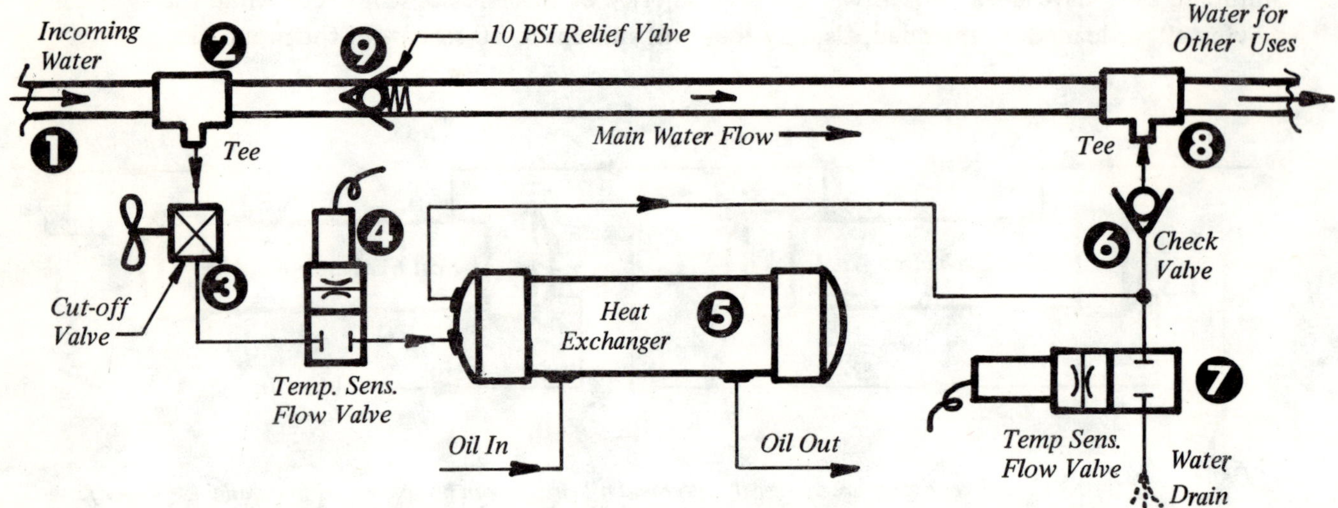

Obtaining a "Free" Source of Cooling Water by Tapping Into Flow Used for Other Purposes.

In the diagram a 10 PSI restriction is placed in the main water line to divert water into the heat exchanger. This can be a low pressure brass or stainless steel relief valve or a check valve with 10 PSI spring.

As long as there is a continuous flow of water through the main line, it will also flow through the heat exchanger. However, if water usage downstream of Point 8 has stopped or is insufficient for heat exchanger needs, Valve 7 will open and discharge the water to a drain. Circuit operation is as follows:

With cut-off Valve 3 open and with water flowing in the main line, part of the water will divert through the heat exchanger then return to the line at Point 8 through check Valve 6. (Valve 6 prevents main line water from discharging to drain if and when Valve 7 opens). In normal operation Valve 4 modulates the water volume to that needed by the heat exchanger, and Valve 7 remains closed. Valves 4 and 7 are the same kind but Valve 7 is set to open at a higher temperature and will open only if Valve 4, by opening fully, is unable to keep hydraulic oil temperature from rising. The sensing elements of both valves can be installed side-by-side in the tank. In one sense, Valve 7 is an emergency valve designed to open only when water usage in the main line is too low to satisfy cooling requirements. Valve 7 could also be a 2-way, N.C. solenoid valve actuated by a temperature switch. The abrupt opening action of a solenoid valve might even be desirable.

Chapter 7

Air Line Filters, Regulators, and Lubricators

CONDITIONING RAW AIR

Raw air, taken directly from an air compressor tank, may be suitable for inflating tires, but is not suitable for operating expensive or precision equipment like air tools, air motors, paint spraying equipment, valves, or cylinders, until suitably processed.

Every air compressor takes in the dust and moisture that is in atmospheric air. The intake filter is relatively coarse and cannot remove all the dust and none of the humidity. Since about 8 cubic feet of free air are compressed into 1 cubic foot of compressed air at shop air pressure, the dust concentration increases

Compressor intake air is usually dirty.

by 8 times and the relative humidity also increases by the same amount. Then to further pollute the compressed air, used oil which by-passes the piston rings in the compressor mixes with moisture and dirt and is subjected to a temperature of about 450° F at the moment of compression. Thus, raw compressed air taken directly from the compressor tank contains dirt, water, abrasive compounds, wet sludges, and varnish-like substances which can cause great damage throughout the system.

Of course a part of the contamination will precipitate in the receiver tank and can be drained off. But a large part of it may find its way into the distribution system and into machinery operating from the distribution system if it is not removed. Suitable air processing components should be placed ahead of every machine using the air to cleanse it, regulate its pressure, and to lubricate it.

Conditioning Equipment The 3-part assembly shown on the next page is known as a "trio" unit, and is installed ahead of every machine using the air. It consists of (1), a filter to remove solid and liquid contaminants, (2), a pressure regulator, usually with a gauge, to adjust and maintain line pressure to the lowest pressure which will give satisfactory operation, and (3), a lubricator (for some installations) to inject finely divided oil droplets into the air stream to be carried into valves, cylinders, air tools, etc., to reduce wear, prevent rust, and to prolong life.

Please note: Each of the three units of a "trio" has been described in detail in "Industrial Fluid Power — Volume 1", and will be mentioned only briefly here.

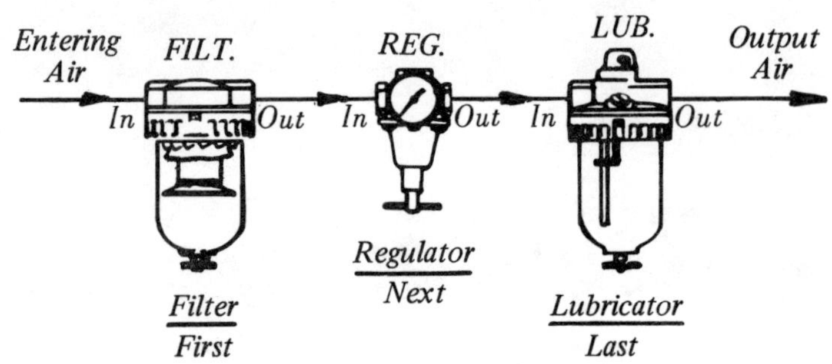

Assemble trio in proper sequence.

Make certain the three units of the trio assembly are plumbed in the right order, with filter first, regulator next, and lubricator last. Be sure each unit is correctly oriented with its inlet port toward the incoming air flow. Filter and lubricator bowls should be visible from a well-traveled aisle.

Filter. For most applications a standard filter with 5 to $10\mu m$ rating will serve quite well. It will remove water mist (finely divided liquid droplets) as well as microscopic dirt. If there is oil mist in the incoming air, probably from the air compressor, the standard trio should be preceded by another filter with an oil absorbing element.

The condensate in the filter should be drained daily, oftener if necessary during high humidity seasons. The filter can have an internal float-type drain valve, or can be connected to an external automatic drain valve which will discharge the condensate into a container or into a sewer line.

Pressure Regulator. Regulators are of two kinds: self-relieving and non-relieving. Self relieving regulators are preferred on industrial fluid power equipment because they have a vent for bleeding off excessive pressure which, on some applications, could back up from the downstream side, or from a leak which may develop in the main poppet. Non-relieving regulators do not have an atmospheric vent. They must be used, rather than the self-relieving type, for handling liquids or for gases other than air. See "Industrial Fluid Power — Volume 1" for a comparison of these types.

Lubricators. While on most applications a lubricator should be used to keep valves and cylinders lubricated, there are certain components in which the manufacturer may recommend no additional lubrication because the component has been pre-lubricated at assembly with a permanent molybdenum disulphide grease. A lubricator would wash out the permanent lubrication.

Two kinds of lubricators are available: standard models break up the oil into droplets of many sizes from about $1\mu m$ to about $50\mu m$. The mist will carry to a distance of about 10 feet down the line before precipitating. If the oil mist must be carried to a distance exceeding 10 feet to reach a cylinder or valve, a lubricator of the atom-mist type should be used. Oil droplets are broken into finer particles no larger than $2\mu m$ and will carry for great distances.

Oil feed rate should be very low on all lubricators, and should be reduced so in normal operation no oil drips out of the exhaust ports of the 4-way control valve. On standard full-feed lubricators, those where all of the drops seen in the sight glass go into the line, a feed rate of one drop every several minutes, or even less than this should be sufficient. On atom-mist types where only one drop in every 20 or 30 seen in the sight glass actually goes into the line, a feed rate of one or two drops every minute should give sufficient lubrication.

CHOOSING A TRIO OF ADEQUATE SIZE

Rule-of-Thumb Sizing. On the majority of compressed air applications, especially where only one cylinder is moving at the same time, a 1/4-inch size trio is more than sufficient. Larger size trios may be desirable where several cylinders are drawing air at the same time or where a single cylinder of larger than 3-inch bore is being operated. The port size of an air cylinder is usually a reliable guide to the maximum size for the trio.

If several cylinders are using air at the same time, a rule-of-thumb is to add piston areas of all cylinders. Use a 1/4" trio if total area is less than 10 square inches; a 3/8" trio if area is less than 20 square inches; a 1/2" trio if area is less than 40 square inches. If area is larger than 40 to 50 square inches, use a 3/4" or 1" size.

Effect of Undersizing. Using a smaller trio than recommended will not affect the amount of cylinder force produced but may cause the travel speed to be less than if a larger size was used. Slow speed can often be increased by opening the speed control valves a little wider. Speed can be increased by raising the pressure but this will also cause the stall-out force of the cylinder to be higher than may be desirable.

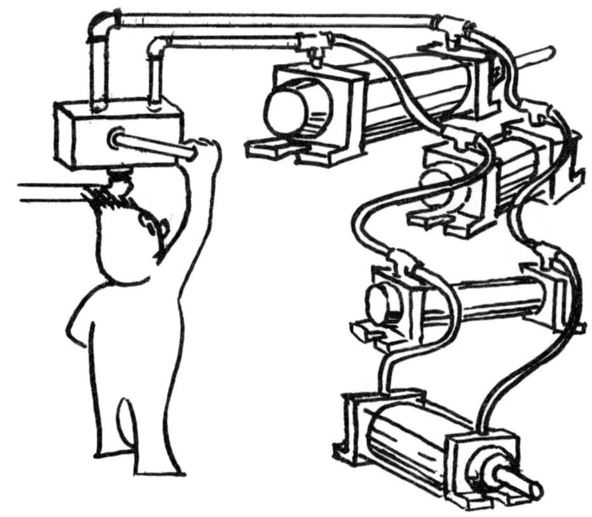

RULE-OF-THUMB

To select trio size for operation of multiple cylinders, add piston area of all cylinders and convert to equivalent area of one large cylinder.

INSTALLATION OF TRIO ASSEMBLIES

Trio units — filters, regulators, and lubricators — are intended to handle single machines. Each machine should have its own trio assembly. Trios, or any individual component, are seldom installed in a distribution system serving several machines. They should be installed as close as practical to the inlet of the 4-way control valve for these reasons:

The filter is designed to pick up water mist as well as dirt. Installing it close to the 4-way valve will reduce the possibility of additional water condensation between the filter and the 4-way valve.

The regulator is adjusted to the lowest pressure which will accomplish the job. It will maintain the adjusted outlet pressure accurately, but if located at a distance from the 4-way valve, the pressure drop in the connecting line will cause the pressure to drop on the 4-way valve as soon as air starts to flow, even though the regulator is delivering full adjusted pressure.

The lubricator delivers small droplets of liquid oil to be carried into valves and cylinders downstream. If located too far from the 4-way valve, some of the oil droplets will precipitate in the line and will not reach the components to be lubricated.

Normal Installations. On most applications the three components in a trio are installed as a complete assembly with brass pipe nipples or slip-joints connecting them together. A "normal" installation is one where all cylinders on the machine can operate at the same pressure level and where the piping distance from the lubricator to the farthest cylinder does not

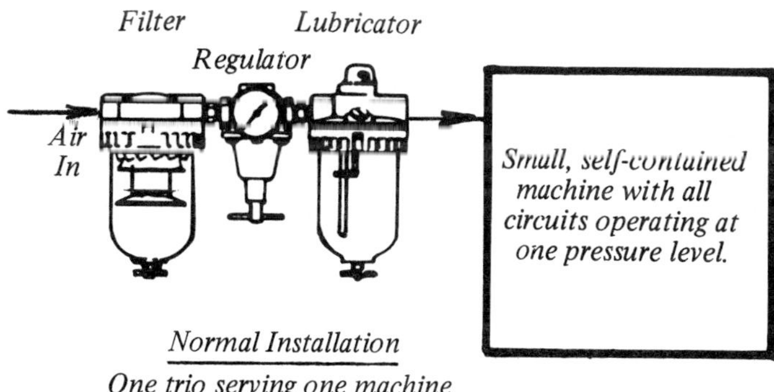

Normal Installation

One trio serving one machine.

exceed 10 to 15 feet. However, by using an atom-mist lubricator and by using large size piping to reduce flow loss, a "normal" system could have longer piping runs up to 100 feet.

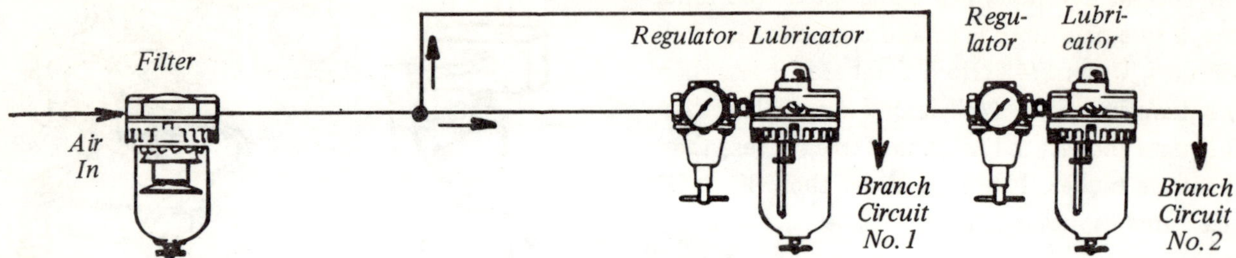

Two branch circuits operating at different pressure levels.

Two Pressure Levels. If more than one pressure level is required on different branches of the same machine, one large filter can serve the entire machine but each branch must be served by a separate pressure regulator followed by a lubricator, installed as close as practical to the 4-way valve inlet port in that branch.

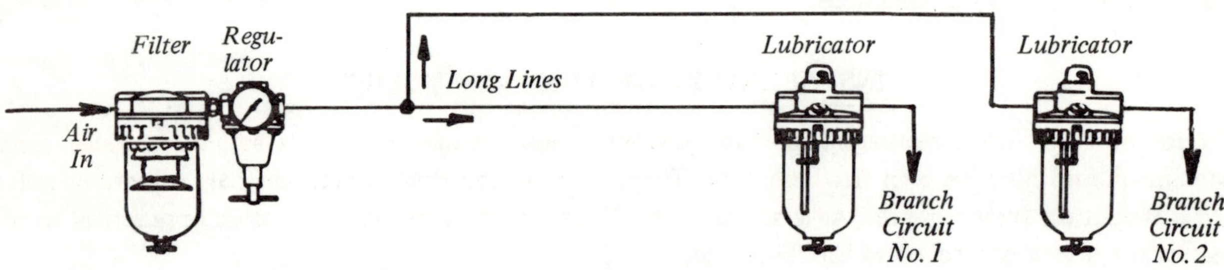

If connecting lines are long, use individual lubricators placed close to valves and cylinders.

Long Plumbing Runs. If all pressure levels in the machine are the same but if the branch circuits are located close to each other but at a long distance from the inlet air supply, one filter and regulator can serve all branches but should be located next to the tee junction where air splits off to the various branches. But, individual lubricators should be used in each branch, located as close as practical to the 4-way valve. However, we suggest that better pressure regulation could be maintained if individual regulators were used in each branch, located just ahead of the lubricators as shown in the top illustration.

Lubrication of Small Cylinders. Air cylinders of small bore and/or short stroke may not receive sufficient lubrication from a lubricator installed upstream of the 4-way valve if the connecting lines between the 4-way valve and the cylinder are long or if the air volume in the lines is about equal to or perhaps greater than the displacement volume in the cylinder. Every time the cylinder reverses direction, the same air simply goes back and forth in the lines and new, lubricated air, cannot reach the cylinder. It may be necessary to install an additional lubricator directly at the rod end cylinder port. This lubricator should be a type which can be adjusted to a very low oil feed. Oil fed into the rod end port will lubricate both the cylinder barrel and the rod bearing.

Many lubricators cannot pass air freely in the reverse direction. Install a check valve around them

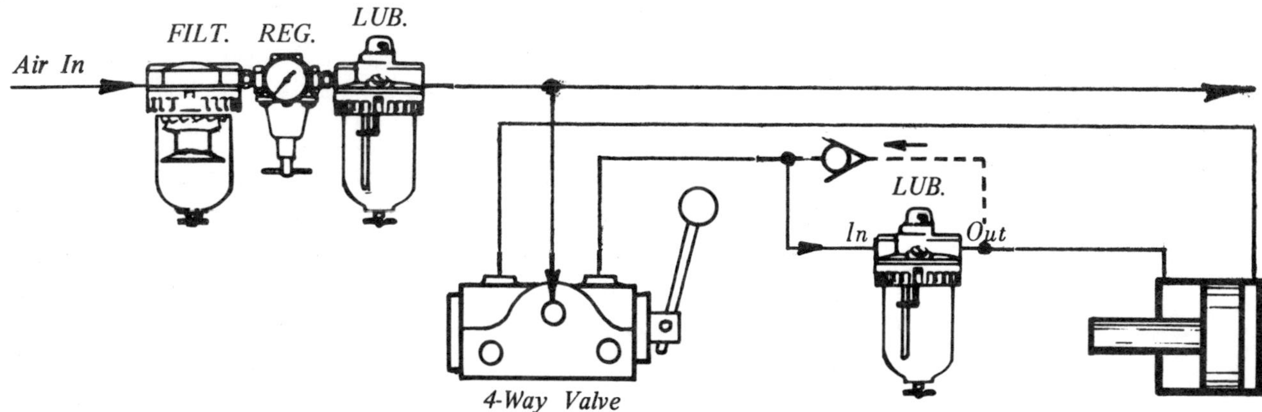

On very small cylinders, an individual lubricator can be installed in a cylinder port.

as shown in the illustration above. Other methods of feeding lubricant into small bore and/or short stroke cylinders are shown on Pages 39 and 40.

Cut-Off Valve. Install a 3-way shut-off valve ahead of each trio so air can be shut off when any unit of the trio or any component in the machine itself must be serviced or replaced. This valve is a safety requirement in some plants. It must be a full ported valve so it will not restrict air flow during normal machine operation. It must be a 3-way type so compressed air trapped in the machine will automatically be vented to atmosphere when the valve is closed. A 2-way valve is not suitable because it would not vent trapped pressure from the machine when closed.

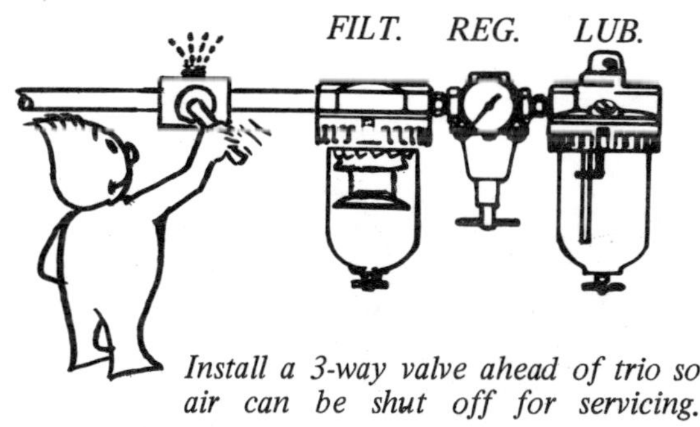

Install a 3-way valve ahead of trio so air can be shut off for servicing.

Install Trio in Accessible Location. Usually a trio should be mounted on the machine which it serves. It should, of course, be located as close as possible to the inlet of the 4-way valve and in such a position that it will be easy and convenient to remove the bowls for cleaning, replenishment, or replacement. Trios which have been located in hard-to-reach places often do not get the attention they should have. If possible, the bowls should be visible from a well-traveled aisle or where the machine operator cannot help seeing them as he operates the machine.

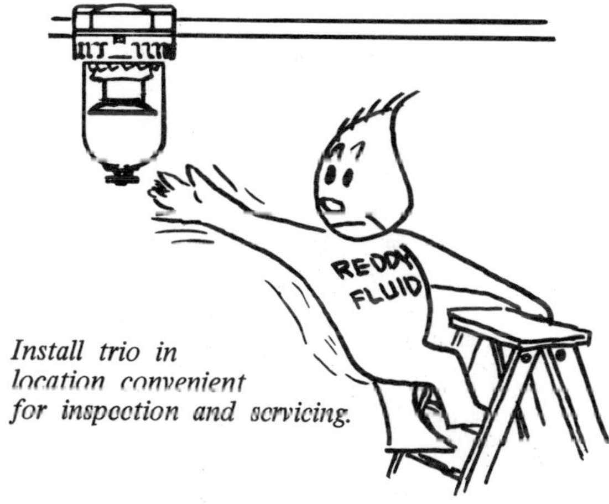

Install trio in location convenient for inspection and servicing.

Metal Bowls for Safety. Because of many industrial accidents, metal bowls, or metal bowl guards have been made available for filters and lubricators. Some bowl manufacturers refuse to supply plastic bowls without bowl guards. They are available for all popular trio brands and should most definitely

be used to protect personnel from these hazards:

(a). High Temperature. If internal or external air temperature exceeds 125° F, plastic bowls lose mechanical strength and could shatter.

(b). Solvents. If solvents are present in the air stream or in the surrounding air, plastic bowls can shatter. Even though protected by a metal guard, solvents will destroy the plastic bowl, and should not be permitted in the vicinity of a trio nor should they ever be used to clean dirt off the unit or surrounding area. When the plastic is exposed to a solvent it will form "craze" marks which will continue to spread until the bowl shatters. If these marks are observed, the bowl should be immediately replaced. Metal bowls should be used on applications near solvent fumes.

Applications possibly harmful to trio units with plastic bowls include injection of various chemicals including odorants, anti-freeze, or fire resistant liquids into air or gas lines. On these applications the safe procedure is to use metal bowls.

(c). High Pressure. Plastic bowls, even though protected by a bowl guard, should not be used on air pressure higher than 150 PSI. Metal bowls should be used for applications at higher pressure.

*Lubricator oil can be obtained
at a local filling station.*

*Before using cooking oil, test
its compatibility with rubber seals.*

Regulator Gauge Port. The gauge port on the regulator can be a convenient extra take-off connection for regulated air. Determine by visual inspection if it a full size 1/8 or 1/4" NPT port.

ROUTINE MAINTENANCE ON TRIO UNITS

What Oil to Use. The precise grade, brand, or viscosity is not extremely critical, but the best oil when available, is non-detergent petroleum oil with viscosity of SAE 5 or 10 or 10W. A higher viscosity should work well if lubricator is working at normal room temperature. Oil with a detergent additive or a viscosity improver additive is usually satisfactory although there is a remote possibility that the additives may not be compatible with plastic bowls. If metal bowls are used, almost any petroleum oil of very low viscosity can be used with no concern for its compatibility with plastic bowls.

A convenient source for small quantities of oil is a nearby filling station. Purchase a non-detergent motor oil of as low viscosity as available. The above information is only a general guide. Always follow recommendations of the trio supplier.

Other Fluids. Air line filters and lubricators are sometimes used for special applications, and when used with fluids other than oil or water, use metal bowls as a safety measure.

Other special applications may include filtration of various liquids or gases, injection of liquids such as anti-freeze into air or gas lines. Trio units may have rubber gaskets or seals which may not be compatible with certain liquids or gases, so compatibility of seals as well as plastic bowls should be checked.

In food processing plants it may be necessary to use cooking oil in air line lubricators because the air may come in contact with food. Caution! Never use unknown oil in a lubricator until you have proven it is harmless to all components in the system.

Oil and rubber compatibility can be checked with simple tests.

Compatibility Test. If no information is available on compatibility of a certain oil or other liquid with seals used in a trio, it may be necessary to make your own test which can be conducted as follows:

The seal sample or plastic bowl should be immersed for at least 72 hours in the liquid in question, and maintained at near 200° F for rubber seals or 175° F for a plastic bowl. An ordinary kitchen double boiler may serve in some cases, with water in the lower section. If the test cannot be continued over night, turn off the heat at night, letting the material stand in the liquid. Then resume the test the next day. When the test has been completed, remove the sample and wash it with soap and water. Give it a visual inspection and a "feel" test for any signs of deterioration. Any stickiness or swelling is a sure sign of incompatibility. Compare it dimensionally with an untreated identical part. If the volumetric swell exceeds 5%, the part is incompatible with that particular liquid. If there is any swelling *at all*, continue the test to see if the swelling increases.

Rubber can be measured more accurately if chilled to sub-zero temperature.

Synthetic rubber hardens slightly at reduced temperatures, so a dimensional check (for swelling) can be made more accurately by chilling both samples in a deep freeze before comparative measurements are made.

Excessive Oil Feed. Oil feed from a lubricator should be reduced if oil drips out or is blown out the exhaust ports on the 4-way control valve. Check again in a day or two to see if the feed should be further reduced.

The oil vapor put into the air by excessive oil feed is a health hazard to personnel working in the area. Also it wastes oil, perhaps causing the system to need

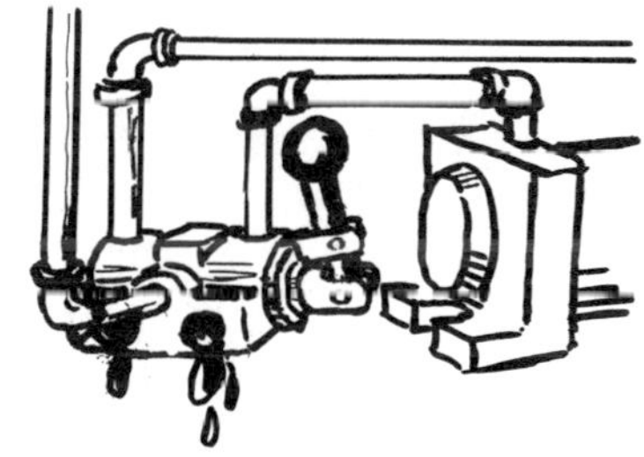

Oil drip from air valve exhaust indicates excessive oil feed.

servicing more often and the excess oil may be harmful to those valves and cylinders which have been permanently pre-lubricated by washing out the permanent lubricant.

Daily Inspections*.* In plants where the air system is in daily use, a routine inspection should be made every day of all trio units to drain water and to inspect or replenish oil level in the lubricator. It is important to make sure that oil is being fed into the system. On systems recently put into operation, the maintenance person should note the oil level every day for a few days in lubricators to be sure oil is being fed, and reduce the feed if it is too much.

Use only dry cloth or soap and water to clean plastic bowls.

Do not use lacquer thinner or other solvents for cleaning plastic bowls.

Cleaning Plastic Bowls*.* If a plastic bowl should become so dirty the oil level cannot be observed, the metal guard should be removed and the plastic cleaned. Use only a damp cloth, with a little soap if necessary. Never use lacquer thinner, gasoline, naptha, or any other solvent because of their destructive effect on the plastic. After cleaning with a solvent, the bowl may look unharmed. But after a while craze marks will appear and the bowl will eventually explode.

While cleaning the bowl of a filter, inspect the element. Blow it out from the inside with an air hose, replace if dirty. Replace it on a routine basis about every six months or as often as required for the particular environment and duty cycle of the machine.

TROUBLESHOOTING TRIO UNITS

Air Leak From Regulator*.* A continuous air leak from the small vent hole in a regulator bonnet may or may not indicate a bad diaphragm or leaky main poppet, depending on whether the regulator is self-relieving or non-relieving.

On a self-relieving type, the one used on the majority of industrial air systems, test for internal leakage by allowing the cylinder to remain stalled at either end of its stroke. If air continues in a steady leak, either the diaphragm itself is cracked or the O-ring seal in the relieving poppet located in the center of the diaphragm is leaking. Remove the diaphragm for examination and replace the O-ring in the center of the diaphragm.

If it is a non-relieving type, any leakage from the vent hole indicates a crack in the diaphragm or a defective gasket seal between main body and the top bonnet. Overhaul kits are available with diaphragm and all seals.

Important! Repair a leaking regulator as quickly as possible after the leak is discovered. A complete rupture of the diaphragm will allow full air pressure downstream to the valve and cylinder. Some solenoid valves, if operated at a pressure higher than their rating may shift by themselves without being energized, causing a cylinder to extend and injuring the operator.

Continuous leak out regulator vent may indicate defective diaphragm.

Pressure Regulator Problems. If adjusting the T-handle fails to reduce outlet pressure, check these possible causes:

(1). The regulator may be connected in backward, with inlet pressure connected to regulator outlet port. It will not work if connected backward.

(2). Regulator improperly connected. The gauge port may have been connected as a main port or used as an outlet port. This is not a full-flow port on some regulators.

(3). Broken parts inside the regulator. If diaphragm is cracked or broken, a steady flow out the vent hole in the bonnet will indicate this as a possible problem.

(4). If the regulated pressure is too low and cannot be raised by screwing down the T-handle, the range can be raised by replacing the main spring in the bonnet with a heavier spring for a higher range.

(5). We recommend a routine overhaul of a regulator once a year even though it appears to be working properly. Use an overhaul kit to replace diaphragm, seals, and gaskets.

Filter Problems. Very few problems are encountered with air line filters provided the element is cleaned or replaced when needed. Here are some suggestions for good filtering action:

(1). Be sure the filter is correctly plumbed, with incoming air connected to the inlet. It has limited filtering ability in the reverse flow direction.

(2). Present day trend is to use a very fine element, 5 to 10 μm rating. Experience has shown how important microscopic dirt removal is for protection of equipment connected downstream.

(3). If the filter element is allowed to become restricted with dirt, the pressure drop through the filter will increase. This will cause cylinders on the machine to operate at a slower speed, although their force at stall will not be affected.

Service the filter before water rises above baffle.

(4). Keep the condensate drained manually or with an automatic drain. If allowed to build up above the baffle plate on the lower end of the element, turbulence inside the bowl may cause dirt to deposit on the element and may allow water mist to pass through the element.

Lubricator Problems. (1). Insufficient Lubrication. If an air cylinder is receiving internal lubrication there should be a very thin oil film on its piston rod. Small-bore and short stroke cylinders may run dry because the volume of air in connecting lines is greater than the volume displaced by the cylinder. Oil mist cannot reach the cylinder. Refer to Page 39 for a discussion and remedies for this problem.

(2). If oil feed cannot be increased with the metering valve, perhaps the viscosity of the oil is too high or the machine is operating in a cold environment.

(3). If lubricator will not feed, be sure it is not plumbed in backward. A standard lubricator, if used on an extremely low flow, may not feed because feed pressure cannot be developed in its throat. Replace it with a smaller size lubricator, bushing down the lines to fit the smaller ports.

(4). We have previously discussed the importance of locating the lubricator close to the 4-way valve inlet. If the oil must be carried to a distance greater than 10 to 15 feet, use an atom-mist type unit which generates an internal oil mist and permits only those oil droplets of 2 μm or smaller to be delivered into the air stream. Larger droplets are returned to the bowl for re-processing.

Chapter 8

Dryers for Compressed Air

Two Forms of Water. In a compressed air system there are two forms of water which may cause trouble. Water *mist* is finely divided droplets of liquid water suspended in the air stream. The mechanical air filter described in the preceding chapter as part of a trio unit, is quite effective in removing water *mist* — up to about 95% removal. However, on some applications described in this chapter even the 5% remaining water could be a problem, and would have to be removed either by a dryer or by an additional mechanical filter. Then there is water *vapor* which is evaporated water and is present in all compressed air systems as *humidity*. This water is in gaseous form and quite invisible. But as long as it remains a *vapor* it will cause no problem — in fact it is beneficial because it will carry fluid power just as well as if it were air. But when a condition is present which causes some of this vapor to condense into *mist*, then it will give trouble.

Relative Humidity. The amount of water *vapor* in air is expressed as "relative humidity". This is the amount of *vapor* in the air compared to the relative amount it is capable of holding at that temperature and pressure. For example, a relative humidity of 50% means that the air (at that particular pressure and temperature) contains 50% of the maximum vapor it could hold before it reaches 100% relative humidity, called saturation. When a condition is present in the air system which would cause the relative humidity to exceed 100%, the excess *vapor* would condense and remain in the air stream as water *mist*.

The air in a container of, say 1 cubic foot volume, will hold more water if heated. That is, its relative humidity will be lower than when it is cooler and it is said to be drier. The same volume of air, if its physical volume is reduced by compression will hold less water than when allowed to expand to a greater volume. Its relative humidity under compression will become higher and it is said to be wetter. The object of an air dryer is to reduce the relative humidity of the air sufficiently that condensation will never occur while the air is passing through the system.

How Water Gets Into the Air Stream. The air compressor takes in air at atmospheric pressure and compresses it to about 1/7th of its original volume. This increases its relative humidity by 7 times. If its relative humidity originally was 30%, its new humidity (after its temperature has returned to its original level) will tend to become 210%. The excess water above 100% humidity or saturation, would drop out, most of it in the bottom of the receiver tank and could be drained off. But the air in the receiver tank would be at 100% humidity as it was fed into the distribution system, and would be subject to further condensation if piped into a cooler area, or if cooling should take place by expansion. Therefore, water can drop at any number of places in an air circuit depending on local conditions. To prevent condensation, the air coming out of the receiver tank at 100% humidity must be processed through a dryer to reduce its humitity to a somewhat lower value. It is important to remember that a regular mechanical filter, no matter how effective for removing water mist and dirt, cannot reduce humidity.

APPLICATIONS FOR DRY AIR

If your process involves any of the following operations, or one of a similar nature, the air must be dried. It may also be necessary to add oil absorbing or special filters to remove microscopic particles or bacteria.

(1). Transferring powdered material through pipes under pressure by flotation. This includes plastic or iron powder, cement, and many other materials.

(2). Paint Spraying. Water droplets are often precipitated at nozzles caused by expansion cooling to cause defects in painted surfaces. Mechanical filters cannot prevent this. Air used by an artist for air brushing must be completely dry.

(3). Cold Locations. If the compressed air is piped outdoors in winter or into a cold storage area, water is likely to precipitate and freeze in the equipment or in the lines.

(4). Instrumentation. Clean, dry air is important because moisture may be corrosive to expensive and delicate instruments.

(5). Air Gauges. For inspection of machined parts, dry air for gauging is most important. Water precipitated from the air can be destructive.

(6). When air is injected into gas mixtures for furnaces, or where jets impinge on or come into contact with hot materials, the air must be dry.

(7). Food Processing. If air comes into contact with food, it must be sterile and dry. To remove bacteria it should also be run through a microalescer filter. Applications like blowing food chips off of weighing scales and into bags require very dry air.

(8). Shot blasting, shot peening, or sand blast applications can tolerate no moisture precipitation.

(9). Packaging. Compressed air is often used on packaging machines to blow bags open, to move products, to blow off crumbs, to preform paper by air blast and many similar applications.

(10). Medical and Dental. It almost goes without saying that air used for aspiration, air drills, blowout, etc., must be sterile and dry.

Paint Spraying

Cold Locations

Instrumentation.

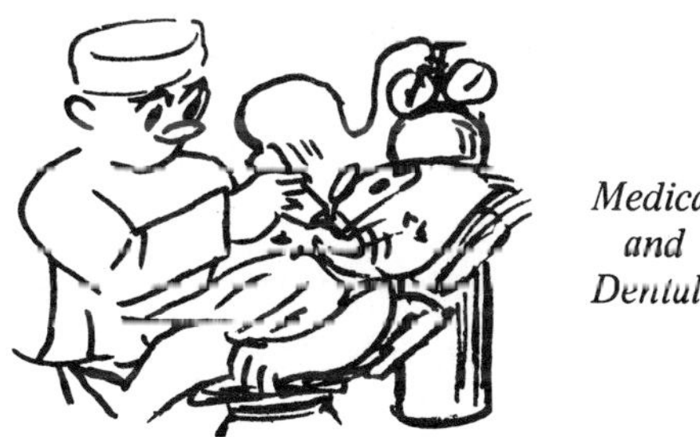

Medical and Dental

In Many Applications Air Must Be Dried as Well as Filtered.

HOW COMPRESSED AIR IS DRIED

Air delivered from the compressor receiver tank into the distribution plumbing is saturated with water *vapor*, that is, its relative humidity is 100%. If its temperature is above room temperature when it enters the distribution plumbing, water will condense along the line as the air becomes cooler. This water will flow along the distribution piping until it reaches a drain leg — a drop line terminating in an automatic drain which will discharge it from the line. The air flow through the distribution plumbing remains saturated with water *vapor* and drops more water if its temperature should fall, even slightly. All of the water vapor need not be removed — just enough so additional temperature drops will not cause more water to drop out. Removing more water vapor than necessary is a waste of power because water vapor is a good conductor of fluid power.

Water vapor removal is done with one of several kinds of air dryers. One large dryer may serve the entire plant, or a smaller dryer may serve only one branch, for example the air supply to the paint spraying room.

△ *Refrigeration-Type Air Dryer*

Twin Tower Heatless Air Dryer ▷

Refrigeration Dryer. These are similar in principle to a home refrigerator but of greater HP capacity. The principle involved is to first cool the air. Reducing its temperature causes some of the water vapor to condense and be drained away. Then the air is warmed back to room temperature and its humidity will be less than 100% because some of its water has been removed.

The dryer is a self-contained freon refrigeration system complete with gas compressor, condenser and evaporator coils and expansion valve. Saturated inlet air coming from the compressor receiver tank is passed in proximity to the evaporator coils and is cooled to about 39° F. As it cools some of its water is forced to condense. At 39° F the air is still saturated but holds much less water because its temperature is lower. Then, the 39° F saturated air is passed in proximity to the condenser coils and warmed back to room temperature before being fed into the distribution plumbing. As it is used in the system, no more water will condense unless its temperature falls to lower than 39° F either because of reaching a colder environment or from its expansion in air system components.

Refrigeration dryers are probably used more often than either of the other types described here. They require virtually no maintenance if equipped with an automatic water drain; they consume more electric power than the other types but they do not waste any of the compressed air; they will not operate at pressure dewpoints lower than 39° F because the condensed water may freeze.

Dessicant Dryer — Twin Tower Heatless Type. Two identical towers are filled with dessicant, usually silica gel, which will adsorb water vapor without chemically combining with it. Water vapor is condensed and held. Later the water can be driven out of the dessicant either by heat or by passing dry air through it. One tower is in service while the other one is being purged. After a timed interval, usually 3 to 5 minutes, the towers are automatically switched so the first one can be purged while the second one is in service.

In this heatless dryer, water is purged from the dessicant during the regeneration cycle by passing through it a small flow of dry air, tapped from the dry air delivered from the opposite tower. The dry air picks up water from the dessicant and blows it out to atmosphere.

The towers are automatically switched by a 4-way solenoid valve controlled from an electric timer. To understand how a small flow of dry air can carry away all the water condensed from a much larger flow, remember that every 1 cubic foot of compressed dry air tapped off the working tower for regeneration, when allowed to expand to atmospheric pressure becomes much drier. For example, when allowed to expand from 100 PSI, its volume will increase by about 7 times and it will become 7 times drier than it was when compressed to 100 PSI. So it can pick up 7 times as much water after being allowed to expand. Thus, 1/7th of the dry air produced by the working tower is used to purge the other tower, and the remaining 6/7ths is fed into the air distribution lines.

Heatless dryers waste about 1/5th to 1/7th of the air for the regeneration process. But these dryers have numerous advantages: (a), they require no regular maintenance; (b), no water drain is required — the condensate is blown out to atmosphere as vapor; (c), they consume only about 50 to 75 watts of electric power, just enough to operate the electric timer and 4-way solenoid valve; (d), they can be shut down at night and will almost immediately produce dry air when started the next morning; (e), they can produce very dry air down to -75° F pressure dewpoint depending on type of dessicant used and volume of air tapped off for regeneration; and (f), they can operate in very cold environments, under arctic conditions if necessary. However, to save power and conserve air, they should be adjusted to a dewpoint no lower than necessary.

Dessicant Dryer — Absorbent Pellet Type. This was the first kind of dryer developed to produce large volumes of dry air for commercial use. They are still used to a limited degree but they are not as popular since the development of refrigeration and twin tower heatless dryers. They are capable of producing very clean and dry air if carefully sized so they are not ever overloaded, and if the incoming air temperature is never allowed to exceed 100° F.

Air is passed through a bed of chemical pellets, several inches to several feet thick depending on dryer size. The pellets pick up water vapor by deliquescing action and slowly dissolve, producing a salt solution which must be regularly drained. The pellet bed must be replenished at intervals of 1 to 3 months depending on dryer duty cycle. Servicing of the pellet bed may be required at more frequent intervals if overly warm inlet air causes the pellets to fuse together.

On some of the more popular brands the pellets are sodium chloride, common salt. A small amount of other chemicals,

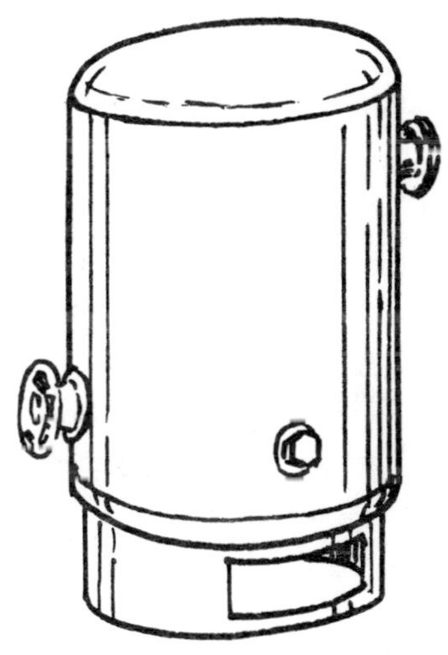

Deliquescent-Type Air Dryer Uses Dessicant in Pellet Form.

possibly 1%, are mixed with the salt to offset its normally corrosive action. The salt mixture is pressed into pellets so there will be spaces between them for air flow. Many other chemicals have the same deliquescing action but salt is one of the best and is cheaper than most.

Inlet air enters the dryer at the bottom, passes through a filter screen, flows upward through the pellet bed, and leaves the dryer at the top through another filter screen which prevents dessicant from being blown out into the distribution line.

These dessicant dryers cost less than the other types but have several disadvantages: (a), pellet cost plus man hours required for servicing and replenishing is usually higher than electric cost for a refrigeration dryer; (b), service life is usually shorter than for the other types because of corrosion caused by the dessicant; (c), salt is sometimes blown out into the distribution lines and causes corrosion in the lines if a maintenance person is in the habit of "blowing out" the air lines before start-up each morning; (d), the solution of condensate and dissolved dessicant cannot be handled reliably by an automatic drain, and the dryer must be manually drained every day; and (e), these dryers are very sensitive to high temperature in the inlet air, and become virtually useless if inlet air temperature should ever exceed 110° F even for a short interval. The dessicant melts together and the services of a maintenance person are required to open up the dryer and break up the pellets.

Other Dryers. For small air flows such as for instrumentation air, small non-regenerative dryers can be used. These are small cartridges filled with a dessicant which will absorb water and hold it until its retention capacity has been reached. The dessicant can then either be discarded and replaced or can be rejuvenated with heat or dry air. There are many chemicals which can absorb water vapor, silica gel being one of the best for most applications. Lower dewpoints can be achieved with certain other dessicants. Dryers of this type are relatively small and not practical for use on the plant main air flow, but could be used in a low-flow branch circuit.

APPLICATION OF COMPRESSED AIR DRYERS

Air Temperature. A basic principle of dryer application is that temperature of incoming air *must* be as low as practical. The lower the temperature, the more effectively it can be dried. For example, inlet air at 110° F will not be as dry after passing through the dryer as air at 100° F. Hot air put into a dryer will be dried to some degree, but when discharged into the distribution line may again reach 100% relative humidity as it cools in the line, and water may condense in the system. Hot air will not harm a refrigeration or heatless dryer but will render it less effective. But hot air will quickly put a pellet type dessicant dryer out of commission by fusing the pellets together. The dryer must then be opened and the pellets replaced or at least broken up. To maintain a low inlet temperature it may be necessary to install an air-cooled or water-cooled aftercooler either between the air compressor and the receiver tank or between the receiver tank and the dryer. The surface of the receiver tank helps to cool the air, so the larger the tank, the less aftercooler is needed or the more effective the dryer.

On all types of dryers, the lower the temperature of the inlet air the drier the outlet air.

Air Pressure. Another basic principle of dryer application is to install the dryer where the air is at high pressure. Dryer capacity is less at lower pressures.

When selecting a dryer, remember that dryer capacity is directly proportional to *absolute* pressure of inlet air. When selecting a dryer from a catalog, if the catalog SCFM flow rating is specified at a pressure different from your pressure, the catalog rating must be adjusted. If the SCFM rating is specified at one pressure, the equivalent rating at any other pressure, higher or lower, can be determined by finding the proportion between the (absolute) pressure values at these two levels. The following example illustrates such a calculation:

Dryer Flow Capacity Is Directly Proportional to Absolute Pressure at Inlet.

Example: If an air dryer has a catalog rating of 25 SCFM (standard cubic feet per minute) at a gauge pressure of 100 PSIG, convert this rating to an inlet pressure of 80 PSIG.

Solution: Remember that dryer flow rating is proportional to *absolute* and not to *gauge* pressure. Convert 100 PSIG and 80 PSIG to absolute pressures by adding 14.7 PSI to each one. Thus, 100 PSIG becomes 114.7 PSIA (absolute), and 80 PSIG becomes 94.7 PSIA. The ratio between these pressures is: 94.7 ÷ 114.7 = .826. The adjusted rating will be 25 x .826 = 20.65 SCFM. The following table can be used in making these conversions:

DRYER RATING CHART

Figures in the chart are percentages. If dryer SCFM flow rating is known for one of the gauge pressures in the left column, adjusted SCFM flow ratings, in percentages of the original, are shown for gauge pressures along the top for 0 PSIG (atmospheric) to 180 PSIG.

Rating Known at:	0 PSIG Atmos.	20 PSIG	40 PSIG	60 PSIG	80 PSIG	100 PSIG	120 PSIG	140 PSIG	160 PSIG	180 PSIG
80 PSIG	15.8%	36.9%	58.0%	79.0%	100%	121%	142%	163%	184%	205%
90	14.3%	33.3%	52.4%	71.4%	90.5%	110%	128%	147%	166%	185%
100	13.0%	30.5%	48.0%	65.4%	82.6%	100%	118%	135%	152%	170%
110	12.0%	28.0%	44.0%	60.0%	76.0%	92.0%	108%	124%	140%	156%
120	11.1%	25.9%	40.7%	55.5%	70.3%	85.0%	100%	115%	130%	144%
130	10.2%	23.9%	37.8%	51.6%	65.4%	79.3%	93.1%	107%	121%	135%
140	9.50%	22.4%	35.4%	48.3%	61.2%	74.1%	87.1%	100%	113%	126%
150	8.92%	21.1%	33.2%	45.4%	57.5%	69.6%	81.8%	93.9%	106%	118%

Use of Chart. The example above can be solved by the chart. Find the percentage of the dryer rating at 100 PSI which will apply if it is used on an 80 PSI inlet. Find in the chart on the 100 PSI line and the 80 PSI column, 82.6% (.826 multiplier). Multiply times 25 SCFM = 20.65 SCFM new rating.

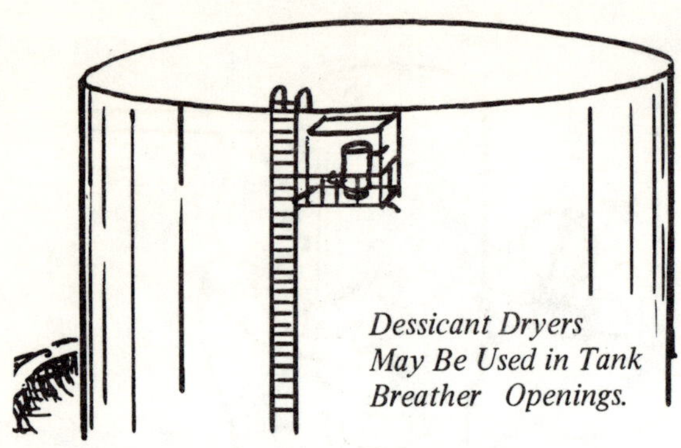

Dessicant Dryers May Be Used in Tank Breather Openings.

CONVERSION FACTORS

For Calculating SCFM Air Flow for Tank Breathing:

Each GPM = 231 CIM (Cu. In./Min.) And, 1728 Cu. In. = 1 Cubic Foot. Therefore, Each GPM = 0.134 CFM.

Dryer Used for Atmospheric Tank Breather. Air dryers can be used on the atmospheric vent of large chemical tanks, to remove water vapor from air which is admitted as the level falls. Dessicant dryers of the silica gel non-regenerative type may be most suitable because no electrical connection is required and they will work at sub-freezing temperatures.

The dryer should be connected with inlet to atmosphere and outlet to the tank. Calculate the SCFM size from the GPM rate at which chemical will be pumped out of the tank.

Calculate the maximum air flow through the dryer from the known GPM discharge rate of the liquid. Use conversion formula shown in the box:

Each GPM requires .134 SCFM dryer capacity.

Use the dryer rating chart on the preceding page to find how much to de-rate a dryer when used at atmospheric pressure. For example, if the dryer catalog gives SCFM rating at 100 PSIG, the chart shows it will have to be de-rated to 13% of its capacity for atmospheric service.

INSTALLATION OF COMPRESSED AIR DRYERS

Location. Since elevated inlet air temperature interferes with operation of any dryer, a favorable location should be selected for its installation.

The dryer cabinet should be shielded from any source of radiant energy. This includes direct sunlight, furnaces, ovens, die-casting machines, etc. If air distribution lines are run close to steam or hot water lines, or space heaters, they should be insulated.

Keep Dryer Away from Sources of Heat Radiation.

A dryer should not be installed in closets or close spaces where there is no air circulation. It should be placed in a location where there is no obstruction to interfere with convection air currents. If possible, locate it near the floor because this location is usually cooler.

Protect dryer from freezing of the condensate. Low ambient temperatures are beneficial to the drying action provided the condensate does not freeze.

A drain connection should be provided near the dryer if needed. Twin tower heatless dryers discharge the water as a vapor into the air and do not need a drain.

Protect Dryer from Freezing of the Condensate.

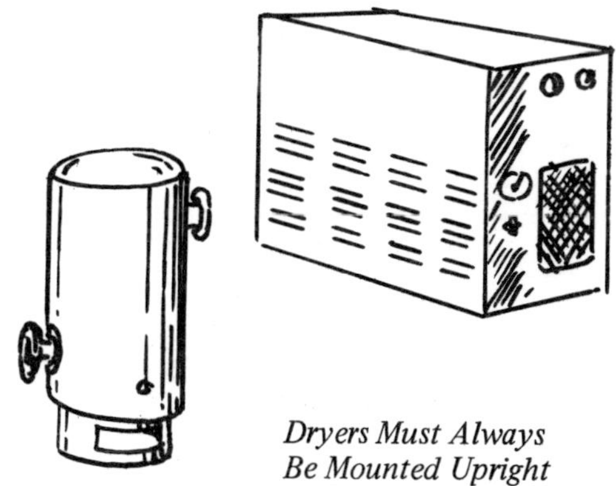

*Dryers Must Always
Be Mounted Upright
So Condensate Can Drain.*

__Mounting Position__. All dryers which discharge condensate must be mounted upright in a vertical position. They should never be mounted horizontally or upside down. Twin tower heatless dryers should be mounted vertically so the gauges are correctly oriented, but if necessary, they can be mounted other than vertically.

To be sure the dryer installation will be effective, inlet air temperature should be checked after installation and periodically thereafter when ambient temperature is most severe. The best way to measure inlet air temperature is to provide a junction in the plumbing close to the dryer inlet for installation of the sensing bulb of a remote reading thermometer. If not possible to use a remote reading thermometer, fasten an ordinary thermometer to the inlet pipe with heat conducting putty. This reading may be as much as 5°F lower than actual air temperature.

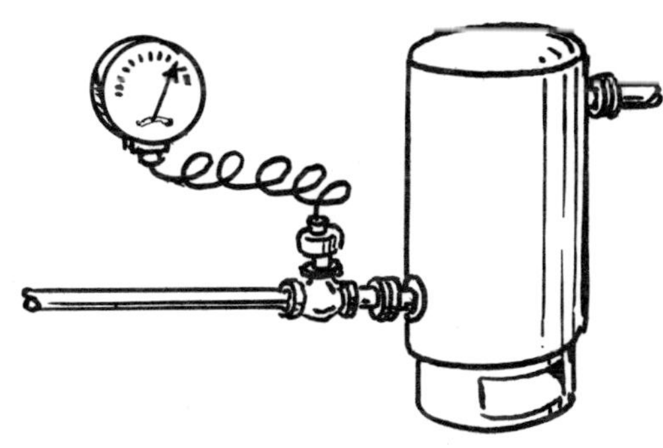

*A Remote Reading Thermometer Is an Accurate
Way of Measuring Dryer Inlet Temperature.*

__Connecting Into the Air Line__. To make a good installation, three full-flow gate or plug valves should be connected in such a way that the dryer could be isolated, disconnected, and removed for maintenance or service. By closing the two series valves and opening the by-pass valve the dryer can be removed and the air system used without a dryer.

__Draining the Dryer__. Install a manual drain valve of same pipe size as drain connection on the dryer. If possible, this should discharge into an open drain so the volume of condensate can be observed.

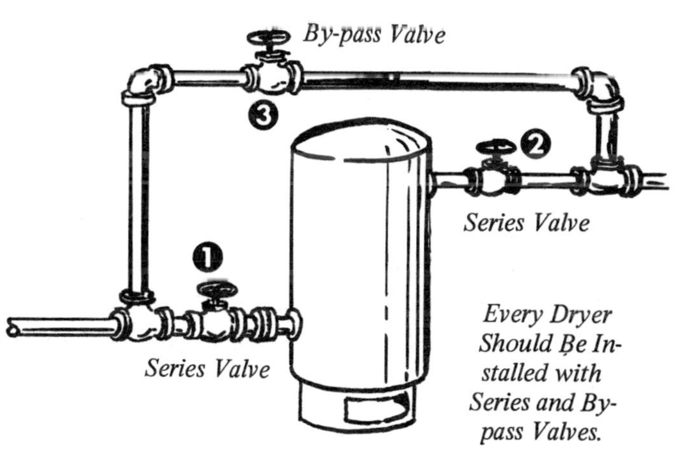

*Every Dryer
Should Be Installed with
Series and By-
pass Valves.*

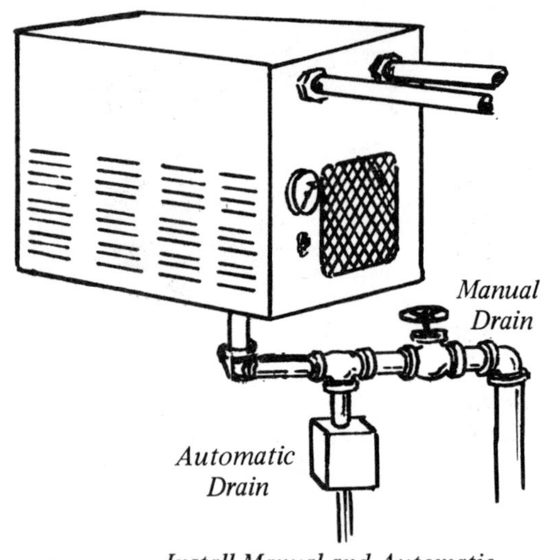

*Install Manual and Automatic
Drain Valves in Parallel.*

On refrigeration dryers an automatic drain valve can be installed in addition to, but not in place of, the manual valve. It should be stubbed off in parallel with the manual valve.

Automatic draining has not been very successful with dessicant dryers of the chemical pellet type because the dissolved salt forms deposits which clog up an automatic drain.

Condensate from a refrigeration dryer is not likely to clog an automatic drain because it is pure water, almost equivalent to distilled water. In fact, some users save it for uses requiring pure water. An occasional inspection should be made, however, to be sure the drain valve is working properly. If it should become clogged, water will back up inside the dryer and interfere with drying action.

Placement of Dryer in System. For most air systems we believe the placement order shown in the diagram below will give the best results, although the connection order can be varied under special circumstances or in special situations.

Compressor Aftercooler. This unit is most effective if located at the highest temperature point in the system, and this is usually in the line connecting newly compressed air into the receiver tank. It can be moved to the outlet side of the receiver tank but will be less effective at that location.

Receiver Tank. This tank serves not only to store the energy of compressed air but performs a valuable function in radiating heat from the stored air. Water which is condensed from the air stream in the aftercooler flows into the receiver tank and can be drained along with the water which condenses in the receiver tank itself.

Air Dryer. It is vital to the drying process that inlet air to the dryer be as cool as practical and hopefully no higher than ambient room temperature. On some installations, additional benefit can be realized by moving the dryer as far downstream as possible, thus providing additional cooling in the line connecting back to the receiver tank. A drop leg should be provided in this connecting line with an automatic drain to prevent condensed water from following the line into the dryer.

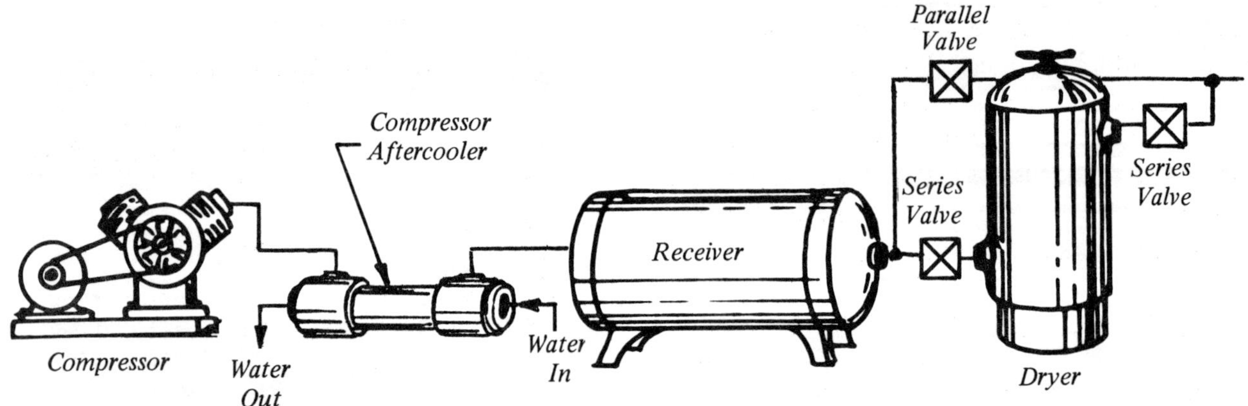

For most installations the compressor aftercooler and dryer should be connected in this order.

Do not install a dryer in the atmospheric intake of an air compressor. Even if the dryer could reduce the humidity of incoming air to 15%, by the time it was compressed to 1/7th of its volume its humidity would again be at 100%. In fact, do not install a dryer in any low pressure part of an air system if it can be installed at another location where the pressure is higher.

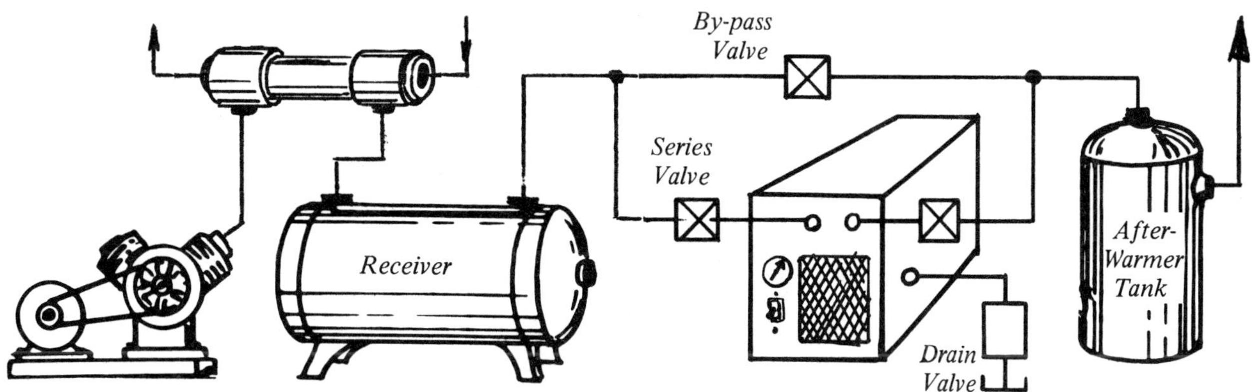

An after-warming tank may increase dryness of air delivered from a refrigeration-type dryer.

Warming Tank. On a refrigeration dryer the outlet air can be piped into another receiver tank (warming tank) before being fed into the distribution system. This lets the air warm up to room temperature. In cold locations an electric heating element can be added if necessary. This reduces air relative humidity because warmer air has lower relative humidity than the same air when cooler. A warming tank will only increase dryness if the dryer is delivering air at below room temperature. Our recommendation for order of placement of components in a refrigeration dryer installation is the same as described on the preceding page.

Blow-Down Valves. Most plant installations have one or more blow-down valves located at the bottom of drop legs. They are used by the maintenance crew to blow water out of the lines before plant start-up.

After installation of an air dryer these blow-down valves should not be necessary for blowing out water, and they should be replaced by smaller valves, 1/4" size which can be opened to test whether the dryer is doing a proper job. Blowing down with a large valve may place a temporary overload on the dryer and permit water to be carried into the system.

When a dryer is installed, blow-down valves must be reduced in size or eliminated.

Mineral-Free Water. Water condensate discharged from the drain of a refrigeration air dryer is virtually free of dissolved minerals and is a good source for small quantities of pure water. Solid particles which may have entered the dryer inlet in the air stream can be removed by filtering.

RULE-OF-THUMB

Tank Volume for Downstream Receivers:

1½ to 2 Gallons per Compressor HP

or

1 Gallon per 2 SCFM Air Flow

Downstream Receiver Tanks. On some systems additional receiver tanks installed at points downstream from the compressor may improve performance.

Tanks installed at the output of a refrigeration air dryer help to warm the dry air which makes it even drier. This was covered on Page 163.

Air-operated machines working at the end of a long distribution line may suffer from low pressure caused by line pressure loss when many other machines upstream are using large volumes of air. A receiver tank installed at this point can maintain air pressure by storing air during intervals when pressure is higher. Provided the plant air compressor has sufficient flow capacity to satisfy all machines, the problem of low pressure can often be solved, or at least improved, by an auxiliary receiver tank at the far end of the distribution system.

Tank volume should be as large as practical, the larger its capacity, the better it can maintain pressure at the end of the line. But we suggest a minimum volume as indicated by the rule-of-thumb based on the plant compressor HP. For a 50 HP air compressor we suggest at least a 75 to 100 gallon receiver tank at the end of the line. This is in addition to the normal receiver tank near the compressor.

RULE-OF-THUMB

Heat to be Removed by Aftercooler
$BTU/hr = SCFM \times 1.08 \times TD$

Compressor Aftercoolers. Most aftercoolers use water for cooling compressor air and these are similar to the shell and tube heat exchangers described in Chapter 6 but with air passing through the inside of the tubes and water circulating through the shell.

The rule-of-thumb can be used for estimating the size of aftercooler which may be required. It is based on the volume of air flow in SCFM and the temperature difference, TD in degrees F between inlet and outlet air. For example, if a compressor is delivering 350 SCFM and the temperature entering the aftercooler is 140° F, find the heat load to be removed to deliver outlet air at 100° F.

From the rule-of-thumb: $BTU/hr = 350 \times 1.08 \times [140 - 100] = 15,120$. Use this as data for specifying aftercooler size. The manufacturer will specify water volume and temperature to remove this amount of heat. If compressor SCFM is not known, figure about 4 SCFM per HP for vane-type or 1-stage piston compressors and about 6 SCFM per HP for 2-stage piston-type compressors.

TROUBLESHOOTING ON A COMPRESSED AIR DRYER

After the installation of a dryer on an existing system, allow several days for the system to stabilize. The inside surfaces of plumbing and components may be wet, or there may be water pockets which take time to dry out. After this stabilizing period, if there is still water in the compressed air, check through these potential problems. Check the entire list because there may be a combination of several problems.

(1). Dryer Undersized. Perhaps the estimate of air flow on which dryer selection was based was in error.

Re-check flow calculations and be sure calculations are based on actual pressure to the dryer. Pressure lower than assumed on the original calculations will result in an under-sized dryer.

The obvious remedy is to add a second dryer in parallel with the first, or add the second dryer in a branch circuit where dry air is particularly important. Each dryer should have its own set of series and by-pass valves. Be sure these valves have full size in-

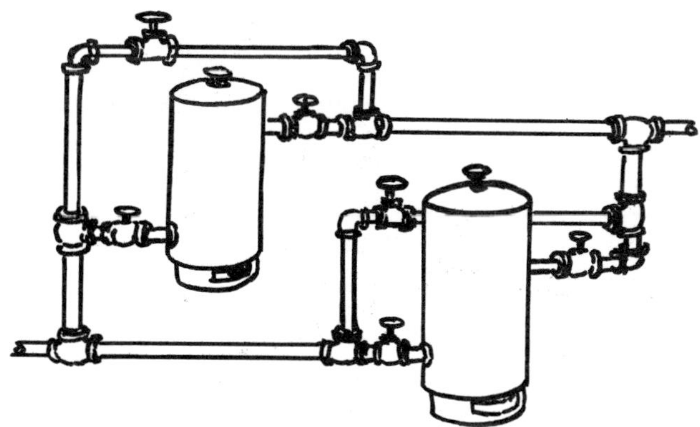

Several Dryers May Be Operated in Parallel.

ternal orifices so they do not restrict the flow. By making each dryer independent, one of them can be cut out for maintenance or servicing (during periods of low air usage) while the other remains in service.

(2). High Inlet Air Temperature. The inlet air temperature may be too high for the dryer to function at its rated capacity. This may occur only at certain times of day, so a 24-hour check should be kept by recording inlet temperature every hour.

First, feel the inlet pipe. If it is above body temperature, the air is too warm for the dryer to deliver its full rated capacity. Measurements of temperature can be made with a recording temperature instrument if one is available. If not, use a bi-metal thermometer installed in the side connection of a tee at the dryer inlet. A remote reading thermometer can be used with its sensing bulb installed in a pipe tee at the dryer inlet. A thermometer placed on the outside surface of the pipe will not give an accurate reading of air temperature inside the pipe.

The remedy for this condition is the addition of a heat exchanger (aftercooler) ahead of the dryer, or in some cases the dryer can be moved further downstream where the inlet air has a chance to cool before entering the dryer.

(3). Radiant Heat. Check to see if the dryer is close to radiant heat sources such as furnaces, steam boilers, or die casting machines.

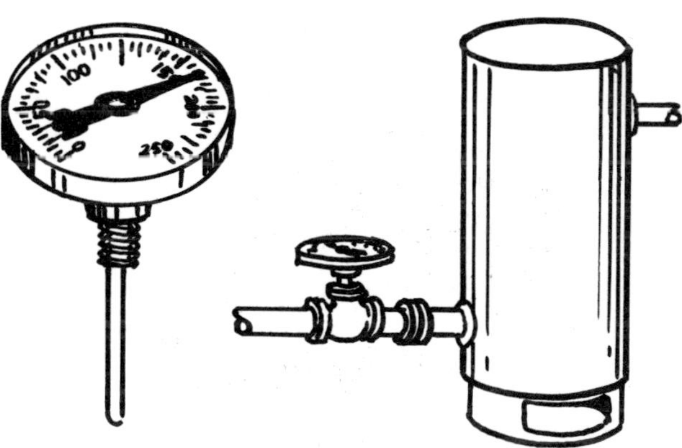

Inlet Temperature May Be Monitored with an Inexpensive Bi-metal Thermometer.

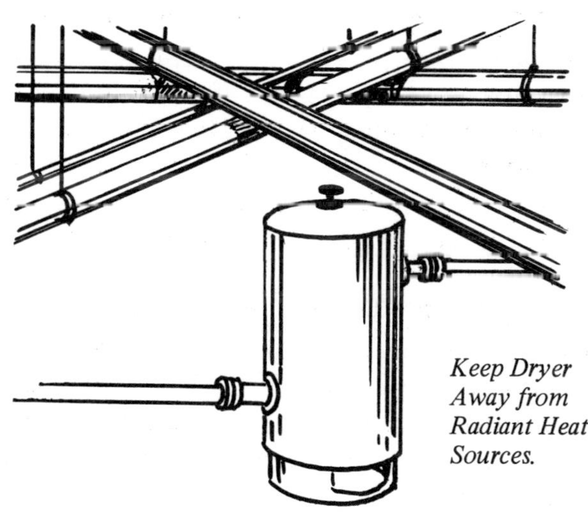

Keep Dryer Away from Radiant Heat Sources.

Adjust Inlet Pressure to Equal or Exceed Pressure for Which Dryer Was Sized.

Blowing Operations May Overload the Dryer.

Radiant heat picked up on the inlet plumbing and sometimes on the dryer housing may impair efficiency of drying. Radiant heat picked up by the outlet plumbing is of no concern because it will not cause water to precipate in the line.

(4). Improper Draining. If the drain line becomes restricted or clogged, water may back up inside the dryer and interfere with dryer action. Check operation of the automatic drain valve.

(5). Inlet Air Pressure Too Low. Drying capacity is directly related to the absolute pressure of the inlet air. The dryer should originally have been sized for the lowest pressure at which it would have to work. Assuming it was originally sized correctly, and if the pressure level later was reduced, the dryer will have insufficient capacity.

The remedy for this condition is to raise the inlet air pressure or to add a second dryer in parallel with the first.

Note: Capacity of an existing dryer can be increased if the inlet air pressure can be increased. This will permit a higher flow of dry air or will give a lower dew point on the same flow of air. The cut-out pressure on the compressor can be raised. This will not affect downstream components which receive air from a pressure regulator. It will, however, increase the cost of air by over-compressing it then reducing it to a lower pressure before use.

(6). Intermittent Surge Flows. Even a momentary overload surge of air through a dryer may carry water into the distribution system. The possibility of surge overloads may not have been considered when dryer size was selected. Look for downstream operations like "blowing" to clean parts, large temporary flows for dusting or drying parts, large blowguns, blow-down valves in drop legs for blowing dirt or water out of the line. If these surge overloads cannot be eliminated, a receiver tank should be added following the dryer to reduce the intensity of the surges.

A dryer can be overloaded when a compressor is first started and before pressure has built up in the distribution lines. This is more likely to occur if there is a large volume in the distribution lines to be filled or if there are other receiver tanks elsewhere in the plant which have to be filled.

(7). Oil Carry-Over into the Air Line. Oil may exist in an air line both in "mist" and "vapor" form. Like water vapor, oil vapor is a gas which cannot be removed with mechanical filters. But fortunately it seldom precipitates in the air system so causes no problem.

On the other hand, oil mist must be excluded from air to be dried. It will foul dessicant dryers and will reduce the efficiency of refrigeration dryers. If oil is suspected, examine some of the dessicant to check for an oil coating.

The air compressor is nearly always the source of oil mist. Rotary compressors may inject oil into the inlet air as a lubricant. Excess lubricating oil coming out with the compressed air must be removed if a dryer is to be effective. Piston compressors may have an adjustable oil feed and this must be adjusted to give adequate lubrication but not so high that excess oil gets into the outlet air. Well-worn piston compressors may feed quite a bit of oil into the outlet air. Oilless rotary compressors give clean oil-free air but are available only in small sizes.

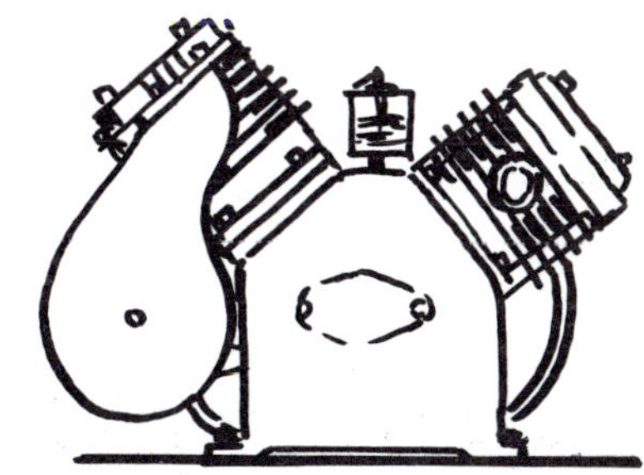

Adjust Compressor Oil Feed to No More Than Manufacturer's Recommendation.

No matter how the oil gets in, it must be removed before it enters the dryer. We believe the most satisfactory arrangement is to pass the air first through a filter with oil absorbing element to remove the majority of oil, then through a "microalescer" filter which will precipitate even microscopic dirt particles and water and oil mist.

PRINCIPLES OF AIR DRYING

A given quantity of air will hold more water vapor at higher temperatures. That is, it becomes drier when raised to a higher temperature. If its temperature is lowered, its relative humidity will increase and will reach the "dew point" when its relative humidity becomes 100%. At this condition it is said to be "saturated".

To be most effective, that is, to deliver the driest air, a dryer should be installed at a point of high pressure. Also it should be installed at a point of lowest temperature. The point in an air system where the best compromise prevails between high pressure and low temperature is following the receiver tank and before entering the distribution system.

On some systems, and we do not recommend this for every system, air can be dried to a satisfactory level without an air dryer. When compressed, its temperature may reach 450° F in the cylinder of an air compressor. Although the water concentration has increased by 7 or 8 times, water will not condense in this area because of the high temperature. But when it enters the receiver tank and cools to a lower temperature, water will condense. By compressing the air to a pressure higher than actually needed in the system, it will drop more of its water as it cools to room temperature. But to be effective, the air must be allowed to cool to room temperature before entering the distribution system. Sometimes, a little extra dryness during periods of weather when outside humidity is very high may be all that is needed, and raising the cut-out pressure on the air compressor might solve the problem.

Chapter 9

Vacuum Systems

Vacuum is another form of fluid power, and is air pressure acting in a negative direction. Vacuum force is produced by the weight of atmosphere above the location at which vacuum is being used, and its pressure will vary from 14.7 PSI at sea level to about 10 PSI at 5000 feet (the elevation of many U.S.A. cities). Due to the limited pressure available, vacuum is limited to low power applications but to those which, because of the nature of vacuum, can be done better, faster, or cheaper than by the higher power of hydraulics or pneumatics. Here are some applications that can be done better by vacuum than by any other medium:

(1). Air bubble removal from wet mixtures such as clay, plastics, plaster molds, cememt, paper-mache', and many similar materials.

(2). Feeding of materials automatically, such as bar stock into screw machines, flat stock into punch presses, paper or card stock into duplicators, collators, and card sorters. Feeding of paper into packaging machines and printing presses.

(3). Suction cups can be used for lifting, transferring, or separating many types of sheet material including paper, sheet metal, cloth, plastic, cardboard, leather, glass and even concrete or stone slabs.

(4). Low pressure bonding or laminating of sheet materials together or to large surfaces. Also shrink packaging of small parts to a cardboard backing for retail sales.

(5). Vacuum chucking of non-magnetic parts or fragile materials for machining. Holding of film in vacuum back cameras, holding of film to sensitized plates for plate making in the printing industry.

(6). Vacuum forming of sheet material, usually hot plastic, into signs, parts for furniture, automobiles, and hundreds of other items.

(7). Transfering sheets from one location to another, with gripping by vacuum and power movement by other means.

These are only some of the uses of vacuum. None of them requires high power but some cannot be done in any other way. Please refer to "Industrial Fluid Power — Volume 1" for sketches and details of the above applications.

In this book we are dealing only with industrial uses of vacuum — hard vacuums are not required and are not desirable. And, for vacuum operated devices to be usable at any elevation from sea level to 5000 feet elevation, vacuum circuits should be able to give full performance at medium vacuums of 10 PSI (about 20'' Hg.). Scientific applications for vacuum often require very hard vacuums, up to a few microns of absolute zero pressure. This book does not cover such applications.

INSTALLATION OF VACUUM EQUIPMENT

Plumbing Materials. As far as compatibility is concerned, any of the materials ordinarily used to plumb compressed air systems may also be used on vacuum systems. However, flexible materials such as hose and plastic tubing must be the kind designed for vacuum to avoid wall collapse. Vacuum hose is built with an internal spiral spring to support the walls.

Take extra care to be sure all joints are leaktight. Leaks are not as obvious as on compressed air, and on small vacuum systems even a small leak can prevent the system from working properly. Pipe sealant should be used to seal pipe threads.

Plumbing Size. Vacuum systems are of two kinds: "Active" systems, such as for moving powdered or granular material through pipe, have a continual air flow at low vacuum pressure; "static" systems quickly pull a vacuum, such as on a suction cup, then hold the vacuum without further air flow. Pipe size should be selected according to the kind of system being plumbed.

For "static" systems, pipe size should be kept very small and as short as possible. By keeping internal pipe volume very small, the response will be quicker and for maintaining the vacuum a very small pipe I. D. is quite adequate.

For "active" systems the pipe size should be as large as practical. Because of the limited vacuum pressure available, flow losses in connecting lines must be kept very small. Use large pipe, bushing down, if necessary, to port size on components.

The table at the foot of this page is offered only as a suggestion for pipe size to carry "active" vacuums a distance of less than 10 feet. For greater distances larger pipe sizes should be used.

On all vacuum systems locate vacuum pump as close to the work as practical to keep piping runs short. As on compressed air systems, plan pipe layout to avoid restrictions and use a minimum number of fittings and the least number of bends. Remember that keeping flow restrictions low is more important on vacuum than on compressed air because of the lower pressure available to operate the system.

All hose used on vacuum must be internally supported to prevent wall collapse.

For "active" vacuum systems, the diameter of all plumbing materials should be larger than would be used for compressed air.

PIPE SIZE FOR VACUUM PLUMBING

Suggested size of standard weight black pipe for runs less than 10 feet.

Free Running SCFM Displ. of Vac. Pump	Approx. Vacuum Input HP	Pipe Size for 10 Ft. Run
0 to 6	3/4	3/8" NPT
7 to 15	1½	1/2" NPT
16 to 30	3	3/4" NPT
31 to 60	5	1" NPT

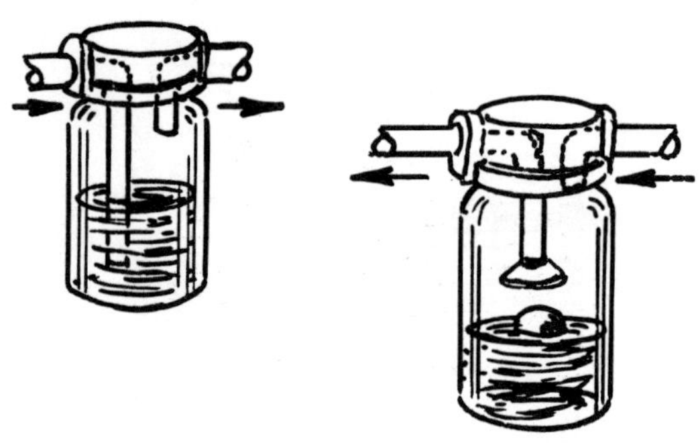

Liquid traps should be used to prevent liquids or solids from being drawn into the vacuum pump.

A lubricated-type vacuum pump should have a filter on both inlet and outlet.

Intake of Liquids and Solids. Liquids or solid particles drawn into a vacuum pump may lock it up and cause it to be destroyed. Also, such liquids may be corrosive to the pump.

Simple devices for excluding liquids are pictured here. They are available from the pump manufacturer. On the left, an in-line trap, with or without a felt filter element, traps contaminants in the bottom of the jar. A felt filter will prevent liquid mist from entering the pump. Care must be used to empty the jar before the level reaches a height where liquid would be drawn into the pump.

A more positive trap is shown on the right. A floating rubber ball blocks the pump intake if the liquid level is allowed to rise too high.

Lubricated or Oilless Pumps. Oilless pumps are convenient because they require no lubrication maintenance. They usually have carbon or graphite vanes or piston rings.

Lubricated models do require replenishment of lubricating oil from time to time, but the oil makes a tight seal between vanes and cam ring and this gives higher efficiency and the capability of producing a slightly higher vacuum. One disadvantage is that there may be a slight discharge of oil into the room. If this is objectionable, the discharge can be directed into an oil reclassifier to precipitate the oil into a jar.

Type of Lubricator. Lubricated models used for "active" systems where air is flowing continuously in a large volume will be equipped with a jar to hold lubricant with an oil wick projecting into the air stream. A standard air line lubricator, described in Chapter 7, can be used in place of the oil wick.

Lubricated models intended for "static" vacuum service must be equipped with a different type of oiling system because most of the time there is little or no air flowing to carry the oil into the pump. These models have an automatic oiler which is activated by circuit vacuum and opens an oil feed line into the pump bearings. Lubricated models can use the same oil recommended for trio units in Chapter 7.

On the other hand, it is important that lubricating oil never get into oilless vacuum pumps. This may cause vanes to stick with partial or complete loss of vacuum. Piston models may be locked up by oil. If oiled accidentally, the oil can be removed by the cleaning procedure on Page 175.

Effect of Back Pressure. Whenever possible, the vacuum pump should be allowed to discharge directly to atmosphere through an exhaust muffler screwed into the pump outlet. But when, for some reason, the exhaust must be piped outdoors or to another area, it is of great importance that the exhaust line be as short as possible and have a large flow area to reduce flow resistance. Flow resistance creates back pressure on the pump outlet. Any back pressure will reduce the vacuum capability of the pump by the amount of back pressure. For example, if the flow resistance through an exhaust

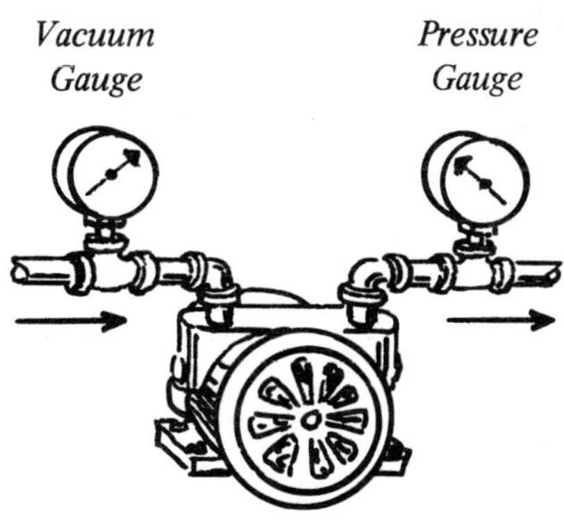

Any back pressure on the exhaust of a vacuum pump can seriously impair the performance.

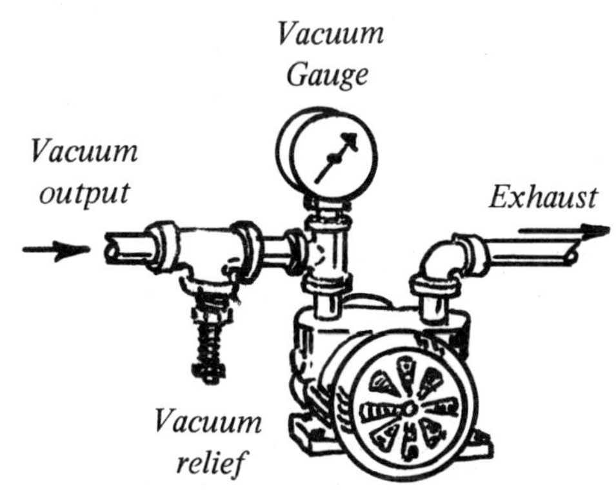

Use a vacuum relief valve to prevent overloading of the driving motor.

line amounts to 1 PSI (2″ Hg), the maximum vacuum which the pump can produce will be reduced by this amount. If the noise is not objectionable, the exhaust muffler can be omitted to reduce back pressure. This should slightly increase the maximum vacuum and should slightly reduce driving HP.

Driving Horsepower. The HP to drive a vacuum pump depends on pump displacement and speed and on the degree of vacuum it is required to develop.

On integral units with electric motor and pump in a package, a vacuum relief valve should be placed across the vacuum line to prevent build-up of excessive vacuum which will overload the motor.

On vacuum pumps driven by a separate motor, drive HP can be determined from specifications in the pump catalog. Protect the motor from overload by using a vacuum relief valve. An ammeter will show if the motor is running at a higher current than shown on its nameplate. Running the pump faster than its rating or exceeding the maximum vacuum will require additional HP and may cause the pump to overheat.

VACUUM CYLINDERS

The force which can be produced by a vacuum cylinder is very small compared to that produced by a compressed air cylinder. For example, operating at a vacuum of 10 PSI (20″ Hg), a cylinder with 2″ bore can produce only 31 pounds, and part of this will be lost in seal friction. The same cylinder on 100 PSI air can produce 300 pounds. Most applications for vacuum cylinders are where only a very small force is required, and in locations where vacuum is the only source of power.

Force from a vacuum cylinder is calculated as it is for air cylinders, but with a variation: Force on the extension stroke is calculated by using the *net* area (full piston area minus rod area) instead of

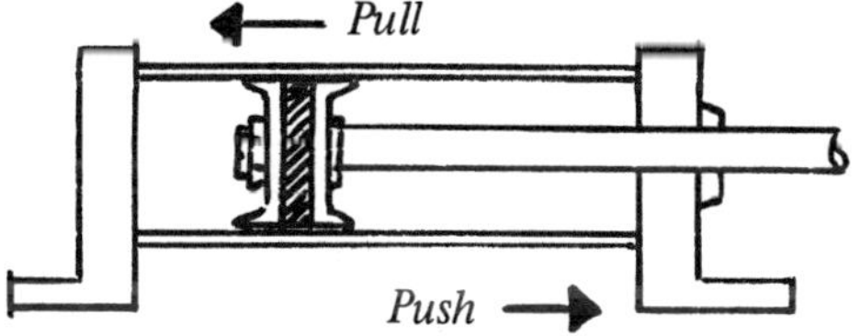

On vacuum cylinders, the "net" piston area (piston area minus rod area) must be used to calculate extension force. The full piston area must be used when calculating retraction force. This is opposite to the method of calculating extension and retraction force of an air cylinder.

the full piston area. Retraction force is calculated using the full piston area because atmospheric pressure (the motivating force) is working not only on the net area inside the cylinder but externally on the rod area. Force is calculated by multiplying vacuum pressure, in PSI, times the areas described. Allow '5 to 10% loss of force in seal friction. Conversion from inches of mercury (''Hg) to PSI is approximately: 2''Hg = 1 PSI.

Choosing a Cylinder for Vacuum. Not all air cylinders are suited for vacuum service. One of the most important requirements is very low seal friction. Piston and rod seals should be cups. Leather cups have lowest friction drag but synthetic rubber cups usually work well. Cylinders with O-ring piston or rod seals are not recommended for vacuum because of high friction drag.

Single-acting air cylinders usually cannot be used on vacuum because the piston seal is sealing in the wrong direction. Vacuum cylinders must be either those built for such service or must be double-acting air cylinders with low seal friction.

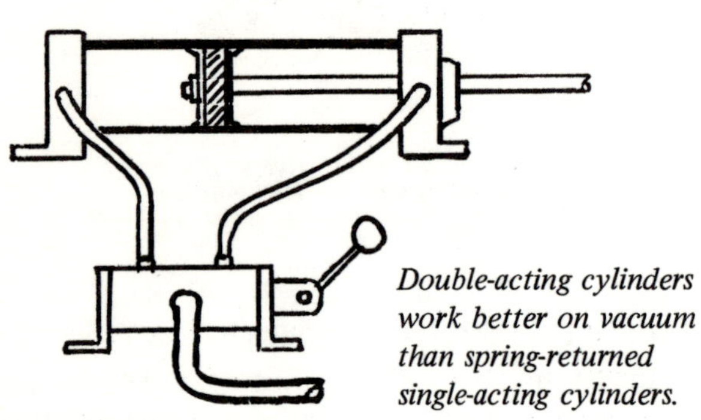

Double-acting cylinders work better on vacuum than spring-returned single-acting cylinders.

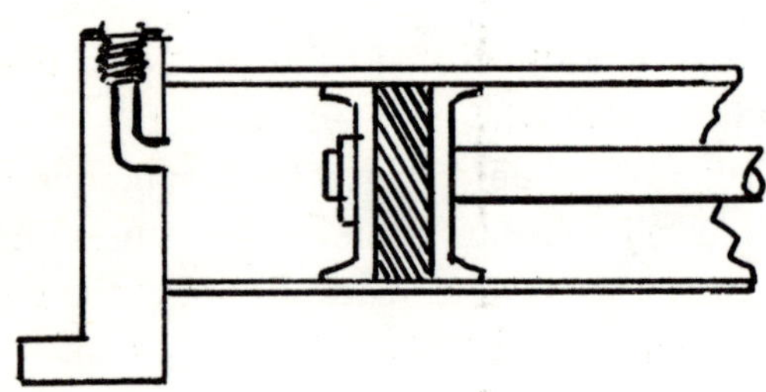

Vacuum cylinders should be designed for minimum drag of the packings.

Single and Double-Acting Cylinders. Single acting air cylinders with internal return spring are not suitable for vacuum because compressing the return spring during the extension stroke consumes about one-half the already limited force.

Select a double-acting cylinder with the smallest rod available in that bore size. The strength of a large rod is not needed on vacuum, and the smaller the rod the greater the force which can be produced while the cylinder is extending.

Vacuum cylinders are not very successful on applications for moving a high friction load at slow speed. They work best on small, fast moving mechanisms supported on bearings.

VALVES FOR VACUUM SERVICE

Directional Control Valves. Some standard 3-way and 4-way compressed air valves can be used on vacuum applications within limitations described in this section.

Double-acting vacuum cylinders can be controlled with 4-way air valves. Single-acting cylinders (without return spring) which depend on reaction from the load to return them to home position, can be controlled with 3-way air valves, or in some cases by 4-way valves with one cylinder port plugged.

Three-way air valves can be used to control vacuum holding or gripping applications such as suction cups, where a single valve is used for both application and release of vacuum. A pair of 2-way valves can be used instead of one 3-way valve on applications where separate valves are needed for application and release of vacuum. Two-way valves can also be used on applications where a release of vacuum is not required at the end of the operation. Limitations on the use of solenoid valves is covered later.

Connecting a 4-Way Valve for Vacuum Service.
Be aware that some brands of air valves cannot be
used on vacuum because spool seals will not seal in
a direction for vacuum. Valves having cup seals
between ports or between the end of the spool and
atmosphere, cannot be used because the seals will
not seal in the right direction. In general, air valves
with O-ring or quad ring seals on the spool or in
body grooves will work on vacuum as well as on
compressed air.

Most air valves are built with dual exhaust ports.
When used on vacuum, the previous inlet port
becomes a common exhaust port for both cylinder
ports, and the two previous exhaust ports must
be teed together for the new inlet. This inlet
should be protected from intake of atmospheric
dirt with an air muffler or filter.

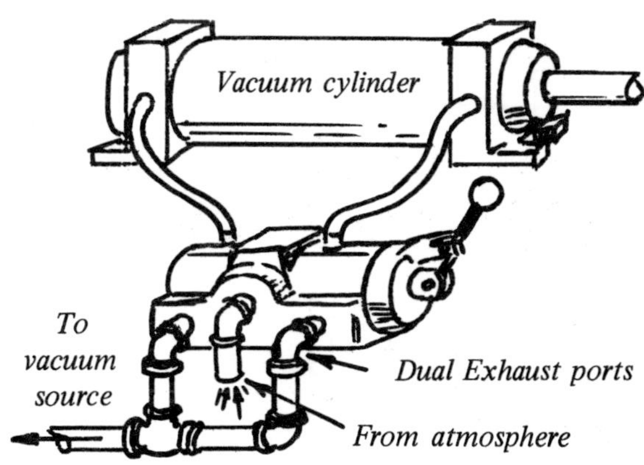

4-way air valve on vacuum service

A general rule for applying air valves to vacuum is to connect the highest pressure in the system to
the port marked inlet or P. This would be atmospheric pressure through a filter or lubricator. Connect
the lowest pressure (the source of vacuum) to the pair of previous exhaust ports marked Ex-A and
Ex-B. Cylinder ports A and B are connected to the vacuum cylinder but in a direction opposite to the
way they would be connected to an air cylinder for piston extension. Air valves with single exhaust
port may be unsuited for vacuum service. Check suitability with their manufacturer.

Lubrication for a vacuum cylinder and its control valve can be provided by installing an air line lubricator at the valve inlet, Port P. If the vacuum pump is an oiless type, the excess lubricant must be
removed from the line before it enters the pump. Install an oil absorbing filter ahead of the pump.
A better arrangement for an oiless pump might be to omit the lubricator and apply permanent
molybdenum disulfide grease to inside surfaces of valve and cylinder.

Directional Valve Size. Keeping flow loss to a minimum in vacuum circuits is extremely important.
Four-way valves should be one or two pipe sizes larger than would be used for compressed air on the
same cylinder. For example, on air cylinders up to 3" bore a full-flow 1/4" valve is quite sufficient.
But on the same cylinder for vacuum service, a 3/8" or 1/2" valve would be better. Use connecting
plumbing the same size as the valve. Bush down at the cylinder to match its port size.

Be sure the directional valve has full flow internal passages. Some miniature valves may have connections of 1/4" NPT but the internal flow passages may be much smaller.

Leaktight directional valves should be used in circuits where a relatively high vacuum must be held
for a period of time. Spool-type valves should have O-rings or other soft seals between all ports.
Poppet-type valves with synthetic rubber seats work very well. Rotary shear seal valves with metal-to-
metal lapped surfaces are tight enough for all but very high vacuum circuits.

Check Valves. Most compressed air check valves are not well suited to vacuum circuits because their
cracking pressure in the free flow direction is too high and consumes too much precious vacuum.
Some of them can be modified by removing the internal spring and mounting the valve body vertically
using poppet weight instead of the spring to keep the poppet normally seated.

Flapper or swing-type check valves are preferred if available. They have a cracking pressure less
than 1 PSI and are probably more reliable than modified air check valves.

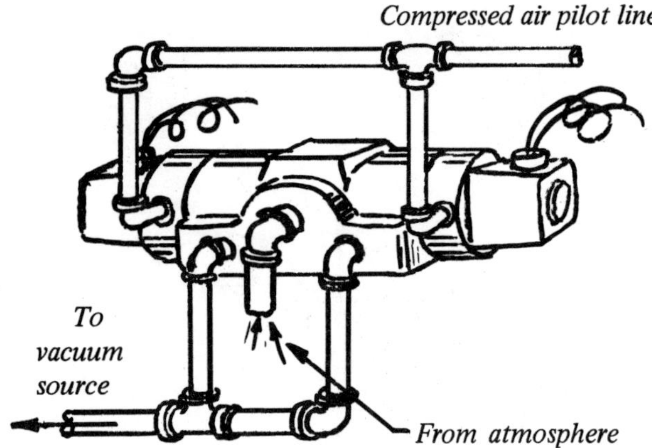

Pilot operated solenoid valves must be connected for external pilot operation.

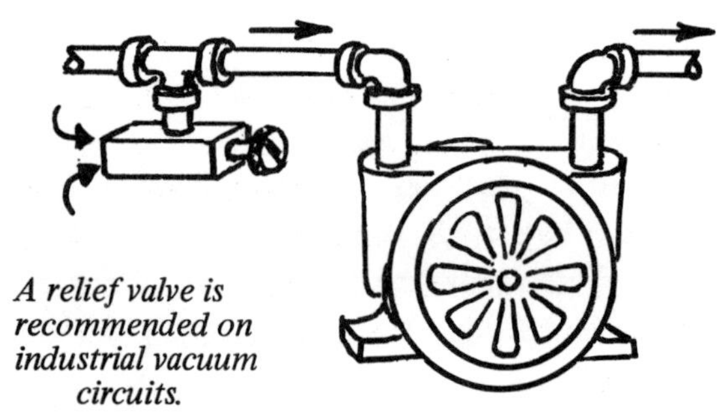

A relief valve is recommended on industrial vacuum circuits.

Button Bleeder Valves. These valves, so popular for non-electrical operation of compressed air cylinders, by their very nature cannot be used on vacuum.

Solenoid Valves for Vacuum. Solenoid valves of the direct-acting type can be used on vacuum.

Solenoid valves of the pilot-operated type cannot be used on vacuum unless they are modified, and most of them are built so they can be modified for vacuum service. They must be converted to "external pilot" operation. A source of external air pressure, 30 to 50 PSI according to valve brand, must be used to shift their main spool; it cannot be shifted by vacuum force. The source of shifting pressure must be connected to the external port provided for this purpose, and the valve must be modified internally to block the internal pilot passage. Some models may have two ports, one on each solenoid, to which air pressure must be connected. Refer to manufacturers catalog for instructions on making the conversion.

Vacuum Relief Valves. We recommend installing a vacuum relief valve on the inlet of the vacuum pump on all applications where the vacuum is "static", that is, where a high vacuum must be held without movement of air. Set the relief valve for the working vacuum level required. A relief valve provides several advantages: (1), it reduces HP input to the pump by limiting the vacuum to no more than is required; (2), heating of the pump is reduced because the input HP is kept to the minimum; (3), air is kept moving through the pump at all times, and this helps to keep it cool and oil can be fed from an air line lubricator; and (4), operation of the machine will be consistent from day to day as barometric pressure may change or if the machine is moved to a location at a different altitude.

Vacuum relief valves are available from the pump manufacturer, or an air pressure relief valve can be used, if both inlet and outlet ports are threaded, by connecting its outlet port to the vacuum pump inlet and leaving its inlet port exposed to atmospheric air pressure through a filter.

Speed Control of Vacuum Cylinders. The flow control valves used for speed control of air cylinders are not suitable for vacuum cylinders because of the relatively high cracking pressure (2 to 5 PSI) of the internal check valves in the "free flow" direction.

Speed control is often not necessary on vacuum cylinders, but when it is, needle valves should be used. Assuming a dual exhaust type of air valve is being used for directional control, the needle valves can be placed in the vacuum inlet lines to the valve. See illustration next page. If a reduction from maximum speed is required in only one direction, use only one needle valve feeding vacuum into the side where speed reduction is needed.

Needle Valves. The same kind of needle valves used for compressed air can also be used in vacuum circuits for metering or for speed control as shown in this illustration.

They should not be used as shut-off valves. Their internal orifice is smaller than their connection size, placing too much flow resistance into the circuit. Also, they may not shut off tightly after being closed a few times if the needle should become scored. For vacuum shut-off, a quarter turn rotary plug valve works out better.

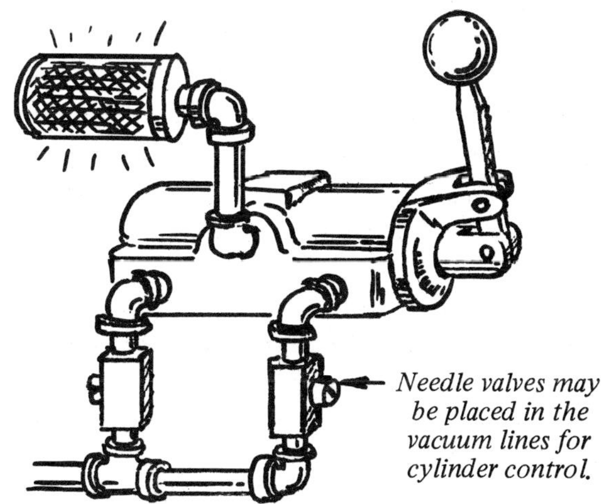

Needle valves may be placed in the vacuum lines for cylinder control.

MAINTENANCE OF VACUUM EQUIPMENT

Lubrication. Maintenance procedure should include inspection of the lubrication system of the pump. If the lubrication system should run out of oil or should fail, the pump may not be able to pull a full vacuum. An oil film on the cam ring is necessary for a tight seal between vane tips and cam ring. It retards rusting and corrosion, and this is particularly important if the application calls for pulling in air of high humidity as from wet mixtures or when atmospheric humidity is high. Also a lack of lubrication will cause accelerated wear inside the pump.

Petroleum oil, without detergent additives, and of SAE 5W or 10W weight is recommended, and can usually be obtained from any filling station. If pump manufacturer specifies a certain kind or viscosity of lubricating oil, his recommendation should be followed.

Oilless models have graphite or plastic vanes or piston rings which need no lubrication. Adding oil to these pumps may cause the vanes to stick, and the pump should be cleaned as described below.

Cleaning Vacuum Pumps. Accumulations of gum or residue may cause vanes to stick or check

Check oil supply frequently on lubricated vacuum pumps.

Pour solvent in very slowly.

valves on piston pumps to leak. Cleaning procedure is to open the inlet plumbing, then with the pump running, pour in *small quantities* of a mild solvent. A temporary filter should be installed on the pump atmospheric outlet to pick up the solvent as it is blown out. The filter element should then be discarded. The solvent must not be harmful to any synthetic rubber seals in the pump. Kerosene, petroleum distillate, or JP jet fuel should not be harmful to the pump. Avoid strong solvents which may be harmful to the pump or may present a fire hazard or a health hazard to personnel. Lacquer thinner, turpentine, gasoline, naptha, etc. are *dangerous* to use. Pour in the solvent in very small quantities. A large slug of liquid could lock up the pump and cause severe damage.

Shaft Seals. A vacuum pump should be returned to the factory service department for replacement of shaft seals. If a field repair is necessary, the body cover should be removed carefully so the gasket is not damaged. The gasket thickness on a vane pump is critical; it provides exactly the right amount of clearance for the vanes. If it must be replaced, measure its thickness and replace with a new gasket of exactly the same thickness. When installing the new shaft seal be sure the lips are turned toward the pressure to be sealed. In the case of a vacuum pump, the highest pressure is atmosphere, so the seal lips should be turned to face the outside.

VACUUM PUMP APPLICATIONS

The information on vacuum applications in this section supplements the more detailed information in "Volume 1 — Industrial Fluid Power".

Lifting or Holding Items by Vacuum. The lifting force, in pounds, under a vacuum pad may be calculated by multiplying square inch area under the pad times the degree of vacuum in PSI. Convert vacuum gauge reading to PSI by dividing by 2 (1 "Hg = approximately 1/2 PSI). The table shows lifting force, in pounds, for circular or square pads up to 12" across.

Control Valve 2 can be a simple 3-way valve. However, we think a 4-way valve may work better on most vacuum pad applications because it will unload the pump between cycles. When the valve is in

LIFTING FORCE OF VACUUM PADS

Circular Pad					Square Pad				
Pad Dia., Ins.	Pad Area Sq.Ins.	Lift at 14" Vac.	Lift at 20" Vac.	Lift at 26" Vac.	Pad Side Dim.	Pad Area, Sq.Ins.	Lift at 14" Vac.	Lift at 20" Vac.	Lift at 26" Vac.
1"	0.785	5.5	7.8	10	1"	1.0	7.0	10	13
2	3.14	22	31	40	2	4.0	28	40	52
3	7.07	50	71	92	3	9.0	63	90	117
4	12.57	88	126	160	4	16.0	112	160	208
5	19.63	137	196	255	5	25.0	175	250	325
6	28.27	200	283	370	6	36.0	252	360	470
8	50.27	350	503	650	8	64.0	448	640	830
10	78.50	550	785	1000	10	100	700	1000	1300
12	113.1	800	1131	1460	12	144	1008	1440	1875

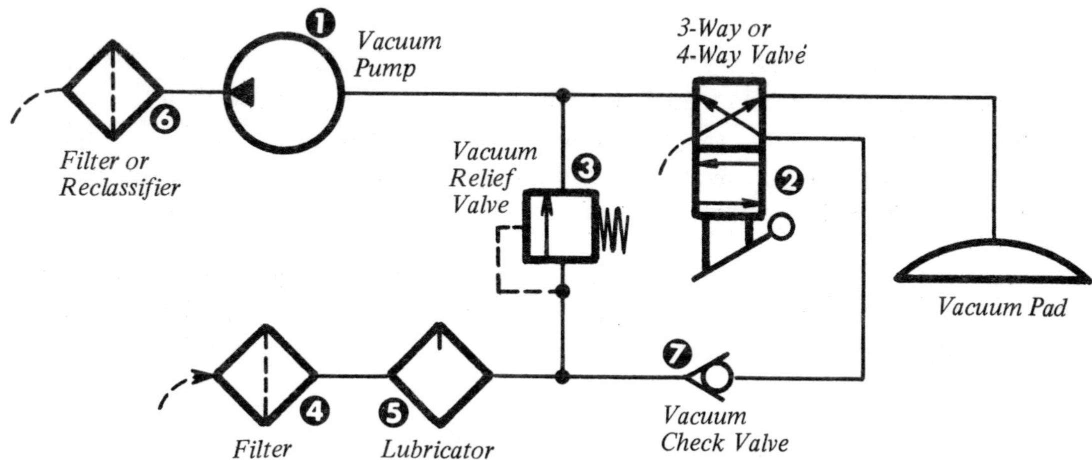

Using a 4-way valve controls the vacuum pad and unloads the vacuum pump between cycles.

the position shown, the vacuum pump is unloaded, and is pulling air through Filter 4, Lubricator 5, check Valve 7, and through 4-way Valve 2 into its inlet and discharging freely to atmosphere. At the same time, the vacuum pad is also vented to atmosphere.

When control Valve 2 is shifted to its other position, a vacuum is being pulled on the pad up to the setting of the vacuum relief valve. Flow of air through Filter 4, Lubricator 5, and relief Valve 3 keeps the vacuum pump lubricated while it is running deadheaded.

Item 6 may be a filter, a muffler, or a reclassifier to condense excess lubrication. Not all these components may be needed on every application. The lubricator, 5, and reclassifier, 6, should be omitted when using an oilless vacuum pump. The lubricator, 5 may be omitted when using a lubricated pump if the pump has a built-in lubrication system.

__Vacuum and Pressure From the Same Pump__. Some applications, packaging machines for example, may require both vacuum and pressure on the same machine. Most vacuum pumps will deliver suction at the inlet and pressure at the outlet, but the combined vacuum and pressure, added together, must not exceed the vacuum rating of the pump. A vacuum pump rated at 24"Hg (12 PSI) maximum will deliver any part of this as vacuum, the remainder as pressure. For example, if a vacuum relief valve

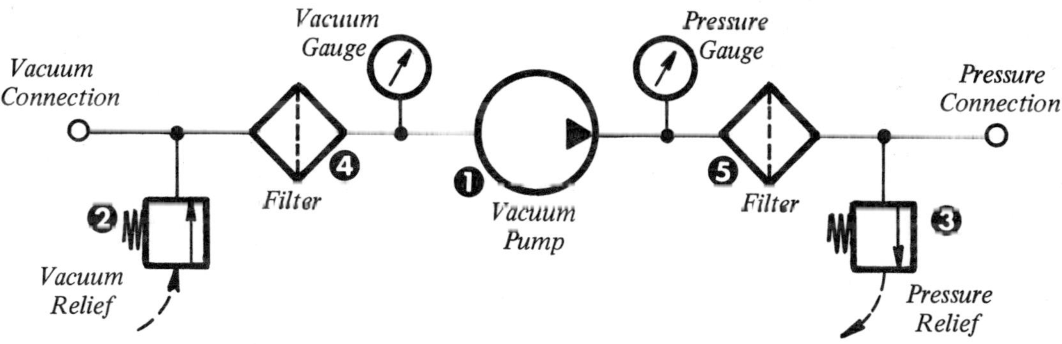

A low pressure and a low vacuum can be obtained at the same time from a vacuum pump.

is connected to the inlet and is set for a vacuum of 10"Hg (5 PSI), then a pressure of 7 PSI will be available at the pump outlet.

The vacuum relief valve is necessary to prevent too much of the pump capacity going into vacuum, leaving insufficient capacity for compressed air. Also it is vital as a path for air into the pump when the pump must produce a *flow* of compressed air. Likewise, the pressure relief valve is necessary to prevent too much of the pump capacity going into compressed air, leaving insufficient capacity for producing vacuum. Filters, lubricators, and reclassifiers can be added to this basic circuit if desired.

Several Vacuum Levels from One Pump. Two or more different degrees of vacuum (or pressure) can be obtained from the same pump using the arrangement in the diagram below. Vacuum levels are shown in this diagram but the circuit can be rearranged to produce several levels of low pressure.

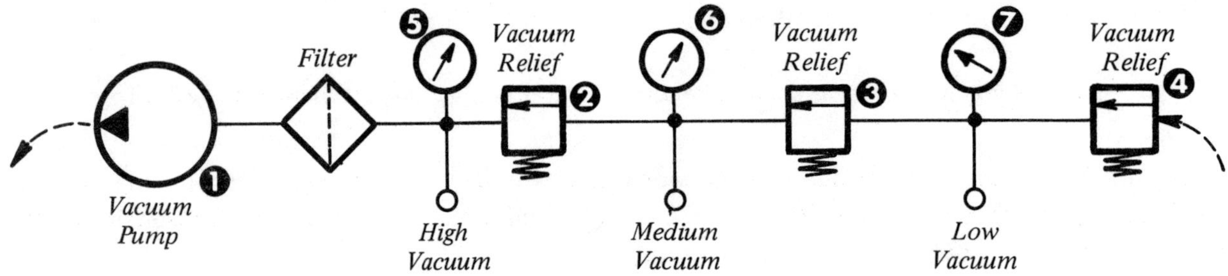

Several degrees of vacuum can be obtained at the same time from one vacuum pump.

Vacuum relief Valves 2, 3, and 4 are in series. Their combined settings must not exceed the maximum vacuum rating of the pump. To adjust the relief valves, install vacuum gauges 5, 6, and 7. Plug the high, medium, and low vacuum inlets then start the pump. Adjust the three relief valves until the desired degree of vacuum is obtained on all three vacuum inlets with the maximum rated vacuum for the pump appearing on Gauge 5.

Filling and Emptying a Closed Vessel. The pump is alternately used as a vacuum pump to draw liquid into the closed vessel, then used as a compressor to expel the liquid from the vessel.

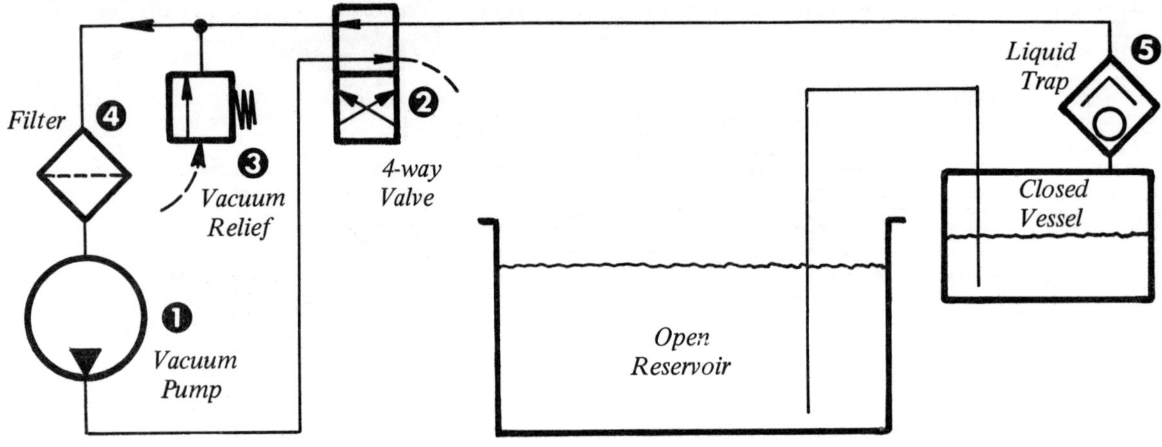

Using a vacuum pump to fill and empty a closed vessel.

Changeover from vacuum to pressure is made by shifting the 4-way valve, 2. A filter, 4, should be installed to prevent ingestion of dirt from the atmosphere. A liquid trap, 5, prevents accidental intake of liquid. As in all vacuum circuits a relief valve, 3, should be used to prevent overloading the pump. A relief valve can be added to the pressure side of the pump for the same purpose.

MISCELLANEOUS VACUUM INFORMATION

Storage Tanks for Vacuum. Receiver tanks for storing vacuum must be specially constructed with internal supports to prevent collapse. Compressed air tanks, even though rated for high air pressure, 350 PSI or higher, are not necessarily suited for vacuum. They are constructed to withstand internal pressure but may not be able to withstand very much external pressure.

The same is true of rubber hose. Compressed air hose may collapse under vacuum unless supported by an internal coil of steel wire or by other means.

Most Efficient Vacuum. We recommend for ordinary industrial uses of lifting, holding, transferring, and the like, a maximum vacuum of about 20 to 24 "Hg. Vacuum pumps become very inefficient as vacuum level increases, and at 28 "Hg most of the input power is going into leakage losses, mechanical friction, or heat generation. Pumps working at these high levels can produce a "static" vacuum but cannot produce much "flowing" vacuum. Beyond 20 to 24 "Hg, the additional increase in vacuum power is usually not worth the extra power input and wear and tear on the pump. On exceptional cases it may be necessary to reach as high a vacuum as possible.

Another good reason for limiting the design vacuum to 20 "Hg is so the machine will perform normally at all altitudes up to about 8,000 feet. If designed to require 26 "Hg vacuum, it would not be able to develop full vacuum power at altitudes above 3000 feet.

Compressed Air Tanks May Collapse Under Vacuum.

Use as Low a Vacuum as Will Do the Job.

Applications using "flowing" vacuum usually require a high displacement pump running at high speed but at a low degree of vacuum. A household vacuum cleaner is such an application.

Flow of Vacuum Through Orifices. A difficult application is a vacuum gripping device having a large number of exposed holes, part of which may be covered by the work during the holding operation, leaving some of the holes open through which vacuum can escape. The vacuum pump to run such a machine must have sufficient displacement to maintain a certain level of vacuum even with some of the holes open for free flow.

The chart below may help in estimating the free running displacement of a vacuum pump for an application having (so many) holes of (such and such) a diameter. The application must be carefully specified as to how many of the holes will be left open, and what degree of vacuum must be maintained both with all holes open to atmosphere and with only a part of the holes left open. Please remember this will give only an estimate because of unknown factors such as orifice shape, etc.

Problem Example. Assume an application in which a vacuum of 6 "Hg must be maintained when 8 holes of 1/16" diameter are left open. From the chart, estimate the SCFM air escape out of these holes at a vacuum of 6 "Hg.

Solution: The chart shows that a 1/16" diameter hole when exposed to a pressure difference of 6 "Hg vacuum will allow .517 SCFM of air to escape. For 8 holes: 8 x .517 = 4.136 SCFM. This is not the free running displacement of a vacuum pump. This is air flow related to a vacuum of 6 "Hg, and the free running displacement of the pump must be determined from graphs in the pump catalog as shown in the next chart.

FLOW OF AIR THROUGH ORIFICES UNDER VACUUM

Figures in body of chart are approximate "free air" flow in SCFM (standard cu. ft. per min.)

Orifice Dia., Inches	Degree of Vacuum Across Orifice									
	4"Hg	6"Hg	8"Hg	10"Hg	12"Hg	14"Hg	16"Hg	18"Hg	20"Hg	24"Hg
1/64	.026	.032	.037	.041	.045	.048	.052	.055	.058	.063
1/32	.100	.128	.148	.165	.180	.195	.208	.220	.231	.250
1/16	.420	.517	.595	.660	.725	.780	.833	.880	.927	1.00
1/8	1.68	2.06	2.37	2.64	2.89	3.12	3.33	3.53	3.71	4.04
1/4	6.74	8.25	9.52	10.6	11.6	12.4	13.3	14.0	14.8	16.2
3/8	15.2	18.5	21.4	23.8	26.0	28.0	30.0	31.8	33.3	36.4
1/2	27.0	33.0	38.5	42.3	46.3	50.0	53.3	56.5	59.2	64.6
5/8	42.2	51.7	59.5	66.2	72.6	78.0	83.3	88.0	92.7	101
3/4	60.6	74.0	85.3	95.2	104	112	120	127	133	145
7/8	82.6	101	116	130	142	153	163	173	181	198
1	108	131	152	169	185	200	213	225	237	258

The above chart shows the approximate flow discharge loss to atmosphere which may be expected from a practical orifice like a drilled hole. The figures are 2/3rds the theoretical discharge through a sharp edge orifice. These values are approximate because the flow characteristic of your orifices can only be determined by actual test under specified conditions.

Note: A multiple-hole gripper works more efficiently at a moderately high vacuum than at a low vacuum. For example, in the chart for a 1/4" diameter hole, the first 6"Hg vacuum causes a discharge loss of 8.25 SCFM, while the last 6"Hg, from 18 to 24"Hg, causes a discharge loss of only 2.2 SCFM. A more efficient design would, therefore, use more smaller holes working at a higher vacuum.

Vacuum Pump Selection. After the SCFM flow requirements are estimated at a certain degree of vacuum, using the chart above, a pump model must be selected by using the performance graphs in the pump manufacturers catalog. A set of performance curves might look like this graph. Pumps 1, 2, and 3 are three models with 4, 6, and 8 SCFM *free running* displacement. The curves drop toward the right because the pumps become less efficient at higher vacuums.

Using the example on the preceding page where a flow of 4.136 SCFM was required at a vacuum of 6"Hg, the graph of Pump No. 1, although it has a free running displacement of 4 SCFM, it will only produce 3 SCFM at a 6"Hg vacuum. So this model is not large enough. Pump No. 2, with a free running vacuum of 6 SCFM will produce 5 SCFM at a 6"Hg vacuum. So this model has sufficient capacity. Pump No. 3 is larger than necessary.

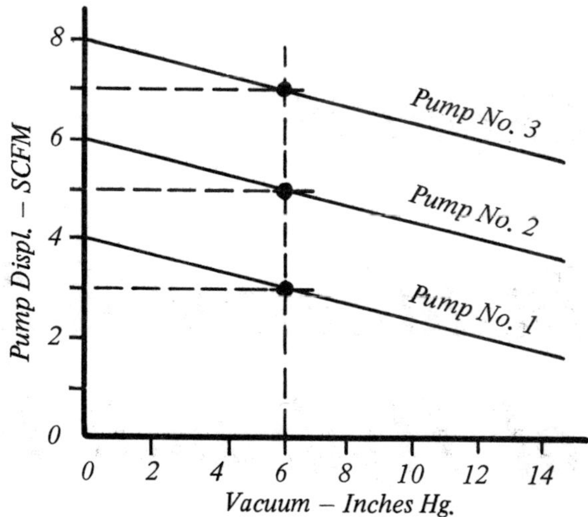

Size Selection Chart from Vacuum Pump Catalog

Chapter 10

Installation & Start-Up of Hydraulic Power Units

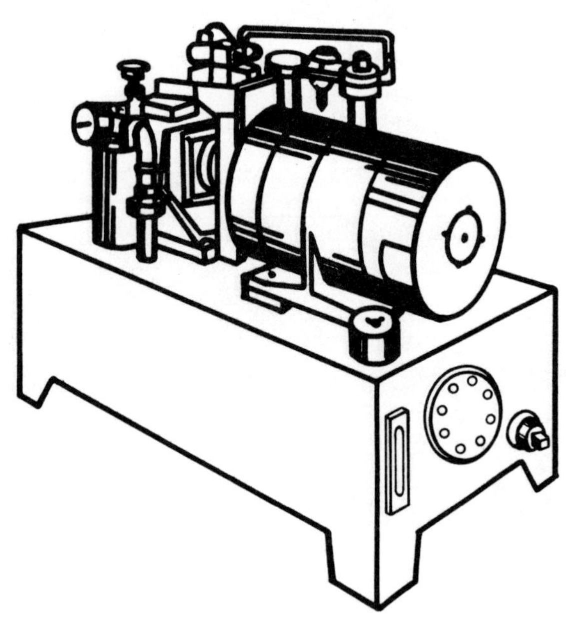

Hydraulic Power Unit

On most hydraulically powered machines the fluid power is generated by a unit assembly called a "hydraulic power unit" or "hydraulic pumping unit". Usually the oil reservoir (tank) is used as a mounting base for electric motor, pump, and sometimes for other components. On machines operating in an industrial plant the hydraulic power unit is designed to set alongside or overhead near the machine, being connected to a cylinder or hydraulic motor on the machine through hose or steel tubing. The installation and start-up of this kind of power unit is the subject of this chapter. See "Volume 1 — Industrial Fluid Power" for construction details and specifications of power units.

Sometimes a hydraulic power unit is designed to be portable so it can be moved around the plant and connected first to one machine then to another. On mobile equipment it is usually more convenient to separate the pump from the rest of the power unit and to mount it in the engine compartment to be driven from the engine power take-off shaft. Still other power units, usually small ones, are self-contained and built so they can be permanently mounted as one component on a large machine — a printing press for example. These smaller power units may not require the detailed installation and start-up instructions given here for industrial power units.

LOCATION AND INSTALLATION OF THE POWER UNIT

Placement. A preferred location in the plant for the installation of a hydraulic power unit is where free air can circulate around all sides, including top and bottom. One of the important functions of the oil reservoir is to radiate heat which may accumulate in the oil. If the reservoir is installed too close to a wall, its heat radiating ability will be impaired and the entire hydraulic system may overheat. An air space of at least 1 foot should be maintained on small and medium size power units and a greater separation on large units. Do not install a power unit inside a cabinet, console, under a table or bench, or other closed space where air cannot freely circulate unless forced air ventilation is provided.

The location should be one where the unit will not be exposed to temperature extremes, either hot or cold. Do not install in direct sunlight without a sunshade, and one which will not restrict vertical air circulation. Do not install near a furnace without interposing a heat shield unless a provision has been made to keep the oil cool with a heat exchanger.

Hydraulic Connections. If the system to which the new power unit is to be connected has not been flushed or otherwise cleaned, we recommend the installation of an in-line micronic filter in the return line to prevent entry of dirt into the power unit reservoir. This filter is particularly beneficial if connecting a new power unit into an older existing system. Accumulations of dirt in the cylinders and lines may be washed back into the power unit reservoir. Connecting hoses, if they have been in service for more than 3 to 5 years, should be replaced at this time. The use of hose instead of rigid plumbing between power unit and cylinders will usually reduce system noise level.

CHARGING THE SYSTEM WITH OIL

Oil Type. Be sure to follow the pump or power unit manufacturers recommendation on type and viscosity of hydraulic oil. The oil recommendation will usually appear on the power unit nameplate or in the instruction manual which accompanies the power unit. If you wish to use a different oil than the one recommended, consult your oil supplier to see if it is equivalent to the one specified. The use of a different type or viscosity than specified may in some cases void the pump warranty, and permission should be obtained from the pump or power unit manufacturer for its use. If there is no oil recommendation, these general rules may be followed:

Most power units, except those built for fire resistant fluids, will operate on petroleum base oil which has been compounded with the proper additives for use in hydraulic pumps, and in the viscosity range of not less than 100 SSU nor more than 500 SSU (These viscosity ratings are at a temperature of 100° F). Mobile hydraulic systems and hydrostatic transmissions quite often operate on high grade automatic transmission fluids which, because of their high V. I. (viscosity index) will operate satisfactorily over a wide range of ambient temperatures.

When connecting a new power unit to an older hydraulic system of valves, cylinders, etc., the old system should be drained of oil unless the oil in the old system is sparkling clean and is known to be of the same brand, type, and viscosity as the oil in the new power unit. It is not a good idea to try to salvage used oil from an old system unless it can be completely re-processed. Simply filtering the old oil may not be good enough. It should be chemically re-refined if it is to be used in a new system.

On power units using pumps which are especially sensitive to dirt, those of the cam plate piston type, an old system should be flushed before adding permanent new oil. Fill the reservoir at least up to the LOW level mark with oil of the same kind which will be used. Put new elements in all micronic filters. Run the system for about an hour without load. Drain and discard the oil. Replace the micronic filter elements.

Quantity of Oil Required. Sufficient oil should be purchased to fill the power unit reservoir up to the FULL mark on the level gauge, plus enough to fill cylinders, plumbing, and heat exchangers, plus enough to fill accumulators to about one-half their total volume, and of course a reserve supply, usually about 25% of reservoir capacity, to make up for leakage or accidental plumbing breaks.

Handling the Oil. As a general rule, any oil added at any time to any hydraulic system should be new oil taken from a sealed factory container. Reserve oil should be stored in original containers, tightly sealed, and stored in an area in which the temperature will not exceed 130° F. Utensils for transferring the oil to power unit reservoir should be kept in a clean place and protected from atmospheric dust. If utensils are found to be dirty they should be cleaned with soap and water then dried before use.

Most hydraulic power units will have a wire or plastic mesh strainer built into the filler opening. If there is no built-in strainer the oil should be poured through lint-free cloth draped over the mouth of a funnel. Experience has shown that factory-sealed containers are not always free of sand, slag, or dirt particles. Keeping additive oil clean is insurance against a future breakdown.

It is very important to immediately replace the filler cap as soon as oil has been added. If the filler cap should be lost, order a replacement immediately and in the meantime, fashion a temporary cover from a piece of screen wire or hardware cloth covered with a shop towel.

If the cylinders on the machine happen to be in their retracted position, fill the reservoir to the FULL mark on the level gauge. If cylinders happen to be in their extended position, fill only to the LOW mark. As the system is put into operation, a little more oil will have to be added to replace oil which fills components. *Caution!* Piston pump and motor cases must be filled with oil before start-up. Some units may be destroyed in a few minutes if started and run without the case being filled completely. See manufacturers instruction manual for filling procedure. Mounting unit with case drain on top will keep the case full.

Watch the reservoir oil level as the system is first operated. When system is fully charged, the oil level should be to the FULL mark on the gauge when all cylinders are in retracted position.

The majority of breakdowns in hydraulic equipment can be traced directly or indirectly to lack of cleanliness in the oil or from allowing the system to run at too high an oil temperature.

ELECTRICAL CONNECTIONS

Proper Voltage. It almost goes without saying that the electrical power source must match the electric motors on a hydraulic power unit as to voltage, phase, and Hertz (cycles per second), and must have sufficient amperage capacity. Motor requirements will be found on motor nameplate.

A 60 Hz induction electric motor connected to a 50 Hz supply will run at only 5/6ths rated speed. This will cause the hydraulic pump to deliver only 5/6ths of its rated flow and will cause the system to be slow. The motor will be unable to deliver full nameplate HP without overheating. A 50 Hz motor connected to a 60 Hz line will run 20% faster than it should and the hydraulic system will operate 20% faster than if operated at correct line frequency. But the motor will not be overloaded and on many systems the operation will be acceptable.

Power units which operate on 3-phase current may be briefly jogged at this point to determine shaft rotation. If incorrect, it will have to be changed before operating the system.

All electrical power wiring should be installed by a qualified electrician and should be in accordance with the National Electrical Code and in some states may also have to meet additional requirements.

A copy of the National Electrical Code handbook may be ordered through any bookstore. Small power units which plug directly into a wall socket do not need the services of an electrician.

The National Electrical Code requires installation of the motor starter within a few feet of the motor, and a cut-off switch with line fuses either at the starter or within line-of-sight not to exceed 25 feet distant. The fuses protect the motor against momentary current overloads such as for starting. To protect the motor against sustained current overloads the motor starter should be equipped with heater coils sized according to the current rating on the motor nameplate. These heater coils do not respond to momentary high current surges but will open the motor circuit if an overload is maintained for a few minutes.

PUTTING THE POWER UNIT INTO OPERATION

Direction of Pump Rotation. After filling the reservoir and the pump and hydraulic motor cases (on piston-type units), and connecting electric power lines and motor starting equipment, the system is ready for start-up. But before starting the electric motor it is a good idea to back off the pump pressure relief valve unless this valve has been set at the factory, by testing, to the correct pressure level.

First, jog the motor to determine direction of shaft rotation. Compare this with markings on pump nameplate or housing. A direction arrow will usually be stamped on the nameplate or may be stamped or cast into the front housing near the shaft. Piston pumps used on hydrostatic transmissions may have a small charge pump inside the rear cover. In this case the rotation marker will be on the rear cover or end cap. If no direction marker can be found, these general rules can be followed:

On gear-type pumps, when looking directly into the shaft, and with pump mounted so shaft and drive gear are on top and idler or driven gear is on the bottom, rotation should be CW (clockwise) if outlet port is on the right side of the housing or to the right side of center (when viewing through the pump), and should be CCW if outlet port is on the left.

On other types of pumps proper rotation cannot be determined from outward physical appearance. These pumps, when jogged will pump air into the reservoir below oil level if running in the wrong direction. This will cause turbulence which can be heard or seen when the breather or filler assembly is removed.

Reversing Electric Motor Rotation. If a 3-phase motor is found to be running in the wrong direction it can be reversed by interchanging any two of its line or load wires either ahead of or following the motor starter, or in the motor junction box. Single-phase motors can be reversed by re-wiring in the motor junction box, reversing polarity of starting winding to running winding. Instructions will be found inside the junction box. Once a single-phase motor is wired to run in a certain direction, it will always be correct even if moved and wired to a different power source.

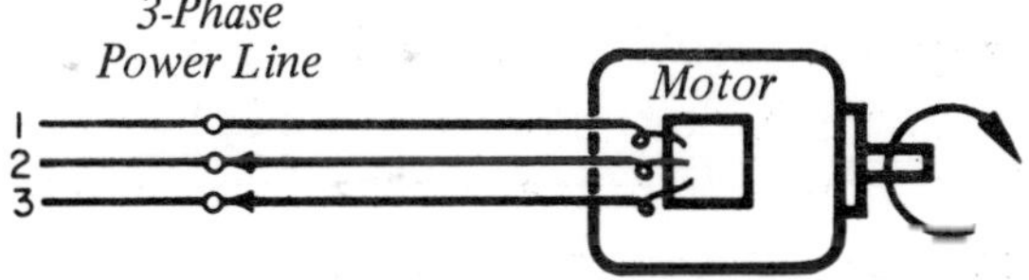

Connections for CW (Clockwise) Rotation

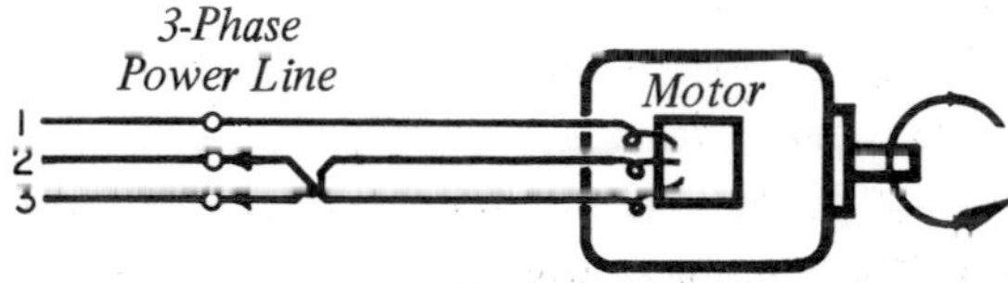

Reverse Wires for CCW (Counter Clockwise) Rotation

Pump Priming. After the motor is running in the right rotation, the next step is to determine if the pump has primed; that is, if it is picking up oil by suction from the reservoir. Do not run the pump except for jogging until you are sure it has picked up the oil. Gear pumps and vane pumps

Jog motor very briefly to prime pump or to check for direction of motor rotation.

will usually prime immediately unless the suction head is too high or the oil is too thick. Suction head is the lift distance from oil surface to the pump inlet port. Piston pumps do not develop as much suction as other kinds and may not immediately prime. They may have to be primed as directed later.

When starting a system it will usually be evident if the pump is working. There may be a slight reading on a pressure gauge in the pump line, or, there should be a change in the sound of the pump as it fills with oil. If there is any doubt, slightly crack a fitting in the pump pressure line to see if oil drips out.

If necessary to prime the pump, the best method and one which works in nearly every case, is to crack a fitting on the pump pressure line to allow the escape of air when oil is picked up. Do not start the electric motor; rotate the pump shaft by hand until oil drips from the cracked fitting. Tighten the fitting and start the motor. On pump and motor combinations too large to be rotated by hand, jog the motor with the motor starter as briefly as possible. Then allow the pump to come to a *complete* stop before jogging again. When oil drips from the cracked fitting, tighten the fitting and you are ready to start the motor. Some pumps may not self-prime if run at high speed by the motor because the mass of the oil prevents them from responding rapidly enough to the suction produced when the pumping elements open. But if rotated slowly by hand the pumps will easily prime in most cases unless the suction lift is too high or if the oil viscosity is too great.

If when once primed, a pump loses its prime when stopped overnight, this almost certainly indicates an air leak into the pump line or a leaking pump shaft seal.

Breaking in the System. After determining that the pump is rotating in the right direction and has primed, the system is ready to operate. There should be pressure showing on a pressure gauge in the pump line but on most systems this may be a very low pressure, 100 to 200 PSI or even less, because the pump is in an unloaded condition. Pressure will appear when the control valve is shifted.

Shift the directional control valve. This should start a cylinder or hydraulic motor into motion. Pump pressure, on some systems, may still be low if the cylinder or motor is free to move unloaded. The system should be operated without load during a break-in period while air is being expelled from lines and cylinders. Cylinders should be reciprocated back and forth without allowing them to bottom out at either end of their stroke or without allowing them to build up high pressure. Hydraulic motors should be run first in one rotation then the other. If allowed to run long enough, all air will be purged back to the reservoir. But the process can be accelerated by cracking fittings, especially those at the cylinder ports. To prevent high pressure build-up during air purging it is a good idea, if practical, to back down the pump relief valve. When operating a system remember that full relief valve pressure will only appear in most systems when cylinders are stalled against an immovable object or against a load too heavy for them to move.

If air purging is not complete the system will be "spongy". When all air is out it should become rigid. If sponginess develops later, look for air leaking into the system.

Operating the System. After the cylinders have been moved back and forth through their full stroke several dozen times, enough air will have been purged so the system can be operated under light loading. Air will not completely purge until the system is operated for awhile under pressure. The remaining air will slowly dissolve in the oil and be carried back to the reservoir where it will effervesce from the oil. The original relief valve setting can now be restored. Run the cylinder to the end of its stroke and let it stall briefly while setting the relief.

Make a final check on reservoir oil level and bring it up to the FULL mark when all cylinders are in their retracted position. Watch the oil level and be sure it does not drop below the LOW mark when all cylinders are extended. If it does, the reservoir is too small for the size of cylinders being operated.

The power unit was run under pressure while being tested after assembly. It was run long enough to pick up and filter out residual dirt. However, when connected to either a new or existing system, there will quite likely be additional dirt, rust, and scale carried out and circulated back to the reservoir. Install a temporary filter in the return line to tank. After a few days when residual dirt should have been picked up, the filter may be removed, or if left in the line its element should be replaced.

System Operates Hot. After the system has been installed and run for several hours or long enough to come up to a maximum levelling-off temperature, the reservoir oil temperature should be measured to see whether a heat exchanger will be needed (if the power unit does not already have one). In-plant systems should level off in the range of 130° F to 150° F maximum. A system installed in winter and working properly may overheat next summer. Overheating may cause accelerated wear in components, deterioration of seals, and chemical contamination of the oil. A heat exchanger should be added to industrial systems which operate above 150 to 160° F.

On mobile hydraulic systems it may be impractical to add a heat exchanger. If a system runs overheated, the entire supply of hydraulic oil should be replaced at the beginning of each season, and perhaps oftener on hard-working systems in unfavorable environments.

START-UP TROUBLE IN A NEW SYSTEM

Troubleshooting procedures in this section are for problems which may be present in a system started for the first time. For additional troubleshooting suggestions refer to Chapter 12.

Pump Wont Build Up Pressure. Remember that a hydraulic system normally places the pump in an unloaded condition while the cylinders are not moving, and pressure in the pump line may be low, 100 to 200 PSI, until the directional valves are shifted to move the cylinder. Even then, until the cylinder stalls, pressure in the pump line will only be proportional to the load resistance. Shift the directional control valve to see if the cylinder will move. If it will, let it stall against a positive stop or its own end cap and watch for pressure to build up.

If cylinder will not move and pressure will not build up when the control valve is shifted, check to be sure the pump has primed. If it was primed before, see if prime has been lost.

Check all mechanical items in the pump drive system such as slipping belts, sheared shaft key or pin, broken shaft (usually inside the pump), broken coupling, or loosened set screw.

Check reservoir for low oil level.

If these checks do not reveal the trouble, the power unit may be tested separately by disconnecting it from the rest of the system, plugging its pressure outlet line and running the pump across its relief valve. Be sure the relief valve is left connected to the pump and do not run pump oil across the relief valve for very long. Further details of this test are given in the next chapter on troubleshooting.

Lack of pressure on systems which use 4-way solenoid valves of the pilot-operated type may be due to a lack of pilot pressure to shift these valves which have tandem center or open center spools out of center neutral position. This problem and suggested remedies are covered in detail in "Industrial Fluid Power — Volume 2".

Lack of pressure may also be due to solenoid valves which fail to shift because of an electrical fault. This problem is covered in Chapter 12 on troubleshooting.

SUGGESTED MAINTENANCE ON HYDRAULIC POWER UNITS

Inspection Schedules. In those plants having a maintenance department, all hydraulic systems should receive regular inspections to prevent an unexpected breakdown at a time when it would be costly. Inspection schedules will vary with plant conditions such as ambient temperature, atmospheric dirt, number of shifts per day, daily hours of operation, and other factors. As starters we offer these suggestions: .

Daily Inspections. Visually inspect for oil leaks, especially around shafts of pumps, cylinders, and hydraulic motors. Leakage often starts at these points just before seal breakdown. Order new seals or overhaul kits at once. Watch the reservoir oil level over a complete cycle of the machine. Be sure the oil level never falls below the LOW mark as cylinders extend. Add oil if necessary.

Weekly Inspections. Once a week a housekeeping inspection should be made to observe general conditions around the machine. Tighten any fittings which leak. Clean up spilled oil. Clean dirt off top of reservoir. Inspect to see that filler and breather caps are in place. Tap a small quantity of fluid from one of the drain valves on the reservoir to see if water is precipitating in the reservoir. If water is found, suspect a water leak from a shell and tube heat exchanger. Water may be caused from condensation on inside walls of the reservoir due to atmospheric temperature changes around the reservoir. If this is the case, it will be necessary to tap off accumulated water every day. Check condition of breather cap; remove accumulations of lint and dirt. Breather must be clean especially at higher altitudes where air pressure is lower.

500-Hour Intervals. Frequency of this maintenance may vary according to plant conditions. Remove and clean pump suction strainers. Replace elements in micronic filters.

Miscellaneous. All hoses should be replaced after being in service approximately 5 years even though they appear to be in good condition. Keep pressure gauges closed off except when readings are to be taken; replace pressure gauges which have been damaged.

Recommended Spare Parts. The need to keep spare parts on hand is governed by the relative cost of lost time on the particular machine. Manufacturers instruction sheets with part numbers of replaceable parts should be kept on file.

Cylinders. Piston and rod seals should always be kept on hand. Overhaul kits can be purchased from cylinder manufacturer which contains all replaceable gaskets, O-rings, and parts which normally wear out. These parts are usually packed in a sealed polyethylene bag. On receipt the bag should be dated. If not used within 2 years a new kit should be ordered to replace it. Rubber deteriorates by oxidation in a few years. If left sealed in an airtight bag it will deteriorate more slowly.

Pumps. Overhaul kits are sometimes available for pumps, and include all soft seals, gaskets, and sometimes bearings. Spare shaft seals should always be kept, and on important machines, entire pumps should be kept as spares. On vane pumps, spare vanes and springs, or complete cartridges, should be kept.

Electrical Parts. Keep replacement coils for solenoid valves, relays, and motor starters. Limit switches, fuses, lamp bulbs, and on important machines, spare relays should be kept.

Filter Elements. Several sets of elements for paper-type micronic filters should be on hand. Wire mesh pump suction strainers seldom need replacing because they can be cleaned in most cases.

Hydraulic Oil. Be sure to have a reserve supply of the same brand and type of oil used in the system. Usually a back-up supply of 10 to 25% of reservoir capacity should be kept as a minimum.

Chapter 11

Installation & Start-Up of Hydrostatic Transmissions

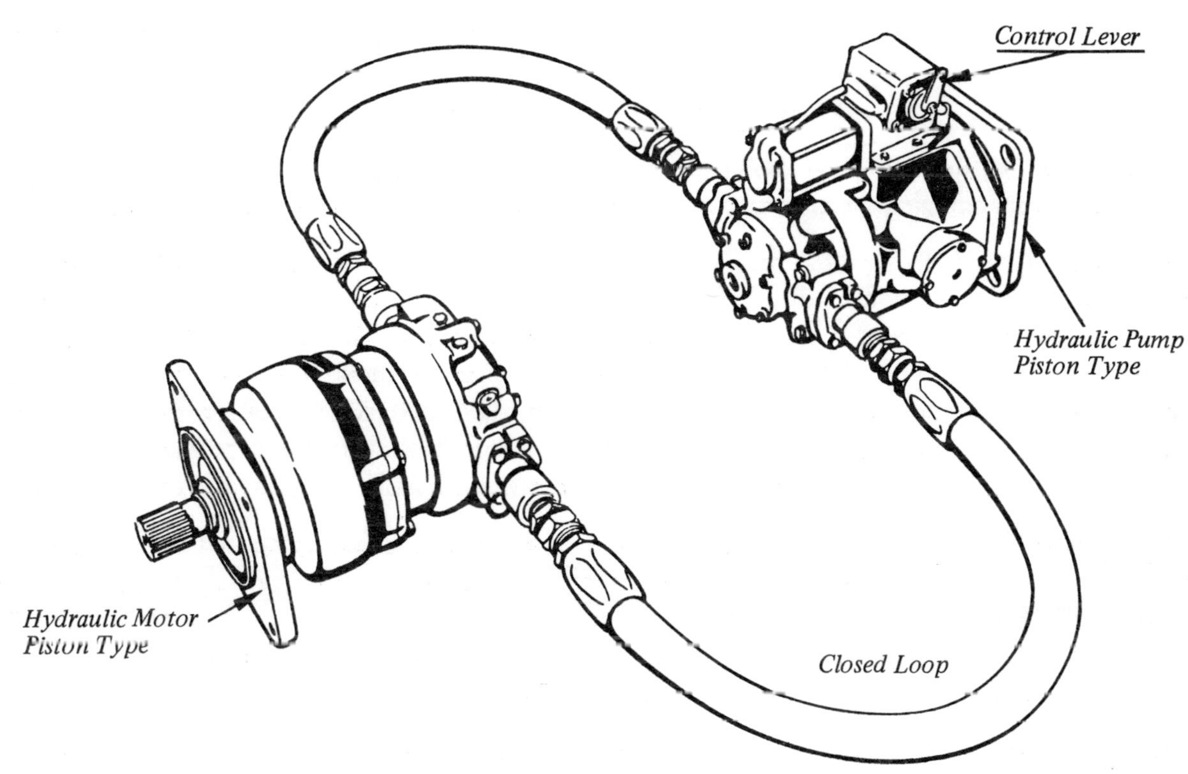

CLOSED LOOP HYDRAULIC PUMP-TO-MOTOR TRANSMISSIONS

Most hydrostatic transimissions use a hydraulic pump driven by an engine or electric motor and connected by means of a closed loop to a hydraulic motor. A closed loop is a circuit in which neither port of pump or motor is vented directly to reservoir. Discharge oil from the motor re-circulates back to the inlet of the pump. A small volume, low pressure, charge pump continually replaces about 10 to 15% of the circulating oil with fresh oil from the reservoir. The replaced oil is returned to the reservoir for filtering and cooling before being placed back in the loop. Pressure is maintained at all times in both sides of the loop, usually about 4000 to 5000 PSI in the working line and from 150 to 500 PSI in the return line depending on the size of the transmission. Hydrostatic transmissions are explained in detail in the Womack textbook "Industrial Fluid Power — Volume 3".

KEEP IT CLEAN!

Piston-type pumps and motors are used in high pressure transmissions. They are precision pieces of equipment, quite expensive, and should be handled as such. They are very sensitive to contamination in the oil. Cleanliness in making the installation and in servicing or trouble-shooting is absolutely essential if the equipment is to have the long and trouble-free life for which it was designed. Do not disassemble any pump, motor, or control under dirty conditions. If necessary to leave a pump, motor, or line open while waiting for repair parts, keep it covered. Protect the ends of hose or pipe which must temporarily remain open. The manufacturers warranty does not cover equipment which has been abused by carelessly failing to take precautions to exclude and remove dirt from the hydraulic system.

PLACEMENT OF THE COMPONENTS

The diagram on the next page shows the basic components found in the usual hydrostatic transmission. Case drain connections may vary from the arrangement shown according to brand.

Main Pump, Item 1. Be sure the rotation of the shaft is in the direction marked either on the front or rear cover. Wrong rotation of the shaft will damage the charge pump. Control linkages or levers for the operator must be connected to the stroking lever on the pump. The case drain line must never be plugged or restricted. It must be full size as specified in the manufacturers literature. An undersize case drain line may cause the pump shaft seal to blow out. The inlet connection to the charge pump (shown on rear cover) must be full size. A restricted line may cause the pump to cavitate. Hydraulic hose for the loop lines must have a working pressure rating equal to the system operating pressure.

Charge Pump, Item 2. The charge pump is usually a gear or gerotor pump housed in the main pump rear cover (occasionally in the front cover). On some installations, if the internal charge pump does not have sufficient volume, the system may have been designed to use a separate external pump. The charge pump keeps a small volume of fresh oil feeding into the closed loop, and furnishes pressure for shifting the pump cam plate when the pump displacement lever is moved from neutral to a side position.

Hydraulic Motor, Item 3. On most installations this will be a fixed displacement piston-type motor of approximately the same displacement as the pump, producing the same maximum speed as the pump driving motor. On some installations the motor may be a slow speed radial piston motor of greater displacement than the pump, giving high torque output at slow speed.
Case drain lines must be full size to avoid blowing out the motor shaft seal.

Charge Pump Filter, Item 4. Oil fed directly into the charge pump inlet port must be filtered to 10μm. This will usually require a paper element filter. The filter should have sufficient filtering area to keep pressure drop very low — to 1 to 2 PSI — when the element is clean. If the filter becomes restricted the charge pump will cavitate. This may cavitate the main loop and may prevent the system from reaching maximum speed. A filter with an electrical or visual indicator is recommended.

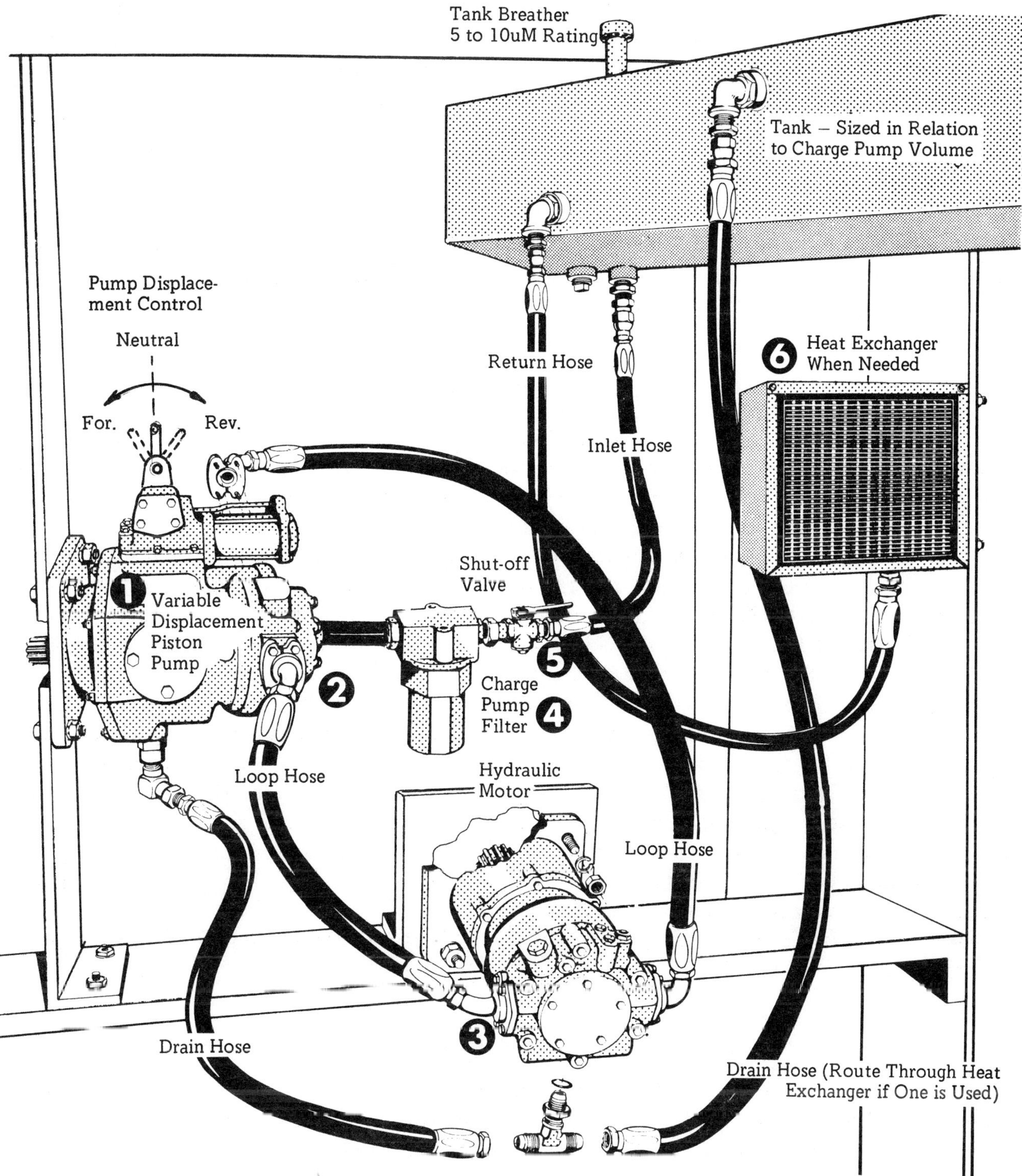

Main Components of a Closed Loop Hydrostatic System

Shut-Off Valve, Item 5. When the reservoir is at a higher elevation than the pump, this shut-off valve is necessary so the oil to the filter can be shut off while the element is being changed. We recommend a quarter turn plug valve which can be padlocked in the open position. Closure of this valve while the system is running might destroy the charge pump and might damage the main pump.

Relief Valves. The high pressure and low pressure relief valves are usually contained in the rear cover of the pump or motor. They have been pre-adjusted and their setting should never be changed except by a maintenance man who is technically qualified to adjust that particular system.

Heat Exchanger, Item 6. Many transmissions will require a heat exchanger. It must be placed in a low pressure location. The only practical location on most transmissions is in the case drain lines from pump and motor. Join case drain flow from pump and motor and run to heat exchanger. The discharge side of the heat exchanger should be piped to reservoir.

Filtering. Although not shown in this illustration, high pressure filters of 3 to 10μm rating are sometimes placed in both loop lines. Special filters are available for this purpose. They have internal check valves to prevent backflow through the filter when the transmission is reversed.

OIL RESERVOIR

On vehicular installations, for the most reliable operation, the reservoir should be sized as for industrial installations, a minimum of twice the volume of oil circulated in one minute. But on some installations a reservoir of this size may not be practical because of limitations in space and weight. In these cases, the reservoir should be as large as space permits.

One of the important advantages of a closed loop system over an open loop system is that a much smaller flow of oil is being circulated into and out of the reservoir, and a smaller reservoir can be used without sacrificing reliability or performance. Capacity of a reservoir for a closed loop system is based on the circulating volume of the charge pump, not the main pump. Charge pump volume is usually only 10 to 20% that of the main pump, so a 20 gallon reservoir on a closed loop will equal the performance of a 100 to 150-gallon reservoir on an open loop system of the same HP rating.

Cleaning of the reservoir during system construction is especially important for a closed loop system. The piston pumps are far more dirt sensitive than the gear and vane pumps used on open loop systems. Some builders sand blast the reservoir to remove weld slag. This is not a good practice unless the reservoir is thoroughly flushed with oil or a solvent to remove all traces of sand. Sand blasting, without flushing, may cause more trouble later than if the reservoir was not cleaned at all.

All hydraulic reservoirs should have one or two drain holes. These are important because they will be used later to drain out water condensate.

Please refer to details of reservoir construction in the Womack textbook "Industrial Fluid Power – Volume 1".

HEAT EXCHANGER

Most closed loop systems of 50 HP or more will require a heat exchanger if they operate at full power for long periods. On vehicular equipment the usual practice is to install an air cooled radiator in front of the engine radiator. For in-plant systems a water cooled shell and tube heat exchanger gives most efficient cooling if a water supply is available. Air cooled radiators, with electric-motor-driven fan are also available. Important characteristics of each type heat exchanger are covered in the Womack textbook "Industrial Fluid Power – Volume 1.

A very important point when selecting a heat exchanger for a closed loop transmission using piston pumps and motors is to make certain the pressure drop through the heat exchanger under the worst condition of cold oil, etc., is very low – 10 PSI or less. Since the heat exchanger is installed in the case drain lines, the pressure drop across it will be reflected back into the pump and motor cases, and if high enough will blow out the seals. Standard piston units will tolerate about 15 PSI case pressure for standard seals or no more than 50 PSI for high pressure shaft seals.

HYDRAULIC OIL AND FILTER RECOMMENDATIONS

Approved Oils. The transmission manufacturer will furnish a list of oil brands and types which can be used in his transmissions. Oil characteristics are more critical for closed loop transmissions because of the higher pressure operation. The use of certain oils, not approved by the manufacturer may void the transmission warranty. If in doubt, contact the Engineering department of the manufacturer and ask for approval.

Changing Oil. On vehicular hydrostatic transmissions it is usually impractical to keep the oil temperature low enough (130°F) to obtain long oil life. The recommended practice is to completely replace the oil periodically. The replacement interval will vary widely according to the operating temperature and the duty cycle. On equipment used occasionally, changing the oil once a year may be often enough. But for hard working equipment in hot weather it may be necessary to replace the oil every month or two. Watch the color of the oil and compare it with a new sample of the same oil. A slight darkening is to be expected, but a change in color to a very dark, perhaps an opaque brown, indicates a great deal of oxidation has taken place and may indicate the necessity for an oil change.

Filters. The piston pumps and motors used on hydrostatic transmissions are very sensitive to dirt, more so than other types operating at lower pressure. Since these units are very expensive, money spent on micronic filters is money well spent. Very careful attention should be given to selecting good filters when the system is installed and to servicing the filters on a regular basis. If a piston pump or motor should fail and if sent to the factory for replacement, the factory will not honor the warranty if traces of dirt are found inside the units.

The minimum filtration which is acceptable for warranty requirements is $10\mu m$ micronic filtration on all oil which enters the loop through the charge pump. This means a $10\mu m$ inlet filter on the charge pump, or, on some transmissions the charge pump oil is brought outside the main pump case for installation of a $10\mu m$ filter on the charge pump outlet. Since a dirt build-up in this filter can cause a disaster inside the main pump, the filter should be one which has a warning signal or light.

In addition to $10\mu m$ filtration in the charge pump circuit, transmission life can be extended by micronic filters of 3 to $10\mu m$ installed in both loop lines. Loop filters are beneficial because some of the contamination is generated by the pump and motor inside the system.

OEM companies manufacturing equipment for resale on which a hydrostatic transmission is used, should install $3\mu m$ filters on the loop during testing, and should let the system run idle at full circulating flow for a period long enough to pick up all contamination. Usually a running period of 30 minutes should be sufficient. Before shipment of the machine the $3\mu m$ filters can be removed and used on the next machine.

Handling the Oil. As a general rule, all oil added to any hydraulic system at any time should be new oil taken from a factory sealed container. Reserve oil should be stored in original containers, kept tightly sealed, and stored in an area which is at or below normal room temperature. Utensils used for transferring the oil should be kept wrapped in plastic bags or other protection and kept in a clean place.

On hydrostatic transmissions the new oil, even though from a factory sealed container, should be poured into the reservoir through a micronic filter. It is important to re-cap the reservoir immediately after the new oil has been added. Be sure a micronic breather is used on the reservoir and that it is in place. The above precautions may seem to be nit-picking, but we have found that dirt has a way of getting into the system in spite of the best precautions, and the oil should be almost surgically clean if the transmission is expected to give long service.

Hydraulic Connections. Hose connections are recommended for connection between units of a hydrostatic transmission because of cleanliness, ease and speed of installation, and reduction in system noise. Steel tubing may be used as an alternate. Black pipe should never be used. Galvanized pipe may be used if internally cleaned.

PUTTING A HYDROSTATIC TRANSMISSION INTO SERVICE

1. Visual Inspection. Unpack all units of the transmission and inspect for physical damage. Make a detailed check against the packing list to be sure you have received all items on the list.

2. Rotation Check. Make a careful comparison of the direction of rotation of your power source with reference to the direction of rotation shown on the front or rear cover of the pump. It is ***absolutely essential*** that the direction of rotation of your power source be the same as that of the pump. No matter whether your power source is diesel, gasoline, steam, or electric motor; no matter if clockwise or counter-clockwise, if its rotation is different from that shown on your pump, ***do not proceed with the installation.***

If the driving source is an electric motor, jog it briefly before coupling to the pump to determine rotation. It can be reversed by interchanging any two of the three feed wires inside the motor junction box, or on the line or load side of the motor starter.

If driving from an engine it will be necessary to contact the factory if pump rotation is not correct. Most transmissions can be altered for the opposite rotation in the field provided the necessary extra parts and instructions can be obtained. On some transmissions the main pump will run equally well in either direction of rotation but the rotation of the charge pump must be changed. On other transmissions the main pump group will not run equally well in the opposite rotation because of off-center timing on the pump cam plate inside the pump. Contact your supplier for exchange of these pumps.

3. Pump Drive Speed. Pump speed should never exceed the rated maximum for that particular model. Even a momentary overspeeding can cause damage to the pump. Engines should be provided with a governor.

4. Mounting and Coupling. Before mounting, check your splined couplings. They must be a slip fit but a snug fit on the pump shaft. Pump shaft must be carefully aligned with the driving source or output load shaft. Alignment may be done with a dial indicator, matching the centerline of the driving source shaft with the bore on the pump or motor mounting bracket. Or, may be done on the two shafts after the transmission pump and motor have been mounted.

When using a flexible or semi-flexible coupling, the alignment must be to the tolerance specified by the coupling manufacturer. If coupling direct, with a rigid coupling, alignment must be to a tolerance of ±.001".

After aligning the shafts, install the complete coupling and test for accuracy of alignment. If a roller chain coupling is used, feel the chain to be sure it is free to slip back and forth. Then rotate the coupling in intervals of 45° to be sure there is no bind at any position. Finally, uncouple the hydraulic motor until the system has been started and run in.

5. Fill the System With Oil. Completely fill the hydraulic reservoir up to the FULL mark on the gauge, using clean practices of straining the oil as it is added. A small amount of additional oil will have to be added as the system components fill with oil.

Important! Fill the pump and hydraulic motor cases with oil. See caution on Page 184. There will usually be alternate case drain ports through which oil can be poured, or, filling can sometimes be done through trunnion ports. Refer to service manual for location of recommended ports for filling.

6. Priming the Charge Pump. When starting a new system or servicing an existing system, the service manual for the transmission must be consulted for location and size of oil ports. Install a gauge to read charge pump pressure. Usually a 1000 PSI range is suitable.

If oil reservoir is at a higher elevation than the pump, crack the fitting at the inlet to the charge pump to release trapped air in the suction line and to allow oil to flow into the inlet filter case and suction line. When oil drips from the fitting, tighten the connection.

If reservoir is below the elevation of the pump, fill the inlet filter case with oil. Prime the charge pump by very briefly jogging the driving electric motor, allowing it to come to a complete stop before jogging again. Be sure the control lever on the main pump is in neutral while jogging the pump. On engine drive systems, jog by de-clutching, or if this is not possible, run the engine at idle speed with main pump control lever in neutral. The pump is primed when a low pressure appears on the gauge. Full pressure may not appear until later, after the system has been purged of air.

7. System Start-Up. Be sure the hydraulic motor is disconnected from the load, or reduce load to a minimum by disconnecting it at some point beyond the motor. Be sure control lever on pump is in neutral.

Start pump into rotation but do not shift the control lever out of neutral until at least a low pressure is present on the charge pump gauge. If no pressure appears on the gauge, go back and re-check for correct pump rotation and re-prime the charge pump if necessary.

Let the pump run for several minutes at the lowest feasible speed. Watch the reservoir oil level during this period, and add oil to bring the level to the FULL mark.

8. First Start. Install high pressure gauges in both sides of the loop. Refer to service manual for location and size of high pressure gauge ports. A gauge range of 0 to 6000 PSI should be right for systems operating at 4000 to 5000 PSI.

Move the pump control lever *a little way* into one side position. Run the system slowly in both directions for 10 to 15 minutes (with no load on the motor) to bleed all air from the system. If hydraulic motor does not rotate in the direction desired for a forward or reverse movement of the control lever, the only way to change this relationship is to interchange the loop hoses on either the pump or the motor.

Gradually increase the hydraulic motor speed in each direction during this run-in period until it is operating at full speed. System noise will gradually decrease as air continues to bleed out of the system.

9. Full-Scale Run-In. Connect the load to the hydraulic motor shaft and operate it under actual working conditions. Test for peak performance under the heaviest loads you may encounter. Check for smoothness of starting in both directions under light and heavy loads. Check for smoothness as control lever is moved through neutral to the opposite direction. Try quick starts and slow starts. Check for steadiness of power flow through the entire speed range.

Maximum pressure in the loop lines cannot be checked until full load is placed on the motor. It may be necessary to lock the motor shaft or to run it against a dead load to determine if maximum system pressure (and torque) can be developed.

If the system does not meet the performance expected, contact the transmission factory for assistance. Be prepared to furnish all pertinent information which will help them diagnose the problem and suggest a solution. Information necessary may include: charge pump pressure both with control lever in neutral and at full throw, and on each side of center neutral; maximum system pressure on both sides of center neutral; approximate case drain flow both with control lever in neutral and at full throw, and on both sides of center neutral; brand, type, and viscosity of oil used.

10. Final Check. Go over all connections inspecting for leaks and tighten as required. Again check oil level in the reservoir. If all the above tests are satisfactory, the system is ready for full scale operation. After running for several hours under typical operating conditions of heavy and light loads, measure temperature of oil in the reservoir. On industrial systems, oil temperature should not exceed 160° F. On mobile systems the oil temperature *should never* exceed 200° F. If higher than these limits, a heat exchanger should be added.

After the system has been in operation for 8 to 15 hours, we suggest replacing the elements on all micronic filters and cleaning all wire mesh strainers.

GENERAL RULES FOR MAINTENANCE

Fluid Level. Check fluid level every 4 to 5 hours the first day of use, then twice a day for the first week, then once a week thereafter.

External Leaks. Keep a surveillance check for leaks. Those which develop suddenly may deplete the reservoir to a dangerously low level.

Hydraulic Oil. If color becomes dark brown to the point of complete opacity, change as soon as possible. In any event, change as often as necessary and at least once a year on systems which run above 160° F temperature.

Filters. Change elements in micronic filters on new systems after 8 to 15 hours of initial operation. Then, after each 600 hours, or more frequently on mobile systems or those operating in a dirty environment. Filters with visual or electrical indicators are recommended. Clean wire mesh strainer elements each 600 hours of operation.

Heat Exchanger. Keep dirt and lint accumulations removed from core of air cooled heat exchangers. On water cooled heat exchangers, inspect and replace zinc anodes at regular intervals. The interval depends on the quality of cooling water. Water with high mineral content consumes the zinc more rapidly.

Oil Reservoir. Check daily for accumulation of condensed water inside the reservoir by cracking a drain valve. Systems exposed to changing ambient temperatures may need daily draining of water from the reservoir. Suspect water leaks in shell and tube heat exchangers.

Chapter 12

Troubleshooting Hydraulic Systems

HYDRAULIC SYSTEM CHECK-OUT FOR OPEN LOOP SYSTEMS

Finding the cause of trouble on any fluid power installation requires some ''know-how'' and good common sense. When trouble develops, don't start taking the system apart. Analyze the problem first. Give it a lot of thought before you act.

Before breaking into the fluid circuit, be sure the trouble is not mechanical or electrical. Jot down a list of symptoms such as: increased noise, difficulty in reaching full pressure, loss of speed, components (especially the pump or hydraulic motor) running very hot, weak or slow performance in only one direction of cylinder movement, sponginess or compressibility in the system, appearance (milkiness) of the oil, did the trouble develop very suddenly or did it develop gradually over a period of time, etc. A careful consideration of these symptoms will often give an important clue to the nature of the trouble.

Information in this chapter is intended primarily for systems which were correctly designed and which have been performing properly. Although it is not intended as a diagnostic check when starting up a new system, still the same principles apply and the step-by-step troubleshooting procedure will sometimes reveal an area where the original design needs to be improved.

SYMPTOMS OF THE TROUBLE

Mechanical Faults — Trouble in the Drive Train

1. Broken or defective shaft coupling.
2. Broken pump shaft.
3. Sheared pin or key on pump shaft.
4. Mis-alignment of pump shaft.
5. Slipping or broken belts on pump drive.
6. Shaft turning in wrong direction (if power unit has just been moved to a new location and re-wired).
7. Broken teeth in a gear pump, broken vanes in a vane pump, broken pistons or valves in a piston pump. These internal mechanical troubles can be checked later after the step-by-step diagnosis.

Electrical Faults

 1. Broken connection in wiring to electric motor or in control circuit. Check for presence of correct voltage throughout the system.

 2. Check fuses on all three phases of a 3-phase electric motor. If the fuse on one phase is blown the motor may run but will overheat and will not produce full power.

 3. If the driving electric motor has just been replaced make sure it is the correct voltage, phase, Hertz, HP, and is running in the correct rotation and is the correct speed for the pump.

 4. Check fuses in the electrical control circuit.

GENERAL SYMPTOMS

Pump Appears to be Delivering too Little or no Flow

 1. Shaft may be rotating in the wrong direction. Shut down immediately. Reversed leads on a 3-phase electric motor are a common cause for wrong rotation.

 2. Pump inlet may be clogged or restricted. Check inlet strainer for dirt. Check for collapsed inlet hose.

 3. Low oil in reservoir. Check oil level with all cylinders extended.

 4. Stuck vanes, valves, or pistons, either from varnish in the oil or from rust and corrosion. Varnish indicates the system is running too hot. Rust or corrosion may indicate water in the oil.

 5. Oil may be too thin either from wrong choice of oil or from thinning out at high temperature. A system with this problem may operate normally the first few hours after start-up, then gradually slow down as the oil becomes overheated and thins out.

 6. Mechanical trouble. Check points listed on the preceding page.

 7. Pump running too slow. Most pumps will deliver a flow at all speeds proportional to RPM. But some vane pumps which depend solely on centrifugal force to extend the vanes into contact with the cam surface will deliver little or no flow at slow speeds such as engine idle speed.

 8. If electric motor has been replaced check for correct speed and rotation.

Pump Noise Has Recently Increased

 1. Cavitation of pump inlet. Check for symptoms of cavitation described later.

 2. Air leaking into system from low oil level or other causes described later.

 3. Mechanical noise caused by loose or worn coupling, loose set screw, badly worn internal parts, and other possible causes listed previously.

 4. System may be running at too high an oil temperature, causing oil viscosity to be too low.

 5. Pump may be running at higher than rated speed.

Unusually Short Pump Life

 1. Pump may be operating above catalog maximum pressure rating, especially if pump must maintain this excessive pressure for a high percentage of its running time.

 2. Oil may be of incorrect viscosity or may be a type not recommended for the system.

 3. The oil temperature may be running too high. A heat exchanger may be needed.

 4. Inadequate filtering. Check oil for dirt contamination and add micronic filtering if necessary.

 5. Oil contamination by water leaking or condensing in the system or by emulsification with air.

 6. Improper maintenance, particularly failure to regularly clean pump inlet strainer.

 7. Mis-alignment of pump shaft with motor or engine shaft. Note: When replacing a bracket mounted pump, leave the bracket and replace only the pump. Chances are that if an exact replacement pump is installed it will not have to be re-aligned with the motor.

8. The pump may be running too fast (or too slow).

9. Inlet cavitation from causes other than listed above.

10. If reservoir lid is not sealed airtight to the tank, contamination will enter the system, possibly causing damage to the pump.

System Overheating

On systems which have recently developed a heat problem:

1. An excessively worn pump will have excessive internal slippage and this will generate heat. The pump will be extremely hot around the shaft.

2. The pump may not be unloading between cycles. This can be checked by installing a pressure gauge in the pump line. However, some circuits using pressure compensated pumps will maintain full pressure in the pump line between cycles even though the pump is unloaded by having its flow reduced to near zero.

3. The pressure relief valve may be set higher than necessary.

4. If a heat exchanger is used, check for adequate water supply if it is a shell and tube type. Also check operation of thermostatic water control valve. On air cooled heat exchangers, clean lint and dirt from between fins.

5. If the duty cycle of the machine has been increased, overheating may not indicate anything wrong with the system but a heat exchanger should be installed to limit oil temperature to 140° F on industrial systems.

STEP-BY-STEP CHECK-OUT OF OPEN LOOP HYDRAULIC SYSTEMS

An open loop system is one in which return oil goes directly to the reservoir, to be picked up again later by the pump. Closed loop system check-out is described later.

This circuit diagram shows the major components in the usual open loop hydraulic system. The step-by-step check-out procedure can be followed as described here, but if the service man has a pretty good idea of the component which has failed it is often quicker to substitute a similar component which is known to be good.

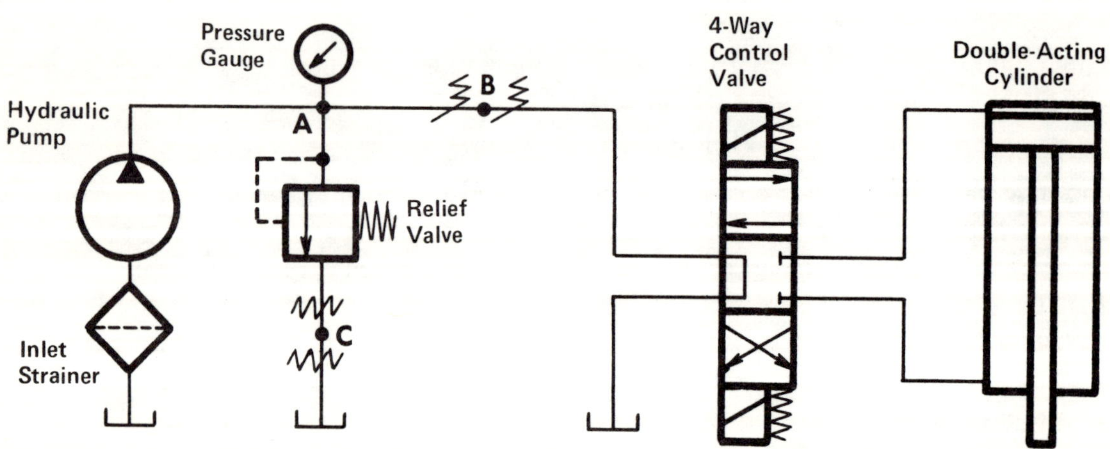

Major Components in an Open Loop Hydraulic System

A pressure gauge is almost a necessity for troubleshooting. A range of 1½ to 2 times the expected pressure should be used. First, the gauge should be installed in the pump pressure line. Later, the same gauge, or another one, can be installed in a branch circuit for further pressure testing.

Many of the failures in a hydraulic system show similar symptoms: a sudden or gradual loss of high pressure, resulting in a loss of cylinder speed and/or force. The cylinder(s) may not move at all, or if they do, may move too slowly or may stall before full force is reached. Often the loss of power is accompanied by an increase in pump noise, especially as it tries to build up pressure against a load.

Of course any major component — pump, relief valve, cylinder, 4-way valve, or filter could be at fault. And in many systems there are additional components such as sequence, by-pass, reducing, and counterbalance valves which could be at fault, but these possibilities are too numerous to be covered in this brief discussion.

By following an organized step-by-step procedure in the order given here, the problem can usually be traced to a general area or to one branch of a multiple branch system. Then, if necessary, each component in that area can either be tested or can be replaced with a similar component known to be good. But it makes good sense to check out the pump area first before going into branch circuits because the majority of system failures occur in the pump and relief valve circuit.

Step 1 — Pump Inlet Strainer

Probably the field trouble encountered most often is cavitation of the hydraulic pump caused by a dirt build-up on the inlet strainer. Not only can this happen on a system which has been in service for some time, it can happen on a new system after only a few hours of operation. It produces the symptoms described above: increased pump noise, loss of high pressure and/or cylinder speed.

Look for the pump inlet strainer. It may either be inside the reservoir below oil level or may be outside the reservoir and next to the pump inlet port. If inside the reservoir it can be removed by uncoupling the inlet line, removing the flanged cover where the line goes into the tank, then withdrawing the strainer. If the hydraulic power unit has been built properly there will be union joints for disconnecting the strainer for removal. On externally mounted strainers, usually the case can be opened for removal of the element without removing the filter body from the line.

Some operators may not be aware of the strainer inside the reservoir and may fail to service it regularly. A dirty strainer restricts flow into the pump and will definitely cause it to fail prematurely.

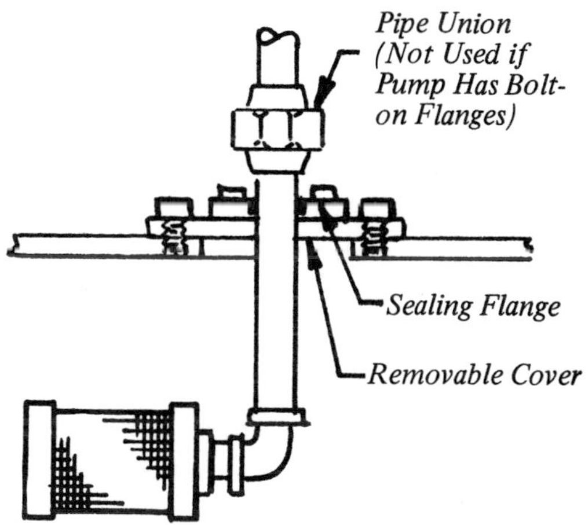

Uncouple inlet line, remove cover plate, and withdraw strainer from reservoir.

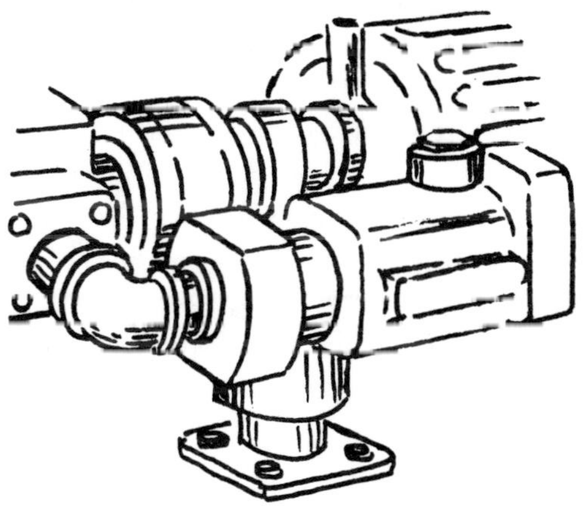

On externally mounted pump inlet strainers, element can be removed without disconnecting filter body from the line.

In any case of system failure, unless the exact cause of the failure is known, the pump inlet strainer should be removed for cleaning as a matter of routine. Clean it whether or not it looks dirty. Some clogging materials are hard to see. Wire mesh strainers can be cleaned with an air hose, blowing from the inside out. If there is any deposit of a brown varnish on the strainer, it should be washed in a solvent, scrubbing with a bristle brush. Caution! Do not use solvents like gasoline, naptha, lacquer thinner, etc., which are explosive and highly flammable. The solvent must be compatible with the fluid in the system. For example, kerosene can be used on hydraulic oil systems. On systems using synthetic or fire resistant fluids, use some of the same fluid for cleaning the strainer. If the varnish deposits are heavy the strainer should be replaced. If there are any holes in the mesh or if there is obvious physical damage, replace the filter element.

Incidentally, when a system is opened, if varnish is found on any component, this is a pretty sure indication that the oil temperature is running too high, and a heat exchanger should be installed.

When re-installing the strainer, inspect all joints in the inlet plumbing for air leaks, particularly at union joints. There must be no air leaks in the inlet line. Check tank oil level to be sure the strainer is immersed at least 2 or 3 inches even at low oil level, which occurs when all cylinders are extended. If it is not well covered, there is danger of a vortex forming above the strainer which may allow air to enter the system when the pump is running.

While cleaning the strainer, notice the condition of the inlet hose (if one is used). A partially collapsed hose or one which has internally swelled due to hot oil has the same negative effect as a clogged inlet strainer.

Some users make a quick check to see if a clogged inlet strainer is the fault by temporarily removing the strainer element and running the pump very briefly to see if the trouble clears up.

Step 2 — Pump and Relief Valve

If cleaning the pump inlet strainer does not solve the problem, isolate the pump and relief valve from the rest of the system and run the pump across the relief valve. Disconnect the plumbing at Point B (see diagram on Page 200), and cap *both* ends of the disconnected lines. This will deadhead the pump into the relief valve. But first, back off the relief valve. Then start the pump and watch the pressure gauge for a pressure build-up as the relief valve adjustment is tightened. If full pressure can be developed, obviously neither the pump nor the relief valve is at fault, and the trouble is further down the line. If full pressure cannot be developed, then either the pump or the relief valve is not working properly. To determine which of these components is at fault, proceed with Step 3.

Caution! To avoid overheating the system, run the pump across the relief valve only briefly, just long enough to observe a gauge indication.

Step 3 — Pump or Relief Valve?

Further testing must be done to determine whether the pump is worn out or if the relief valve is malfunctioning.

The volume of oil discharged from the relief valve must be observed. If practical, disconnect the tank port of the relief valve at Point C of diagram on Page 200. Attach a short length of hose to the tank port and hold the open end of the hose over the tank filler opening so the rate of flow can be observed. Start the pump and run the relief valve adjustment up and down while observing the relief valve discharge. If the pump is bad, a full stream of oil may be observed when the relief valve is backed down to its lowest setting, but this stream will diminish and probably stop as pressure builds up. In this case, the oil flow is slipping back to the pump inlet and is not passing across the relief valve. If a flowmeter is available the discharge flow can be measured and compared with pump catalog rating

If a flowmeter is not available, the flow can be estimated by discharging the flow into a container, say for 15 seconds. The discharged volume can be measured and multiplied by 4 to arrive at a figure for the GPM flow rate. However, even without measuring the flow, a bad pump is indicated if discharge varies widely as the relief valve adjustment is run up and down. The discharge flow should be fairly constant at all pressure levels, although it normally will drop off slightly at higher pressures.

If the relief valve tank port cannot be disconnected to measure flow, the mechanic can remove the filler/breather assembly and reach inside the tank. By placing his hand near the end of the tank return pipe he can judge whether the discharge volume is varying widely as the relief valve is adjusted.

If the flow decreases markedly at higher pressures, but still a moderate, but not full pressure, can be reached, this probably indicates a badly worn but not completely inoperative pump. Proceed to Step 4.

During this test if gauge pressure does not rise above a low value, say 100 to 200 PSI, and if a full discharge is obtained at all relief valve adjustments, the relief valve may be at fault and should be cleaned or replaced as instructed in Step 5.

Step 4 — Pump

If a full stream of oil from the relief valve is not obtained in Step 3, or if the stream diminishes markedly as the relief valve setting is raised, the pump is probably worn out. Assuming that the inlet strainer has been cleaned and the inlet plumbing inspected for air leaks and collapsed hose, the pumped oil is slipping inside the pump from outlet back to inlet instead of going across the relief valve. Either the pump is worn too badly to be able to perform properly or the oil is too thin. High temperature in the oil will cause oil viscosity to decrease, sometimes so much that it cannot be pumped to high pressure. High slippage in the pump will cause it to run much hotter than the oil in the tank. Under normal conditions, with a good pump, the pump case may run 20 to 30° F higher than tank temperature. If greater than this, excessive pump slippage may be the cause.

Check also for slipping belts, sheared shaft key or pin, broken shaft, broken coupling, loosened set screw, and other possible mechanical causes.

Step 5 — Relief Valve

If Step 3 has shown the pump to be OK, the relief valve may be at fault, and the quickest proof may be to temporarily replace it with one known to be good. The faulty valve may later be disassembled and cleaned. Pilot-operated relief valves have small internal orifices which can become blocked with dirt. Use an air hose to blow through all passages and pass a small wire through orifices. Check also for free movement of the spool or poppet. Valves with pipe thread connections in the body may distort if the pipes are screwed too tightly into the body. The distortion may bind the spool. If possible, check for spool binding before unscrewing threaded connections, or, while testing on the bench, screw pipe fittings tightly into the port threads.

Step 6 — Cylinder

If the pump will develop full pressure while deadheaded into the relief valve in Step 2, both of these components can be assumed to be good. Test cylinder piston seals as described on Pages 207 and 208.

Step 7 — Directional (4-Way) Valve(s)

Do not test for excessive leakage on 4-way valve spools until the cylinder piston seals have been tested and found to be reasonably tight. It is rare that a valve spool becomes worn so badly that it will slip the entire volume of the pump under pressure, but it could happen. As valve spools become worn, there may be a loss of cylinder speed, especially under a heavy load. There may be difficulty in building up full system pressure even with the relief valve adjusted to a higher setting. Valve spool leakage is a more serious problem when using a pump of small displacement operating at very high pressure. Symptoms will usually develop gradually over a long period of time. Valve spool leakage can be checked by the method described on Page 208.

Other Components

If the above procedure does not reveal the trouble, check other components individually. Usually the quickest and best troubleshooting procedure is to replace suspected components, one at a time, with ones known to be good. Pilot-operated solenoid valves which will not shift out of center position may have insufficient pilot pressure available. See more information later in this chapter.

TROUBLESHOOTING ON HYDRAULIC PUMPS

Some of the points summarized here were mentioned briefly earlier in this chapter or in other chapters. Pumps and hydraulic motors are subject to more mechanical wear than other components and the ones more likely to wear out. The following checklist is for pumps but some of the points may also apply to hydraulic motors.

On systems where the pump seems to wear out too soon or has to be rebuilt more often than seems necessary, one or more of the following problems may be present:

Pump Cavitation

Cavitation is the inability of a pump to draw a full charge of oil either because of air leaks or restrictions in the inlet line. When a pump starts to cavitate its noise level increases and it may soon become very hot around the shaft and front bearing. Other symptoms of cavitation are erratic movement of cylinders, difficulty in building up full pressure, and perhaps a milky appearance of the oil in the reservoir. If cavitation is suspected, check these points:

1. Check condition of pump inlet strainer. Clean it even if it does not look dirty. Follow procedure described in preceding pages. Varnish deposits in the wire mesh may be almost invisible but may be restricting oil flow. If brown varnish deposits are found on internal surfaces of pumps, motors, or valves, this is a positive indication that the system is running with too high an oil temperature. On industrial systems a heat exchanger should be added. On mobile systems take whatever remedies are available and these are described in previous chapters, especially Chapter 4.

Cavitation is not as common a problem with hydraulic motors but if present will cause accelerated wear. The most common cause of motor cavitation is overrun by the load, causing a partial vacuum at the motor inlet. Overrun may occur while the motor is decelerating to a stop, or, as in the case of a winch motor, when the load is overrunning the motor faster than oil is being supplied by the pump. This calls for a re-design of the circuit with counterbalance or by-pass valves. The Womack book "Industrial Fluid Power — Volume 3" deals at length with this subject.

<u>2</u>. Check for restricted or clogged pump inlet plumbing. If hoses are used, be sure they are not collapsed or internally swelled by hot oil. Hose is not usually a good plumbing medium for the pump inlet line. If used, only hose designed for vacuum service should be used. It has an internal wire braid to prevent wall collapse.

<u>3</u>. Be sure the air breather on the reservoir is not clogged with dirt or lint. On systems where the air volume (above the oil) in the tank is very small, the pump could cavitate during the extension stroke of a cylinder if the breather became clogged.

<u>4</u>. Oil viscosity may be too high for that particular pump running at that speed. Some pumps cannot pick up prime on heavy oil or may run continuously in a partially cavitated condition.

Cold weather start-up is especially damaging to a pump. Running a pump for several hours across its relief valve to warm up the oil can severely damage the pump if it is running in a cavitated condition during this time. On outdoor equipment use an oil with as high as possible V. I. (viscosity index, not viscosity). This minimizes the viscosity *change* from cold to hot oil operation and reduces cavitation on a cold start-up.

<u>5</u>. Check pump inlet strainer size. Be sure the original strainer has not been replaced with a smaller one. Increasing its size (number of square inches filtering surface) may help on some systems where the original size selection may have been marginal.

<u>6</u>. The use of higher quality oil, and a grade which is especially designed for hydraulic systems, may reduce varnish formation and cavitation.

<u>7</u>. Determine recommended maximum speed of pump. Check pulley or gear ratios. Be sure the original electric motor has not been replaced with one which runs at a higher speed.

<u>8</u>. Be sure the original pump has not been replaced with one which delivers a higher flow which might overload the inlet strainer. Increase strainer size if possible.

Air Leaking Into the System

The air in a newly assembled system will purge itself after a short period of operation. The system should be run for perhaps 15 to 30 minutes under very low pressure. Air will dissolve in the oil, a little at a time and be carried to the reservoir where it can escape. The air purging process can, of course, be accelerated by bleeding trapped air from high points in the system by cracking fittings, especially on cylinders.

Air which keeps entering the system from continuous air leaks will emulsify with the oil and cause it to have a milky appearance, but the oil will usually become clear in about an hour after shut-down. To find where air is entering, check these points:

<u>1</u>. Be sure the oil reservoir is filled to its normal level, and that the pump intake is well below the minimum oil level. The NFPA reservoir specifications call for the highest point on the strainer to be at least 2 to 3 inches below minimum oil level.

Check oil level when all cylinders are extended to be sure it is above the LOW mark on the level gauge, but be careful about over-filling because the reservoir may overflow when the cylinders retract.

<u>2</u>. Air may be entering around the pump shaft seal. Gear and vane pumps which are drawing oil from the reservoir by suction will have a slight vacuum behind the shaft seal. When this seal becomes badly worn, air may be pulled by suction through the worn seal. Piston pumps (and motors) usually have a small positive pressure, less than 15 PSI, inside their cases and behind the shaft seal. Air is unlikely to enter around the shaft of such a pump.

3. Check for air leaks in the pump inlet plumbing, especially at union joints. A defective O-ring or lack of an O-ring, in an SAE port flange, will result in a leak. Check for leaks in hoses used in the inlet line. An easy way to check for air leaks is to squirt a few drops of oil around the pump shaft seal or over a suspected leak. If the pump noise dimishes, you have found your leak.

Check also around the inlet port of the pump to see if port threads have been damaged by cross-threading. Screwing a tapered pipe fitting into a straight thread porthole will damage the threads, usually beyond repair.

4. Air may be entering through the rod seal of a cylinder. This is more likely to happen on a cylinder mounted with rod up, and not properly counterbalanced. On the downstroke, the gravity load may draw a partial vacuum in the rod end of the cylinder and suck in air. Cylinder rod seals are not usually designed to keep pressure out, so even a good rod seal can leak under these conditions.

Air may leak into a cylinder through defective barrel gaskets or O-rings when the cylinder is operated under the above conditions.

5. Be sure the main tank return pipe inside the reservoir discharges below oil level and not above it. On new designs it is good practice to enlarge the diameter of the tank return line for a few feet before it enters the reservoir. This reduces oil velocity and minimizes turbulence in the oil.

Water Leaking Into the System

Water leaking into the system will emulsify with the oil when the pump is running and after it has circulated through the system at least once. After the system is shut down the oil will usually clear in an hour or so. The presence of water can be checked after the oil has cleared by tapping a fluid sample from the reservoir drain port. Water usually enters a system in one of these ways:

1. A leak in a shell and tube water cooled heat exchanger may leak water into the oil if water pressure is higher than oil pressure. It will leak oil into the water if oil pressure is higher.

2. Water may condense on the interior surfaces of the reservoir which are above oil level. This water will drain down and collect in a layer on the bottom of the reservoir. Condensation is nearly unavoidable on systems operating in environments where surrounding air temperature changes from day to night. During periods when reservoir walls are cooler than surrounding air, condensation may take place if ambient humidity is fairly high. The best practical solution to this problem is to daily or weekly tap off a small quantity of fluid and inspect it for water. This should be done before start-up or during a period when the system has been shut down for an hour or two.

3. Be sure that any pipe or tubing which carries cold water to the interior of the reservoir enters and leaves below oil level. Cold pipe inside the reservoir may condense a large volume of water.

Oil Leakage Around the Pump

1. Leakage Around the Shaft. On piston pumps and on other pumps which take inlet oil from an overhead reservoir, there is usually a slight internal pressure behind the shaft seal. As the seal becomes well worn, external leakage may appear. This will usually be more pronounced while the shaft is rotating and may disappear while the pump is stopped.

Other pumps such as gear and vane types usually run with a slight vacuum behind the seal. A worn-out seal may allow air to leak into the oil while the pump is running and oil to leak out after the pump has been stopped.

Prematurely worn shaft seals may be caused by excessive oil temperature. At an oil temperature of 200° F or higher, a rubber shaft seal will have a very short life.

Abrasives in the oil may wear out shaft seals quickly, and may also produce circumferential scoring on the shaft. If abrasives are present they will settle out of an oil sample drawn from the reservoir, after the sample has been allowed to stand an hour or more. Check all crevices and cracks in the reservoir where dust could enter. The most common point of entry is through the air breather. In extreme cases where the system is operating in a mill, gin, or foundry, the reservoir may have to be sealed airtight and a slight air pressure of no more than 1 PSI maintained in the reservoir air space. This pressure can be taken from the shop air supply through a precision-type, low range pressure regulator.

2. Leakage Around a Pump Port. Sometimes leakage at these ports may be caused by damaged threads from cross-threading or from trying to screw a tapered pipe fitting into a straight thread port. Once the threads have been damaged it is very difficult ever to obtain an airtight connection.

Check tightness of fittings in the ports. If dryseal (NPTF) pipe threads are used there should seldom be a need for any kind of sealant. If a sealant is used we recommend Teflon Sealant paste. We do not recommend Teflon tape. Beware of screwing taper pipe threads too tightly into a pump or valve body casting. In the past this has been the cause of many cracked housings.

3. Small Cracks. If leakage is from a small crack in a casting, this has most likely been caused by over-tightening a taper pipe fitting, or from operating the pump in a system where either the relief valve has, at some time, been adjusted too high, or where high pressure spikes have been generated as a result of shocks. While it is possible that the casting may originally have been defective, this has rarely turned out to be the cause.

LEAKAGE TESTING OF CYLINDERS AND VALVES

Air Cylinders. Piston and rod seals eventually become worn to the point where cylinder performance and efficiency are effected. Leakage around the rod seal can easily be detected and the seal can be replaced. At that time the rod bearing which is usually brass, bronze, or plastic, should also be replaced, as also should the rod wiper. On commercially built cylinders, kits are available which contain a rod bearing, wiper, rod seal, and barrel seals. Sometimes rod bearing and seals are available in a cartridge and can be replaced as a single unit. Some cylinders are now being constructed so the rod cartridge can be replaced without removing the cylinder from its mounting and without removing the end cap. After the piston rod has been uncoupled from the load, the cartridge can be unbolted or unscrewed and pulled out without further dis-assembly. Continuous air leaks in whatever part of the system, should be repaired without undue delay. Even a small air leak can waste power at the rate of 1 HP or more even when the machine is not cycling

Leakage of piston seals is not as easy to detect but will usually show up as a continuous escape of air from one of the exhaust ports on the 4-way control valve while the cylinder is stopped or stalled against a positive stop with full pressure on one side of the piston. Leakage from a 4-way valve exhaust port can also indicate a badly worn spool in the 4-way valve, and procedure for determining whether the leak is in the cylinder piston or in the control valve will be described later.

Air cylinders and many air valves are designed with soft seals for leaktight operation, and there should never be an appreciable leakage from valve exhausts. But a small amount of spool leakage is normal for those valves using metal-to-metal spool to body fit.

In the majority of cases of valve exhaust seepage, cylinder piston seals are the cause, so they should be tested first.

Leak Testing of an Air Cylinder. In the next figure the leak paths through the cylinder and 4-way valve are shown. To determine whether the leak is in the cylinder or in the valve, proceed as follows:

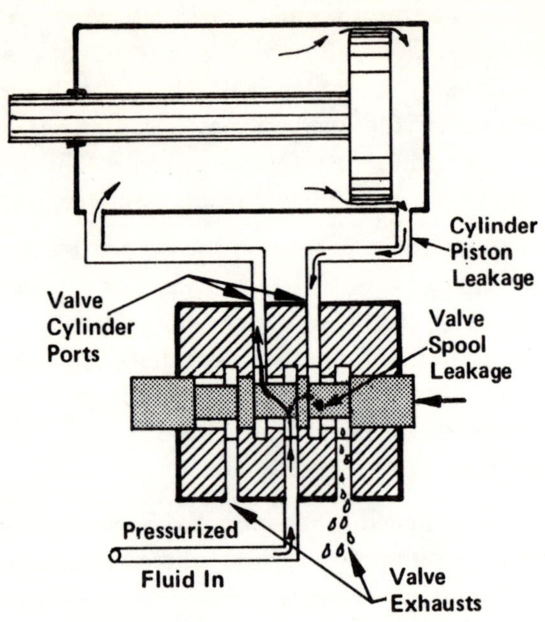

Leakage Paths Through a Cylinder and Valve.

With the air pressure shut off, remove one of the connecting lines between cylinder and valve. Plug the open port on the valve but leave the cylinder port open. Re-connect the air pressure and *carefully* shift the valve in the direction to move the cylinder to the end of its stroke. When the cylinder stalls, air which comes out of the open cylinder port is air which leaked past the piston seals. There should be no leakage if the seals are in good condition, although a *very slight* leakage may be acceptable on large-bore cylinders. If leakage is greater than a whisper, the piston seals should be replaced.

Re-connect the cylinder line and repeat the above test with the opposite line. A cylinder may leak in one direction and be tight in the other direction because a separate sealing lip is active in each direction. Occasionally a cylinder may be leaktight at both ends of its stroke but may leak at some intermediate position, possibly due to an irregularity such as a scratch or dent in the barrel. This will cause a loss of power when the piston reaches this position. If an intermediate piston leak is suspected, block the piston rod in this position and repeat the above test. Once in a while, usually on a large-bore cylinder, a piston seal may leak intermittently. This may be caused by a soft seal or O-ring seal moving slightly, rolling, or twisting as the piston strokes.

Hydraulic Cylinders. The leak test described for air cylinders can also be used for testing hydraulic cylinders. A small amount of by-pass flow (leakage) is normal for many cylinders, and a substantial by-pass flow may be observed on cylinders with metal piston ring seals.

Air Valve Testing. Refer to the figure above for flow path of air which leaks past the valve spool. To test an air valve, shut off the air supply and disconnect both lines running to the cylinder. Plug the open ports on the valve. Turn on the pressure and shift the valve spool back and forth between side positions. Air coming out either exhaust port is spool leakage.

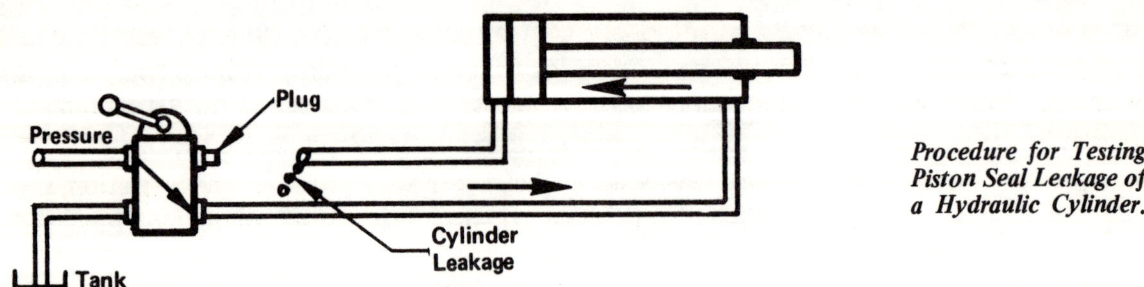

Procedure for Testing Piston Seal Leakage of a Hydraulic Cylinder.

Hydraulic Valves. If the tank return line can be disconnected so leakage flow can be observed, a hydraulic valve can be tested as described above. If the tank return line cannot be disconnected to observe leakage flow, the valve can be temporarily replaced with another valve known to be good.

SOLENOID VALVE PROBLEMS
Checklist for Recurrent Solenoid Burn-Out of A-C Valves

Coil burn-out is more common on A-C than on D-C valves because of the high inrush current. Until the armature can pull in completely and close the air gap in the magnetic loop, the inrush current may be up to 5 times as high as the steady state, or holding, current after the armature has seated. On a D-C valve, the inrush current is about the same as the holding current.

Coil Does Not Match Operating Voltage. Improper match between electrical source and the coil will cause a coil to burn out. Check these possible causes:

<u>1.</u> Voltage Too High. The operating voltage should not be more than 10% higher than the coil voltage rating. Excessive voltage causes excessive current draw which may overheat the coil.

<u>2.</u> Voltage Too Low. Voltage should be no less than 10% below coil rating. Low voltage reduces mechanical force of the solenoid. It may continue to draw inrush current without being able to pull in completely.

If low voltage is suspected as the cause of coil burn-out, a voltmeter should be used to measure voltage directly on the coil wires while the solenoid is energized *and with its armature blocked open* so it will continue to draw inrush current. Energize the coil just long enough to get a voltage reading. Also, take a voltage reading on the wires which feed the coil but with the coil disconnected. A difference of more than 5% between these two measurements indicates excessive resistance in the machine electrical circuit, or, insufficient volt-ampere capacity in the control transformer which is supplying current to the solenoids.

<u>3.</u> Frequency. Operation of a 60 Hz coil on a 50 Hz line causes the coil to draw more than normal current. Operation of a 50 Hz coil on 60 Hz causes the coil to draw less than normal current and it may burn out from inability to pull the armature closed.

Overlap in Energization. On some double solenoid valves, if both solenoids are energized at the same time and held in this state for a short time, the last coil to be energized will burn out from drawing excessive current because it cannot pull its armature in. This burn-out condition only exists on those double solenoid valves where the two solenoids are yoked to opposite ends of a common spool. On valves where each solenoid is free to immediately close its air gap without interference from the opposite solenoid, neither will burn out if both are energized at the same time.

Careful attention should be given to electrical circuit design to make certain it is not possible for the machine operator, through accident or on purpose, to energize both solenoids of any double solenoid valve at the same time. Electrical interlocking contacts on relays or pushbuttons may be employed to keep one solenoid from being energized while the opposite solenoid is energized.

Even with correct design and interlocking circuits, a relay with sticking contacts or which releases too slowly, could be responsible for a momentary overlap of energization on each cycle. Heat will accumulate in one coil over a period of time and it may eventually burn out.

Too Rapid Cycling. Since inrush current may be up to 5 times the holding current, a standard A-C coil on an air-gap solenoid may overheat and burn out if required to cycle too frequently. The extra heat generated during the inrush periods cannot dissipate fast enough. The gradual build-up of heat inside the coil winding may, in time, damage the coil insulation.

High cycling applications can be roughly defined as those requiring the solenoid to be energized more than 5 to 10 times every minute. On those applications, the electrical circuit can be designed to operate with D-C solenoid valves, or A-C oil immersed solenoid valves can be used; the coil will not overheat as easily because heat is conducted more efficiently to the surface of the solenoid cover.

__Dirt in Oil or Atmosphere.__ A small solid particle lodging under the solenoid armature may keep it from fully seating against the iron core, causing coil current to remain higher than normal during the holding period. Be sure, on older solenoid valves, that solenoid dust covers remain tightly in place to keep out atmospheric dust.

Small dirt particles in the oil may lodge on the surface of the spool, glued there by "varnish" circulating in the oil, or the varnish itself may cause the spool to drag or stick in the bore. This will cause solenoid current to be high and will generate greater than normal heat.

Warning! Varnish deposits inside a system mean the oil is running at too high a temperature.

__Environmental Conditions.__ Abnormally high or low ambient temperatures to which a solenoid is exposed for an extended time may cause a coil to burn out.

__1.__ High Temperature. Coil insulation may be damaged and one layer of wire may short to the next layer. A heat shield between the valve and heat source may give some protection against radiated heat. Oil immersed solenoids give the best protection against heat conducted either through metal surfaces or from surrounding high temperature air.

__2.__ Low Temperature. Cold ambient temperatures cause oil to become more viscous, possible overloading solenoid valve capacity. Mechanical parts of the valve or solenoid structure may distort, causing the valve spool to stick and burn out the coil. Change to an oil more suitable for the low temperature or use oil immersed solenoids.

__Atmospheric Moisture.__ High humidity, along with frequently changing ambient temperature, may form corrosion on metal parts of the solenoid structure, causing the armature to drag or the spool to stick. Humidity also tends to deteriorate standard solenoid coils, causing shorts in windings.

Change to molded coils or oil immersed solenoids. Keep solenoid protective covers tightly in place, and perhaps seal the electrical conduit openings after the wiring is installed.

__Dead End Service.__ Fluid circulating through a solenoid valve carries away a great deal of the heat generated by electric current. Some valves depend on fluid flow to keep excessive heat from accumulating, and if used on dead end service, where the solenoid may remain energized for long periods when there is no fluid flow, the coil can burn out from this effect, possibly in combination with other problems.

Pilot-Operated Solenoid Hydraulic Valves

This type of solenoid valve is described in "Volume 2 — Industrial Fluid Power" textbook. A problem sometimes encountered with these valves, especially when putting a newly designed system into operation for the first time, is their failure to shift out of center position when the START button is pressed. Use the manual override buttons (located on the end of solenoid covers) to determine whether the problem is lack of electric current to the solenoids or a lack of sufficient hydraulic pilot pressure at the valve inlet for shifting the main spool.

With the pump running, press the override button on the solenoid which is supposed to be energized. Pressing the override button manually shifts the solenoid pilot spool. If the valve main spool shifts and the cylinder starts forward, chances are that the problem is in the electrical circuit, and an electrical troubleshooting procedure should be started. If the valve main spool does not shift and the cylinder does not start forward when the override button is pressed, the problem may be that there is not sufficient hydraulic pressure at the inlet port of the valve to shift the main spool. Usually a pilot pressure of at least 75 PSI is required. In this case, refer to the textbook "Industrial Fluid Power — Volume 2" for changes in the fluid circuit to provide sufficient pilot pressure.

TROUBLESHOOTING ON CLOSED LOOP HYDRAULIC TRANSMISSIONS

Most closed loop systems use a variable displacement piston pump, driven by an electric motor or engine, and connected to a hydraulic motor through two loop lines, both of which are under pressure at all times. For CW rotation of the hydraulic motor, one loop line is at high pressure, usually 4000 to 5000 PSI, and the other loop line, the return line, is usually at 150 to 500 PSI depending on the size and brand of the transmission. For CCW rotation, the pressure in the two lines is reversed and the first loop line becomes the return line. Troubleshooting on a closed loop system is approached a little differently than on an open loop system.

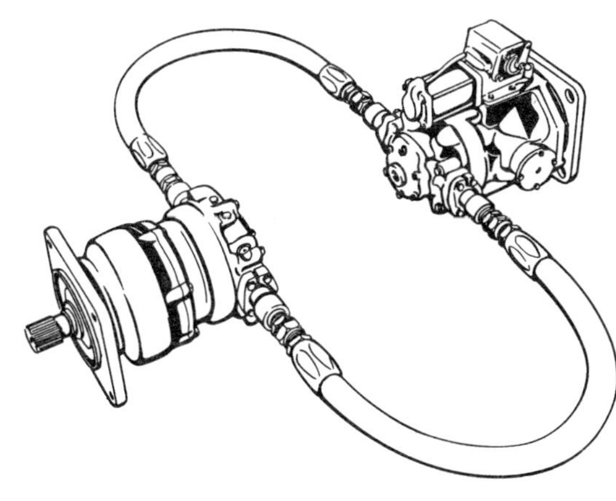

The serviceman who is not familiar with these systems should refer to the textbook "Industrial Fluid Power — Volume 3" for a detailed description.

Getting Ready for Servicing. Before going to the job a serviceman should obtain a copy of the factory service manual for the transmission to be serviced. Find location and size of gauge ports. Determine which components are located in pump and motor covers. Although there are variations, the charge pump on most transmissions is located in the pump rear cover. Some systems may use an external charge pump driven separately.

Equip yourself with pressure gauges and port adaptors for connecting the gauges. You will need the usual mechanics tools and may need extra fittings and short lengths of high pressure and low pressure hose. The main loop lines usually have SAE split flange port pads and other connections are UN straight thread with O-ring seal. Be very careful not to screw taper pipe threads into straight thread portholes. Most transmissions have no taper pipe threads on them.

Pressure gauges should include one with a range up to 6000 or 10,000 PSI for loop pressure and high pressure relief valve gauging. A 500 PSI gauge will be needed to measure charge pump pressure.

One important observation will be the volume of oil coming out the case drain port during various tests. The flow can be observed by disconnecting the case drain line and connecting a short length of low pressure hose to the drain port and discharging the flow into the reservoir in such a way that its volume can be observed. When observing the drain flow from one unit, leave the drain connected to the other unit. Do not ever plug the case drain. Measure case drain flow separately for pump and motor. Do not measure combined flow from both units; this will not reveal useful information. The actual rate of flow is not as important as how much it changes as the pump displacement lever is moved or as load is applied to the system.

Mechanical Inspection. Before starting tests in the hydraulic system, make a visual inspection of the installation for possible mechanical damage such as broken shafts or couplings, slipping belts, etc. Also very important, *see if malfunction occurs equally in both forward and reverse.* If it does, certain troubles can immediately be ruled out such as leaking or sticking high pressure relief valves, leaking charge pump check valves, etc. In systems using two or more identical pumps or motors, if two similar units can be switched, this may help to pinpoint whether the trouble is in a pump or a motor, or it may not be in the hydraulic system at all.

Loss of Power. The malfunction encountered most often is a partial or total loss of power and/or speed in the hydraulic motor. The motor may run when unloaded but will not produce full torque or speed. After making a general inspection, proceed with tests in this order:

1. Measure Charge Pump Pressure. Install a low range gauge in one of the specified ports where this pressure can be measured. The location and size of these gauging ports will have to be determined from a service drawing of the pump or motor. The correct value for charge pump pressure will also have to be determined from the manual. Charge pump pressure puts the pump on stroke, and if it is low, the pump will not go on full stroke and the hydraulic motor cannot develop full speed.

On a system which is working correctly, the charge pump pressure will be close to the value specified in the manual and will hold fairly steady while the control lever on the pump remains in neutral. And it should remain fairly steady as the control lever is moved into forward or reverse, although there will be a momentary fluctuation when the lever is first moved. This momentary drop occurs while the charge pump oil is filling the power cylinder for shifting the cam plate. When full stroke is reached, however, the charge pump pressure should hold steady although it may be very slightly less than with control lever in neutral.

The actual pressure of the charge pump oil is determined by the setting of the low pressure relief valve. This valve may be contained in either the pump or motor cover, or there may be one in each. Location will have to be determined from the service manual. On small transmissions, less than 75 HP, the charge pump pressure will usually be 150 to 200 PSI. On very large transmissions it may be up to 500 PSI.

Note: Do not change the low pressure relief valve setting on transmissions which were previously running normally. If the charge pump pressure is not correct, this is symptomatic of other troubles.

2. Interpret Charge Pump Action. If there is little or no charge pump pressure at any position of the control lever, the charge pump may be at fault. Suspect a broken drive shaft or coupling to the charge pump. To confirm the diagnosis, remove the charge pump cover and inspect for broken parts. Complete loss of charge pump pressure could also be the result of spring breakage, damage, or dirt in the low pressure relief valve.

Fluctuating Charge Pump Pressure. If the charge pump pressure is erratic with control lever both in neutral and in both side positions, with pressure level a little lower than it should be, this may indicate cavitation of the charge pump either from low oil level in the reservoir, from a collapsed suction hose, or from a dirty inlet filter. Check each of these conditions and make repairs if needed.

Charge Pump Pressure Drops When Control Lever is Shifted. If the charge pump pressure appears to be about normal when control lever is in neutral, with about the right pressure, then falls rapidly as the control lever is shifted to *either* side position, this may indicate the charge pump is worn out, but usually indicates that the pumped oil is escaping through clearances which open up as the control lever tries to put the pump on stroke. The most likely fault is a scored valve plate. When this plate gets scored, oil pressure gets under it and tends to lift it from tight contact with the cam plate. This allows some of the high pressure oil together with some or all of the charge pump oil to escape into the pump case. This diagnosis can be confirmed by observing the volume of case drain flow. A scored valve plate is indicated if the drain flow increases heavily at the same time the charge pump pressure falls. If the valve plate is indeed scored, the charge pump pressure may fall very low, too low to stroke the pump to more than a small cam angle. The result is that the hydraulic motor cannot develop full speed when the torque load is high. The remedy for this problem is a complete factory overhaul of the pump.

3. Measure High Pressure. If the charge pump pressure seems to respond normally as the control lever is shifted, perhaps dropping slightly but not a great amount, install a high range gauge in one of the gauge ports for the loop lines. Install the gauge in the side of the loop which is malfunctioning. If the system is malfunctioning in both forward and reverse, use either port, and move the control lever in a direction which should produce high pressure on that port. The motor shaft must be blocked to obtain a maximum pressure reading. On vehicle drive systems block the drive wheels, set the brake, or stall the vehicle against a solid wall.

4. Interpret High Pressure Gauge Readings. On a properly operating system the pressure in the loop, with the control lever in neutral, will vary with the application. For example, if operating against a non-reactive load, with the brakes set, the gauge will probably read very little above charge pump pressure. If the load is reacting against the hydraulic motor, the loop pressure could be any value between charge pressure and near full operating pressure. However, as the control lever is moved to forward or reverse, the pressure should immediately pick up to the value required by the load. If the vehicle brakes are set, the pressure would immediately pick up to the compensator setting.

No High Pressure Can Be Developed. If there is little or no rise in loop pressure when the control lever is moved out of neutral, one of the high pressure relief valves (usually located in the motor rear cover) may be stuck or damaged. Since there are usually two of these valves, one for each direction of operation, try moving the lever to both forward and reverse. If pressure can be developed in one direction and not in the other, one of the relief valves may be damaged. If pressure cannot be developed in either direction, the relief valves are probably not at fault. If relief valves are suspected, the cartridges can be removed and cleaned or replaced with good cartridges.

Do not change the setting of the high pressure relief valves unless you have the instruments and the factory instruction manual showing how to re-set them. They must be set to about 500 PSI higher than the pressure compensator setting in the manner outlined in the manual.

Full High Pressure Cannot Be Developed. If only partial pressure is obtained on the high pressure gauge when the control lever is moved out of center, one of the following faults may be the cause. This is assuming charge pump pressure has been measured and found to be normal when the control lever is shifted to a side position.

(a). First consider the possibility that the fault may lie elsewhere than in the hydraulic system. Stripped gears, pins, or keys in a gear box attached to the pump could be the cause. Investigate other mechanical transmission items in the power train before proceeding with the next tests:

(b). One or both check valves which feed charge pump oil into the loop may be damaged, stuck, or leaking. The physical location of these valves must be determined from the service manual. Usually they can be removed for checking without dis-assembling the main pump. Oil from the main pump may be back-flowing into the charge pump circuit and escaping across the low pressure relief valve. This would cavitate the loop. Symptoms of this fault would be fairly normal operation of the hydraulic motor when unloaded, but inability to build up high torque for heavy loads. A system with this fault might also show signs of overheating. If the motor will not build up torque in either direction, the check valves are not a likely fault unless by coincidence both check valves started leaking at the same time, and this is hardly possible. If motor fails to build up torque in only one direction, investigate these check valves.

(c). Leakage across the loop lines, from high pressure side to low pressure side, could occur across a faulty shuttle valve. Location of this valve must be determined from the service manual.

(d). Excessively worn piston shoes may cause excessive leakage of high pressure oil into the case to seriously reduce motor torque. Symptoms would be similar to leaky check valves or shuttle valve. With this fault, performance would be effected equally in both directions. A major factory overhaul is necessary if piston shoes are worn out.

(e). The pressure compensator internal parts may be jammed, dirty, or damaged. If parts of the compensator are removed, it must be re-set to its original setting by service manual procedure.

5. Interpreting Case Drain Flow. Observation of case drain flow is of great value for determining the condition of the cam plate inside the pump or motor. Measure drain flow of pump and motor separately to determine which unit is defective. Total flow from drains of both pump and motor (plus low pressure relief valve discharge if separate) should equal charge pump flow rating unless charge pump is cavitating due to a restricted inlet.

Pump Drain Flow. In a system operating normally the case drain flow should not exceed the flow given in the specifications. If it greatly exceeds the manual specifications, the main pump may be worn so badly that a part of the high pressure oil is slipping into the case. The pump rotating group may be due for replacement.

In a malfunctioning system an excessive amount of drain flow, particularly if the flow increases suddenly when the control lever is moved out of neutral, usually indicates very serious damage, either to the swash plate, the valve plate, or the piston shoes. If charge pump operation is normal, and the drain flow increases in a short duration spurt as the control lever is moved out of neutral, this may indicate loop oil leaking into the pump case, usually through bad charge pump check valves.

Motor Drain Flow. A sharply increased drain flow from the motor case as the pump control lever is moved out of neutral may indicate a damaged valve plate or badly worn piston shoes in the motor.

Appendix
Reference Information

GRAPHIC SYMBOLS FOR FLUID POWER DIAGRAMS

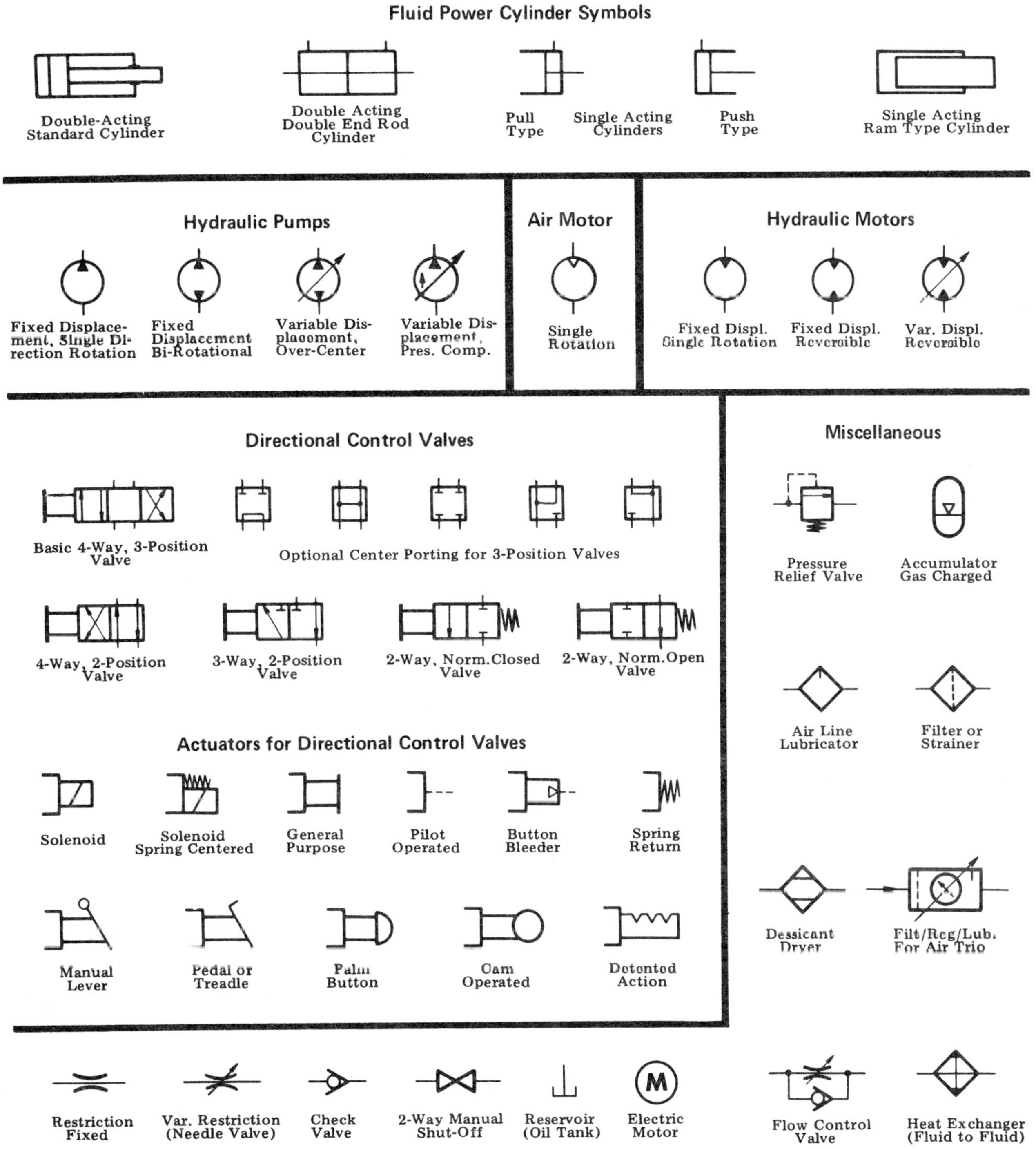

Fluid Power Cylinder Symbols

Double-Acting
Standard Cylinder

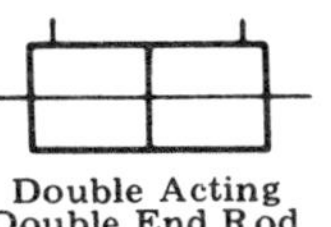

Double Acting
Double End Rod
Cylinder

Pull
Type

Single Acting
Cylinders

Push
Type

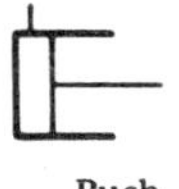

Single Acting
Ram Type Cylinder

Hydraulic Pumps

Fixed Displace-
ment, Single Di-
rection Rotation

Fixed
Displacement
Bi-Rotational

Variable Dis-
placement,
Over-Center

Variable Dis-
placement,
Pres. Comp.

Air Motor

Single
Rotation

Hydraulic Motors

Fixed Displ.
Single Rotation

Fixed Displ.
Reversible

Var. Displ.
Reversible

Directional Control Valves

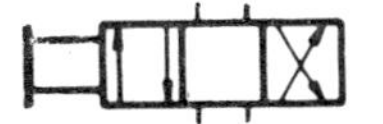

Basic 4-Way, 3-Position
Valve

  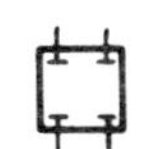 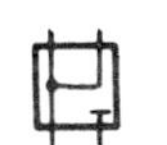 

Optional Center Porting for 3-Position Valves

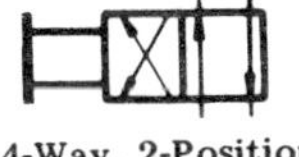

4-Way, 2-Position
Valve

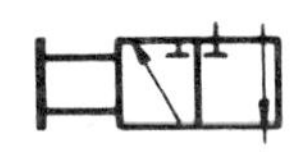

3-Way, 2-Position
Valve

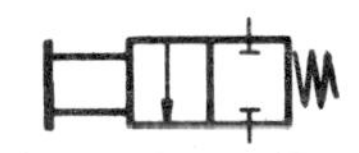

2-Way, Norm. Closed
Valve

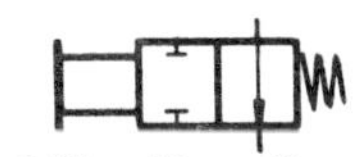

2-Way, Norm. Open
Valve

Actuators for Directional Control Valves

Solenoid

Solenoid
Spring Centered

General
Purpose

Pilot
Operated

Button
Bleeder

Spring
Return

Manual
Lever

Pedal or
Treadle

Palm
Button

Cam
Operated

Detented
Action

Miscellaneous

Pressure
Relief Valve

Accumulator
Gas Charged

Air Line
Lubricator

Filter or
Strainer

Dessicant
Dryer

Filt/Reg/Lub.
For Air Trio

Flow Control
Valve

Heat Exchanger
(Fluid to Fluid)

Restriction
Fixed

Var. Restriction
(Needle Valve)

Check
Valve

2-Way Manual
Shut-Off

Reservoir
(Oil Tank)

Electric
Motor

STEEL TUBING — PRESSURE AND FLOW RATINGS FOR HYDRAULIC PLUMBING

Pressure ratings are for annealed steel tubing "hydraulic grade" having a tensile strength of 55,000 PSI (the kind most often used for plumbing hydraulic systems). See notes at foot of next page for adjustment to other steels.

Pressure ratings are shown for S.F. (safety factors) from 2 to 6. A factor of 4 is recommended for general service. On shockless systems a smaller factor is sometimes used.

GPM ratings are shown for flow velocities from 10 through 30 f/s (feet per second). A general guideline is to use 10 f/s on systems operating at a maximum pressure of 1000 PSI; 15 f/s on systems having maximum pressures from 1000 to 2000 PSI; 20 f/s if system pressure is in the range of 2000 to 3500 PSI; and a velocity of 30 f/s on systems designed for pressures higher than 3500 PSI.

Wall, Inches	.025	.032	.035	.042	.049	.058	.065	.072	.083	.095	.109
5/16" O.D. TUBING — 55,000 PSI Tensile											
PSI @ S.F. = 6	1470	1880	2050	2460	2880	3400	3800	- - - -	- - - -	- - - -	- - - -
PSI @ S.F. = 5	1760	2250	2460	2960	3450	4080	4580	- - - -	- - - -	- - - -	- - - -
PSI @ S.F. = 4	2200	2820	3080	3700	4310	5100	5720	- - - -	- - - -	- - - -	- - - -
PSI @ S.F. = 3	2930	3750	4100	4930	5750	6800	7630	- - - -	- - - -	- - - -	- - - -
PSI @ S.F. = 2	4400	5630	6160	7390	8630	10,200	11,440	- - - -	- - - -	- - - -	- - - -
GPM @ 10 f/s	1.69	1.51	1.44	1.28	1.13	0.95	0.82	- - - -	- - - -	- - - -	- - - -
GPM @ 15 f/s	2.53	2.27	2.16	1.92	1.69	1.42	1.23	- - - -	- - - -	- - - -	- - - -
GPM @ 20 f/s	3.37	3.02	2.88	2.56	2.25	2.25	1.89	- - - -	- - - -	- - - -	- - - -
GPM @ 30 f/s	5.06	4.54	4.32	3.83	3.38	2.83	2.45	- - - -	- - - -	- - - -	- - - -
3/8" O.D. TUBING — 55,000 PSI Tensile											
PSI @ S.F. = 6	1220	1560	1710	2050	2400	2840	3180	3520	- - - -	- - - -	- - - -
PSI @ S.F. = 5	1470	1880	2050	2460	2880	3400	3800	4220	- - - -	- - - -	- - - -
PSI @ S.F. = 4	1830	2350	2570	3080	3590	4250	4770	5280	- - - -	- - - -	- - - -
PSI @ S.F. = 3	2440	3130	3420	4100	4790	5670	6350	7040	- - - -	- - - -	- - - -
PSI @ S.F. = 2	3670	4690	5130	6160	7190	8500	9530	10,550	- - - -	- - - -	- - - -
GPM @ 10 f/s	2.59	2.37	2.28	2.07	1.88	1.64	1.47	1.31	- - - -	- - - -	- - - -
GPM @ 15 f/s	3.88	3.55	3.42	3.11	2.82	2.46	2.20	1.96	- - - -	- - - -	- - - -
GPM @ 20 f/s	5.17	4.74	4.56	4.15	3.76	3.29	2.94	2.61	- - - -	- - - -	- - - -
GPM @ 30 f/s	7.76	7.11	6.84	6.22	5.64	4.93	4.40	3.92	- - - -	- - - -	- - - -
1/2" O.D. TUBING — 55,000 PSI Tensile											
PSI @ S.F. = 6	920	1170	1280	1540	1800	2130	2380	2640	3040	- - - -	- - - -
PSI @ S.F. = 5	1100	1400	1540	1850	2150	2550	2860	3170	3650	- - - -	- - - -
PSI @ S.F. = 4	1380	1760	1920	2310	2700	3190	3580	3960	4560	- - - -	- - - -
PSI @ S.F. = 3	1830	2350	2570	3080	3590	4250	4770	5280	6090	- - - -	- - - -
PSI @ S.F. = 2	2750	3520	3850	4620	5390	6380	7150	7920	9130	- - - -	- - - -
GPM @ 10 f/s	4.96	4.65	4.53	4.24	3.96	3.61	3.35	3.10	2.73	- - - -	- - - -
GPM @ 15 f/s	7.43	6.98	6.79	6.35	5.93	5.41	5.03	4.65	4.10	- - - -	- - - -
GPM @ 20 f/s	9.91	9.31	9.05	8.47	7.91	7.22	6.70	6.20	5.46	- - - -	- - - -
GPM @ 30 f/s	14.9	14.0	13.6	12.7	11.9	10.8	10.1	9.30	8.19	- - - -	- - - -
5/8" O.D. TUBING — 55,000 PSI Tensile											
PSI @ S.F. = 6	730	940	1030	1230	1440	1700	1910	2110	2435	2790	3200
PSI @ S.F. = 5	880	1130	1230	1480	1720	2040	2290	2530	2920	3340	3840
PSI @ S.F. = 4	1100	1400	1540	1850	2160	2550	2860	3170	3650	4180	4800
PSI @ S.F. = 3	1470	1880	2050	2460	2870	3400	3810	4220	4870	5570	6390
PSI @ S.F. = 2	2200	2800	3080	3700	4320	5100	5720	6340	7300	4360	9600
GPM @ 10 f/s	8.10	7.71	7.54	7.17	6.80	6.34	6.00	5.66	5.16	4.63	4.06
GPM @ 15 f/s	12.1	11.6	11.3	10.7	10.2	9.52	9.00	8.50	7.38	6.95	6.08
GPM @ 20 f/s	16.2	15.4	15.1	14.3	13.6	12.7	12.0	11.3	10.3	9.26	8.11
GPM @ 30 f/s	24.3	23.1	22.6	21.5	20.4	19.0	18.0	17.0	15.5	13.9	12.2
3/4" O.D. TUBING — 55,000 PSI Tensile											
PSI @ S.F. = 6	610	780	860	1030	1200	1420	1590	1760	2030	2320	2660
PSI @ S.F. = 5	730	940	1030	1230	1440	1700	1900	2110	2430	2790	3200
PSI @ S.F. = 4	920	1170	1280	1540	1800	2130	2380	2640	3040	3480	4000
PSI @ S.F. = 3	1220	1560	1710	2050	2400	2840	3180	3520	4060	4640	5330
PSI @ S.F. = 2	1840	2340	2560	3080	3600	4260	4760	5280	6080	6960	8000
GPM @ 10 f/s	12.0	11.5	11.3	10.9	10.4	9.84	9.41	8.99	8.35	7.68	6.93
GPM @ 15 f/s	18.0	17.3	17.0	16.3	15.6	14.8	14.1	13.5	12.5	11.5	10.4
GPM @ 20 f/s	24.0	23.0	22.6	21.7	20.8	19.7	18.8	18.0	16.7	15.4	13.9
GPM @ 30 f/s	36.0	34.6	34.0	32.6	31.2	29.5	28.2	27.0	25.1	23.0	20.8

(This Table is Continued on the Next Page)

(This Table is Continued From the Preceding Page)

Wall, Inches	.025	.032	.035	.042	.049	.058	.065	.072	.083	.095	.109
7/8″ O.D. TUBING — 55,000 PSI Tensile											
Inside Area	.5346	.5166	.5090	.4914	.4742	.4525	.4359	.4197	.3948	.3685	.3390
PSI @ S.F. = 6	520	670	730	880	1030	1210	1360	1500	1740	1990	2280
PSI @ S.F. = 5	630	800	880	1060	1230	1460	1630	1810	2090	2390	2740
PSI @ S.F. = 4	790	1000	1100	1320	1540	1820	2040	2260	2600	2990	3420
PSI @ S.F. = 3	1050	1340	1470	1760	2050	2430	2720	3020	3480	3980	4570
PSI @ S.F. = 2	1580	2000	2200	2640	3080	3640	4080	4520	5200	5980	6840
GPM @ 10 f/s	16.7	16.1	15.9	15.3	14.8	14.1	13.6	13.1	12.3	11.5	10.6
GPM @ 15 f/s	25.0	24.2	23.8	23.0	22.2	21.2	20.4	19.6	18.5	17.2	15.9
GPM @ 20 f/s	33.3	32.2	31.7	30.6	29.6	28.2	27.2	26.2	24.6	23.0	21.1
GPM @ 30 f/s	50.0	48.3	47.6	46.0	44.3	42.3	40.8	39.2	36.9	34.5	31.7
1″ O.D. TUBING — 55,000 PSI Tensile											
Inside Area	.7088	.6881	.6793	.6590	.6390	.6138	.5945	.5755	.5463	.5153	.4803
PSI @ S.F. = 6	460	590	640	770	900	1060	1190	1320	1520	1740	2000
PSI @ S.F. = 5	550	700	770	920	1080	1280	1430	1580	1830	2090	2400
PSI @ S.F. = 4	690	880	960	1150	1350	1590	1790	1980	2280	2610	3000
PSI @ S.F. = 3	920	1170	1280	1540	1800	2130	2380	2640	3040	3480	4000
PSI @ S.F. = 2	1380	1760	1920	2300	2700	3180	3580	3960	4560	5220	6000
GPM @ 10 f/s	22.1	21.4	21.2	20.5	19.9	19.1	18.5	17.9	17.0	16.1	15.0
GPM @ 15 f/s	33.1	32.2	31.8	30.8	29.9	28.7	27.8	26.9	25.5	24.1	22.5
GPM @ 20 f/s	44.2	42.9	42.4	41.1	39.8	38.3	37.1	35.9	34.1	32.1	29.9
GPM @ 30 f/s	66.3	64.3	63.5	61.6	59.8	57.4	55.6	53.8	51.1	48.2	44.9
1¼″ O.D. TUBING — 55,000 PSI Tensile											
Inside Area	1.131	1.105	1.094	1.068	1.042	1.010	.9852	.9607	.9229	.8825	.8365
PSI @ S.F. = 6	365	470	515	615	720	850	950	1060	1220	1390	1600
PSI @ S.F. = 5	440	560	610	740	860	1020	1140	1270	1460	1670	1920
PSI @ S.F. = 4	550	700	770	920	1080	1280	1430	1580	1820	2090	2400
PSI @ S.F. = 3	730	940	1030	1230	1440	1700	1900	2110	2430	2790	3200
PSI @ S.F. = 2	1100	1400	1540	1840	2160	2560	2860	3160	3640	4180	4800
GPM @ 10 f/s	35.3	34.4	34.1	33.3	32.5	31.5	30.7	29.9	28.8	27.5	26.1
GPM @ 15 f/s	52.9	51.7	51.2	49.9	48.7	47.2	46.1	44.9	43.2	41.3	39.1
GPM @ 20 f/s	70.5	68.9	68.2	66.6	65.0	63.0	61.4	59.9	57.5	55.0	52.2
GPM @ 30 f/s	106	103	102	100	97.4	94.5	92.1	89.8	86.3	85.5	78.2
1½″ O.D. TUBING — 55,000 PSI Tensile											
Inside Area	1.651	1.612	1.606	1.575	1.544	1.504	1.474	1.444	1.398	1.348	1.291
PSI @ S.F. = 6	305	390	430	515	600	710	795	880	1014	1161	1332
PSI @ S.F. = 5	370	470	510	610	720	850	950	1050	1210	1390	1600
PSI @ S.F. = 4	460	590	640	770	900	1060	1190	1320	1520	1740	2000
PSI @ S.F. = 3	610	780	850	1030	1200	1420	1590	1760	2030	2320	2660
PSI @ S.F. = 2	920	1180	1280	1540	1800	2120	2380	2640	3040	3480	4000
GPM @ 10 f/s	51.5	50.2	50.1	49.1	48.1	46.9	45.9	45.0	43.6	42.0	40.2
GPM @ 15 f/s	77.2	75.4	75.1	73.6	72.2	70.3	68.9	67.5	65.4	63.0	60.4
GPM @ 20 f/s	103	101	100	98.2	96.3	93.8	91.9	90.0	87.2	84.0	80.5
GPM @ 30 f/s	154	151	150	147	144	141	138	135	131	126	121
2″ O.D. TUBING — 55,000 PSI Tensile											
Inside Area	2.986	2.944	2.926	2.883	2.841	2.788	2.746	2.705	2.642	2.573	2.490
PSI @ S.F. = 6	230	295	320	385	450	530	595	660	760	870	1000
PSI @ S.F. = 5	270	350	390	460	540	640	710	790	910	1040	1200
PSI @ S.F. = 4	340	440	480	580	670	800	890	990	1140	1300	1500
PSI @ S.F. = 3	460	590	640	770	900	1060	1190	1320	1520	1740	2000
PSI @ S.F. = 2	680	880	960	1160	1340	1600	1780	1980	2280	2600	3000
GPM @ 10 f/s	93.1	91.8	91.2	89.4	88.6	86.9	85.6	84.3	82.4	80.2	77.6
GPM @ 15 f/s	139	138	137	135	133	130	128	126	123	120	116
GPM @ 20 f/s	186	184	182	180	177	174	171	169	165	160	233
GPM @ 30 f/s	279	275	274	270	266	261	257	253	247	241	233

For steels with tensile strength other than 55,000 PSI, pressure rating is proportional to tensile strength.

Barlow's Formula *(P = 2t x S ÷ O)* was used for pressure calculations, in which P = burst pressure, PSI; t = wall thickness, inches; S = tensile strength of material, in PSI; and O = outside diameter of tube, in inches. Working pressure = burst pressure ÷ safety factor.

The formula: *GPM = V x A ÷ 0.3208* was used for calculating flow velocity, in which V = velocity in feet per second, and A = inside area, in square inches.

PRESSURE AND FLOW RATINGS OF IRON PIPE

Pressure ratings are for wrought steel, butt welded pipe, the kind used most often in hydraulic plumbing. It is a low carbon steel with 40,000 PSI tensile strength. High carbon steel pipe, with tensile ratings up to 60,000 PSI is also available. Its pressure rating is higher in proportion to the increase in its tensile rating.

Schedule 40 is "standard weight" pipe. Schedules 80 and 160 have the same O. D. for a given size, but have heavier walls and a smaller internal flow area. Double extra strength pipe is also available in sizes of 1/2" and larger where higher pressure ratings are necessary.

Taper pipe threads, if used, should be NPTF (dryseal type) to minimize leakage around the crest of the threads.

Abbreviations used in the table are: S.F. is safety factor on the pressure ratings; f/s is flow velocity in feet per second.

1/4" NPT PIPE — 40,000 PSI Tensile

Pipe Schedule →	40	80	160
PSI @ S.F. = 6	2170	2940	- - - -
PSI @ S.F. = 5	2610	3530	- - - -
PSI @ S.F. = 4	3260	4410	- - - -
PSI @ S.F. = 3	4340	5880	- - - -
PSI @ S.F. = 2	6520	8820	- - - -
GPM @ 10 f/s	3.00	2.40	- - - -
GPM @ 15 f/s	4.50	3.60	- - - -
GPM @ 20 f/s	6.00	4.80	- - - -
GPM @ 30 f/s	9.00	7.20	- - - -

3/8" NPT PIPE — 40,000 PSI Tensile

Pipe Schedule →	40	80	160
PSI @ S.F. = 6	1800	2490	- - - -
PSI @ S.F. = 5	2160	2990	- - - -
PSI @ S.F. = 4	2700	3730	- - - -
PSI @ S.F. = 3	3590	4980	- - - -
PSI @ S.F. = 2	5390	7470	- - - -
GPM @ 10 f/s	6.00	4.20	- - - -
GPM @ 15 f/s	9.00	5.30	- - - -
GPM @ 20 f/s	12.0	8.40	- - - -
GPM @ 30 f/s	18.0	11.0	- - - -

1/2" NPT PIPE — 40,000 PSI Tensile

Pipe Schedule →	40	80	160
PSI @ S.F. = 6	1730	2330	2980
PSI @ S.F. = 5	2080	2800	3580
PSI @ S.F. = 4	2600	3500	4480
PSI @ S.F. = 3	3460	4670	5970
PSI @ S.F. = 2	5190	7000	8950
GPM @ 10 f/s	9.50	7.20	5.33
GPM @ 15 f/s	14.0	11.0	8.00
GPM @ 20 f/s	19.0	14.0	10.6
GPM @ 30 f/s	29.0	22.0	16.0

3/4" NPT PIPE — 40,000 PSI Tensile

Pipe Schedule →	40	80	160
PSI @ S.F. = 6	1430	1950	2780
PSI @ S.F. = 5	1720	2350	3340
PSI @ S.F. = 4	2150	2930	4170
PSI @ S.F. = 3	2870	3910	5560
PSI @ S.F. = 2	4300	5870	8340
GPM @ 10 f/s	16.0	15.0	9.33
GPM @ 15 f/s	25.0	22.0	14.0
GPM @ 20 f/s	33.0	30.0	18.6
GPM @ 30 f/s	50.0	44.0	28.0

1" NPT PIPE — 40,000 PSI Tensile

Pipe Schedule →	40	80	160
PSI @ S.F. = 6	1350	1810	2530
PSI @ S.F. = 5	1620	2180	3040
PSI @ S.F. = 4	2020	2720	3800
PSI @ S.F. = 3	2700	3630	5070
PSI @ S.F. = 2	4040	5440	7600
GPM @ 10 f/s	27.0	22.2	16.0
GPM @ 15 f/s	41.0	33.3	24.0
GPM @ 20 f/s	55.0	44.0	32.0
GPM @ 30 f/s	83.0	66.0	48.0

1¼" NPT PIPE — 40,000 PSI Tensile

Pipe Schedule →	40	80	160
PSI @ S.F. = 6	1120	1530	2000
PSI @ S.F. = 5	1350	1840	2410
PSI @ S.F. = 4	1690	2300	3010
PSI @ S.F. = 3	2250	3070	4020
PSI @ S.F. = 2	3370	4600	6020
GPM @ 10 f/s	47.0	40.2	32.6
GPM @ 15 f/s	70.0	60.0	49.0
GPM @ 20 f/s	94.0	68.0	65.0
GPM @ 30 f/s	140	120	98.0

1½" NPT PIPE — 40,000 PSI Tensile

Pipe Schedule →	40	80	160
PSI @ S.F. = 6	1020	1400	1970
PSI @ S.F. = 5	1220	1680	2370
PSI @ S.F. = 4	1530	2100	2960
PSI @ S.F. = 3	2030	2810	3940
PSI @ S.F. = 2	3050	4210	5920
GPM @ 10 f/s	65.0	55.2	43.3
GPM @ 15 f/s	95.0	83.0	65.0
GPM @ 20 f/s	130	110	87.0
GPM @ 30 f/s	190	166	130

2" NPT PIPE — 40,000 PSI Tensile

Pipe Schedule →	40	80	160
PSI @ S.F. = 6	865	1220	1930
PSI @ S.F. = 5	1040	1470	2320
PSI @ S.F. = 4	1300	1840	2900
PSI @ S.F. = 3	1730	2450	3860
PSI @ S.F. = 2	2590	3670	5790
GPM @ 10 f/s	105	91.2	68.6
GPM @ 15 f/s	156	138	103
GPM @ 20 f/s	210	185	137
GPM @ 30 f/s	312	276	206

2½" NPT PIPE — 40,000 PSI Tensile

Pipe Schedule →	40	80	160
PSI @ S.F. = 6	940	1280	1740
PSI @ S.F. = 5	1130	1540	2090
PSI @ S.F. = 4	1410	1920	2610
PSI @ S.F. = 3	1880	2560	3480
PSI @ S.F. = 2	2820	3840	5220
GPM @ 10 f/s	120	132	111
GPM @ 15 f/s	222	198	166
GPM @ 20 f/s	300	265	220
GPM @ 30 f/s	444	396	332

3" NPT PIPE — 40,000 PSI Tensile

Pipe Schedule →	40	80	160
PSI @ S.F. = 6	820	1140	1670
PSI @ S.F. = 5	985	1370	2000
PSI @ S.F. = 4	1230	1710	2500
PSI @ S.F. = 3	1650	2290	3340
PSI @ S.F. = 2	2470	3430	5000
GPM @ 10 f/s	225	206	167
GPM @ 15 f/s	345	310	250
GPM @ 20 f/s	450	412	334
GPM @ 30 f/s	690	620	500

DIMENSIONS AND FLOW AREAS OF PIPES

Nom. Pipe Size	Outside Diam., Inches	Circum- ference, Inches	Schedule 40		Schedule 80		Schedule 160		Schedule XXS	
			Inside Diam., Inches	Inside Area, Sq. Ins.	Inside Diam., Inches	Inside Area, Sq. In.	Inside Diam., Inches	Inside Area, Sq. Ins.	Inside Diam., Inches	Inside Area, Sq. Ins.
1/8 NPTF	.405	1.27	.269	.057	.215	.036	- - - -	- - - -	- - - -	- - - -
1/4 NPTF	.540	1.70	.364	.104	.302	.072	- - - -	- - - -	- - - -	- - - -
3/8 NPTF	.675	2.12	.493	.191	.423	.141	- - - -	- - - -	- - - -	- - - -
1/2 NPTF	.840	2.64	.622	.304	.546	.234	.464	.169	.252	.050
3/4 NPTF	1.05	3.30	.824	.533	.742	.432	.612	.294	.434	.148
1" NPTF	1.32	4.13	1.05	.866	.957	.719	.815	.522	.599	.282
1¼ NPTF	1.66	5.22	1.38	1.50	1.28	1.29	1.16	1.06	.896	.631
1½ NPTF	1.90	5.97	1.61	2.04	1.50	1.77	1.34	1.41	1.10	.950
2" NPTF	2.38	7.46	2.07	3.37	1.94	2.96	1.69	2.24	1.50	1.77
2½ NPTF	2.88	9.03	2.47	4.79	2.32	4.23	2.13	3.56	1.77	2.46
3" NPTF	3.50	11.0	3.07	7.40	2.90	6.61	2.62	5.39	2.30	4.15

FLOW AREAS OF STEEL TUBING

Figures in the body of this chart are internal flow areas, in square inches, of steel tubing. When connecting from pipe into steel tubing, use this chart to find equivalent internal flow area to match areas of pipe in the chart above.

Tube O.D.	Wall Thickness, Inches										
	.025	.032	.035	.042	.049	.058	.065	.072	.083	.095	.109
5/16	.0541	.0485	.0462	.0410	.0361	.0303	.0262	- - - -	- - - -	- - - -	- - - -
3/8	.0830	.0760	.0731	.0665	.0603	.0527	.0471	.0419	- - - -	- - - -	- - - -
1/2	.1590	.1493	.1452	.1359	.1269	.1158	.1075	.0995	.0876	- - - -	- - - -
5/8	.2597	.2472	.2419	.2299	.2181	.2035	.1924	.1817	.1655	.1486	.1301
3/4	.3848	.3696	.3632	.3484	.3339	.3157	.3019	.2884	.2679	.2463	.2223
7/8	.5346	.5166	.5090	.4914	.4742	.4525	.4359	.4197	.3948	.3685	.3390
1	.7088	.6881	.6793	.6590	.6390	.6138	.5945	.5755	.5463	.5153	.4803
1¼	1.131	1.105	1.094	1.068	1.042	1.010	.9852	.9607	.9229	.8825	.8365
1½	1.651	1.612	1.606	1.575	1.544	1.504	1.474	1.444	1.398	1.348	1.291
2	2.986	2.944	2.926	2.883	2.841	2.788	2.746	2.705	2.642	2.573	2.490

COPPER TUBING TO SCHEDULE 40 PIPE — EQUIVALENT FLOW CAPACITY

Copper tubing is a good plumbing medium for compressed air but is not recommended for hydraulic oil plumbing. This chart shows the size tubing which should be used to connect into components which have NPTF pipe thread portholes. For example, if the porthole size is 1/4" NPTF, 3/8" O.D. tubing must be used if full flow capacity is to be maintained. Usually brass fittings or braze-type fittings are recommended for permanent installations. The SAE flare angle of 45° is most often used rather than the JIC angle of 37° as used on steel tubing for hydraulics. Ferrule type or plastic fittings are not recommended for permanent plumbing.

If Pipe Size is:	1/8" NPT	1/4" NPT	3/8" NPT	1/2" NPT	3/4" NPT	1" NPT
Nearest Equivalent Tubing Size Is:	1/4" or 5/16"	3/8" O.D.	1/2" O.D.	5/8" O.D.	3/4" O.D.	1" O.D.

HOSE — EQUIVALENT PIPE AND TUBING SIZES

Hose is not standarized to the same extent as pipe and tubing. There is a wide variation in dimensions, pressure ratings, and availability between manufacturers.

Hose is specified by its inside diameter, overall length fitting to fitting, type and size of end fittings, pressure rating, composition, and temperature range. This table is limited to showing characteristics of one popular brand of low pressure, small diameter hose. Space does not permit listing the wide variety of sizes and types which are available. Consult catalogs of hose manufacturers.

Hose I.D.	Hose O.D.	Minimum Bend Radius	Working Pressure	Min. Burst Pressure	Equivalent Pipe Size	Nearest Equiv. Tubing Size
1/4"	.50"	4"	400 PSI	1250 PSI	1/8" NPT	5/16" O.D.
3/8"	.63"	4"	300 PSI	1000 PSI	1/4" NPT	3/8" or 1/2"
1/2"	.78"	6"	150 PSI	750 PSI	3/8" NPT	1/2" or 5/8"
5/8"	.91"	6"	140 PSI	700 PSI	1/2" NPT	5/8" or 3/4"

CIRCUMFERENCES, AREAS, & DECIMAL EQUIVALENTS OF CIRCLES

Circumferences calculated from πD or $2\pi R$ — Areas calculated from πR^2 or $\pi D^2 \div 4$

R = Radius of Circle — D = Diameter of Circle

Diam., Inches	Decimal Equiv.	Circum., Inches	Area, Sq.Ins.
1/16	.06250	.19635	.00307
3/32	.09375	.29452	.00690
1/8	.12500	.39270	.01227
5/32	.15625	.49087	.01917
3/16	.18750	.58905	.02671
7/32	.21875	.68722	.03758
1/4	.25000	.78540	.04909
9/32	.28125	.88357	.06213
5/16	.31250	.98175	.07670
11/32	.34375	1.0799	.09281
3/8	.37500	1.1781	.11045
13/32	.40625	1.2763	.12962
7/16	.43750	1.3744	.15033
15/32	.46875	1.4726	.17257
1/2	.50000	1.5708	.19635
17/32	.53125	1.6690	.22166
9/16	.56250	1.7671	.24850
19/32	.53975	1.8653	.27688
5/8	.62500	1.9635	.30680
21/32	.65625	2.0617	.33824
11/16	.68750	2.1598	.37122
23/32	.71875	2.2580	.40574
3/4	.75000	2.3562	.44179
25/32	.78125	2.4544	.47937
13/16	.81250	2.5525	.51849
27/32	.84375	2.6507	.55914
7/8	.87500	2.7489	.60132
29/32	.90625	2.8471	.64504
15/16	.93750	2.9452	.69029
31/32	.96875	3.0434	.73708
1	1.0000	3.1416	.78540
1-1/16	1.0625	3.3379	.88660
1-1/8	1.1250	3.5343	.99400
1-3/16	1.1875	3.7306	1.1075
1-1/4	1.2500	3.9270	1.2272
1-5/16	1.3125	4.1233	1.3530
1-3/8	1.3750	4.3197	1.4849
1-7/16	1.4375	4.5160	1.6230
1-1/2	1.5000	4.7124	1.7671
1-9/16	1.5625	4.9087	1.9175
1-5/8	1.6250	5.1051	2.0739
1-11/16	1.6875	5.3014	2.2365
1-3/4	1.7500	5.4978	2.4053
1-13/16	1.8125	5.6941	2.5802
1-7/8	1.8750	5.8905	2.7612
1-15/16	1.9375	6.0868	2.9483
2	2.0000	6.2832	3.1416
2-1/16	2.0625	6.4795	3.3410
2-1/8	2.1250	6.6759	3.5466
2-3/16	2.1875	6.8722	3.7583
2-1/4	2.2500	7.0686	3.9761
2-5/16	2.3125	7.2649	4.2000
2-3/8	2.3750	7.4613	4.4301
2-7/16	2.4375	7.6576	4.6664
2-1/2	2.5000	7.8540	4.9087
2-9/16	2.5625	8.0503	5.1572
2-5/8	2.6250	8.2467	5.4119

Diam., Inches	Decimal Equiv.	Circum., Inches	Area, Sq. Ins.
2-11/16	2.6875	8.4430	5.6727
2-3/4	2.7500	8.6394	5.9396
2-13/16	2.8125	8.8357	6.2126
2-7/8	2.8750	9.0321	6.4918
2-15/16	2.9375	9.2284	6.7771
3	3.0000	9.4248	7.0686
3-1/16	3.0625	9.6211	7.3662
3-1/8	3.1250	9.8175	7.6699
3-3/16	3.1875	10.014	7.9798
3-1/4	3.1250	10.210	8.2958
3-5/16	3.3125	10.407	8.6179
3-3/8	3.3750	10.603	8.9462
3-7/16	3.4375	10.799	9.2806
3-1/2	3.5000	10.966	9.6211
3-9/16	3.5625	11.192	9.9678
3-5/8	9.6250	11.388	10.321
3-11/16	3.6875	11.585	10.680
3-3/4	3.750	11.781	11.045
3-13/16	3.8125	11.977	11.416
3-7/8	3.8750	12.174	11.793
3-15/16	3.9375	12.370	12.177
4	4.0000	12.566	12.566
4-1/16	4.0625	12.763	12.962
4-1/8	4.1250	12.959	13.364
4-3/16	4.1875	13.155	13.772
4-1/4	4.2500	13.352	14.186
4-5/16	4.3125	13.548	14.607
4-3/8	4.3750	13.744	15.083
4-7/16	4.4375	13.941	15.466
4-1/2	4.5000	14.137	15.904
4-9/16	4.5625	14.334	16.349
4-5/8	4.6250	14.530	16.800
4-11/16	4.6875	14.726	17.257
4-3/4	4.7500	14.923	17.728
4-13/16	4.8125	15.119	18.190
4-7/8	4.8750	15.315	18.665
4-15/16	4.9375	15.512	19.147
5	5.0000	15,708	19.635
5-1/16	5.0625	15.904	20.129
5-1/8	5.1250	16.101	20.629
5-3/16	5.1875	16.297	21.135
5-1/4	5.2500	16.493	21.648
5-5/16	5.3125	16.690	22.166
5-3/8	5.3750	16.886	22.691
5-7/16	5.4375	17.082	23.221
5-1/2	5.5000	17.279	23.758
5-9/16	5.5625	17.475	24.301
5-5/8	5.6250	17.671	24.850
5-11/16	5.6875	17.868	25.406
5-3/4	5.7500	18.064	25.967
5-13/16	5.8125	18.261	26.535
5-7/8	5.8750	18.457	27.109
5-15/16	5.9375	18.653	27.688
6	6.0000	18.850	28.274
6-1/8	6.1250	19.242	29.465
6-1/4	6.2500	19.635	30.680
6-3/8	6.3750	20.028	31.919

Diam., Inches	Decimal Equiv.	Circum., Inches	Area, Sq. Ins.
6-1/2	6.5000	20.420	33.183
6-5/8	6.6250	20.813	34.472
6-3/4	6.7500	21.206	35.785
6-7/8	6.8750	21.598	37.122
7	7.0000	21.991	38.485
7-1/8	7.1250	22.384	39.871
7-1/4	7.2500	22.776	41.282
7-3/8	7.3750	23.169	42.718
7-1/2	7.5000	23.562	44.179
7-5/8	7.6250	23.955	45.664
7-3/4	7.7500	24.347	47.173
7-7/8	7.8750	24.740	48.707
8	8.0000	25.133	50.265
8-1/8	8.1250	25.525	51.849
8-1/4	8.2500	25.918	53.456
8-3/8	8.3750	26.311	55.088
8-1/2	8.5000	26.704	56.745
8-5/8	8.6250	27.096	58.426
8-3/4	8.7500	27.489	60.132
8-7/8	8.8750	27.882	61.862
9	9.0000	28.274	63.617
9-1/8	9.1250	28.667	65.397
9-1/4	9.2500	29.060	67.201
9-3/8	9.3750	29.452	69.029
9-1/2	9.5000	29.845	70.882
9-5/8	9.6250	30.238	72.760
9-3/4	9.7500	30.631	74.662
9-7/8	9.8750	31.023	76.589
10	10.000	31.416	78.540
10-1/4	10.250	32.201	82.516
10-1/2	10.500	32.987	86.590
10-3/4	10.750	33.772	90.763
11	11.000	34.558	95.033
11-1/4	11.250	35.343	99.402
11-1/2	11.500	36.128	103.87
11-3/4	11.750	36.914	108.43
12	12.000	37.699	113.10
12-1/4	12.250	38.485	117.86
12-1/2	12.500	39.270	122.72
12-3/4	12.750	40.055	127.68
13	13.000	40.841	132.73
13-1/4	13.250	41.626	137.89
13-1/2	13.500	42.412	143.14
13-3/4	13.750	43.197	148.49
14	14.000	43.982	153.94
14-1/4	14.250	44.768	159.48
14-1/2	14.500	45.553	165.13
14-3/4	14.750	46.338	170.87
15	15.000	47.124	176.71
15-1/4	15.250	47.909	182.65
15-1/2	15.500	48.695	188.69
15-3/4	15.750	49.480	194.83
16	16.000	50.265	201.06
16-1/4	16.250	51.051	207.39
16-1/2	16.500	51.836	213.82
16-3/4	16.750	56.622	220.35
17	17.000	53.407	226.98

TAP DRILL SIZES — U. S. THREADS

To determine correct tap drill size, find threads per inch in left column. Move horizontally to per cent of full threads desired. The figure is double thread depth. Subtract from tap diameter. This is the size tap hole to drill.

Example: Find tap drill for 1/4-20 for 75% thread depth. Find 20 in left column. Move across to 75% column. The figure of .0486'' is double thread depth. Subtract from 1/4'' = .2014 drill diameter or No. 7 drill.

Most machine threads are tapped from 50 to 85% full depth. Rarely more than 85% except on very soft material. For most work, 65 to 75% is adequate. Choose next smaller drill since twist drills tend to drill oversize.

Double Thread Depth, Inches

Th'ds per Inch	90% Full Th'd	85% Full Th'd	80% Full Th'd	75% Full Th'd	70% Full Th'd	65% Full Th'd
2-1/4	.5196	.4908	.4618	.4330	.4040	.3752
2-3/8	.4922	.4648	.4376	.4100	.3846	.3572
2-1/2	.4676	.4416	.4156	.3896	.3636	.3376
2-5/8	.4454	.4207	.3959	.3712	.3464	.3217
2-3/4	.4250	.4014	.3778	.3542	.3306	.3070
2-7/8	.4066	.3840	.3614	.3388	.3162	.2936
3	.3896	.3680	.3464	.3246	.3030	.2814
3-1/4	.3597	.3397	.3197	.2998	.2798	.2598
3-1/2	.3340	.3154	.2968	.2784	.2598	.2412
4	.2922	.2760	.2598	.2436	.2272	.2110
4-1/2	.2598	.2454	.2310	.2165	.2021	.1877
5	.2338	.2208	.2078	.1948	.1818	.1688
5-1/2	.2126	.2008	.1890	.1772	.1653	.1535
6	.1949	.1840	.1732	.1624	.1516	.1407
7	.1670	.1576	.1484	.1392	.1298	.1206
8	.1460	.1380	.1298	.1218	.1136	.1054
9	.1298	.1226	.1154	.1083	.1010	.0938
10	.1170	.1104	.1040	.0975	.0910	.0844
11	.1062	.1002	.0944	.0884	.0826	.0766
12	.0974	.0920	.0864	.0810	.0756	.0704
13	.0900	.0850	.0800	.0750	.0700	.0650
14	.0834	.0788	.0742	.0696	.0648	.0602
15	.0778	.0736	.0692	.0648	.0606	.0562
16	.0730	.0690	.0648	.0609	.0568	.0526

Th'ds per Inch	90% Full Th'd	85% Full Th'd	80% Full Th'd	75% Full Th'd	70% Full Th'd	65% Full Th'd
18	.0650	.0614	.0576	.0538	.0504	.0466
20	.0584	.0552	.0520	.0486	.0454	.0422
22	.0530	.0500	.0472	.0442	.0412	.0384
24	.0488	.0460	.0432	.0406	.0378	.0352
26	.0450	.0424	.0400	.0374	.0350	.0324
28	.0416	.0394	.0370	.0348	.0324	.0300
30	.0390	.0368	.0346	.0324	.0302	.0282
32	.0364	.0344	.0324	.0304	.0284	.0264
34	.0344	.0324	.0304	.0286	.0266	.0248
36	.0324	.0306	.0288	.0270	.0252	.0234
38	.0308	.0290	.0272	.0256	.0238	.0222
40	.0290	.0274	.0258	.0242	.0226	.0210
42	.0278	.0262	.0248	.0232	.0216	.0202
44	.0266	.0250	.0236	.0222	.0206	.0192
46	.0254	.0240	.0224	.0210	.0196	.0184
48	.0242	.0228	.0218	.0202	.0188	.0174
50	.0234	.0220	.0208	.0194	.0182	.0168
52	.0224	.0212	.0200	.0186	.0174	.0162
56	.0208	.0196	.0184	.0174	.0162	.0150
60	.0194	.0182	.0172	.0162	.0150	.0140
64	.0182	.0172	.0160	.0150	.0140	.0132
68	.0172	.0162	.0152	.0144	.0134	.0124
72	.0162	.0152	.0144	.0134	.0126	.0116
80	.0146	.0138	.0128	.0122	.0112	.0106

Standard Twist Drill Sizes — 3/64 Through 1-Inch

Drill Size	Diameter	Drill Size	Diameter	Drill Size	Diameter	Drill Size	Diameter	Drill Size	Diameter	Drill Size	Diameter	Drill Size	Diameter
3/64	.0938	27	.1440	12	.1800	D	.2460	Q	.3320	15/32	.4688	3/4	.7500
41	.0960	26	.1470	11	.1910	1/4,E	.2500	R	.3390	31/64	.4844	49/64	.7656
40	.0980	25	.1495	10	.1935	F	.2570	11/32	.3438	1/2	.5000	25/32	.7812
39	.0995	24	.1520	9	.1960	G	.2610	S	.3480	33/64	.5156	51/64	.7969
38	.1015	23	.1540	8	.1990	17/64	.2656	T	.3580	17/32	.5312	13/16	.8125
37	.1040	5/32	.1562	7	.2010	H	.2660	23/64	.3594	35/64	.5469	53/64	.8281
36	.1065	22	.1570	13/64	.2031	I	.2720	U	.3680	9/16	.5625	27/31	.8438
7/64	.1094	21	.1590	6	.2040	J	.2770	3/8	.3750	37/64	.5781	55/64	.8594
35	.1100	20	.1610	5	.2055	K	.2810	V	.3770	19/32	.5938	7/8	.8750
34	.1110	19	.1660	4	.2090	9/32	.2812	W	.3860	39/64	.6094	57/64	.8906
33	.1130	18	.1695	3	.2130	L	.2900	25/64	.3906	5/8	.6250	29/32	.9062
32	.1160	11/64	.1719	7/32	.2188	M	.2950	X	.3970	41/64	.6406	59/64	.9219
31	.1200	17	.1730	2	.2210	19/64	.2969	Y	.4040	21/32	.6562	15/16	.9375
1/8	.1250	16	.1770	1	.2280	N	.3020	13/32	.4062	43/64	.6719	61/64	.9531
30	.1285	15	.1800	A	.2340	5/16	.3125	Z	.4130	11/16	.6875	31/32	.9688
29	.1360	14	.1820	15/64	.2344	O	.3160	27/64	.4219	45/64	.7031	63/64	.9844
28	.1405	13	.1850	B	.2380	P	.3230	7/16	.4375	23/32	.7188	1	1.000
9/64	.1406	3/16	.1875	C	.2420	21/64	.3281	29/64	.4531	47/64	.7344		

DRIVE HORSEPOWER FOR A HYDRAULIC PUMP

Figures in the body of this table show the horsepower needed to drive a hydraulic pump having an efficiency of 85%. Most positive displacement pumps (gear, vane, piston) fall in the range of 80% to 90% efficiency so this chart should be accurate to within 5% for almost any pump. The table was calculated from the formula:

$$HP = PSI \times GPM \div [1714 \times 0.85]$$

For pumps with other than 85% efficiency, the formula can be used, substituting actual efficiency, in decimals, in place of 0.85.

Using the Table. The range of 500 to 5000 PSI covers most hydraulic systems, but power requirements can be determined for conditions outside the range of the table, or for intermediate values. For example, power at 4000 PSI will be exactly 2 times the figure shown for 2000 PSI. At 77 GPM, power will be the sum of the figures shown in the 75 and 2 GPM lines, etc. For systems operating below 500 PSI, horsepower calculations tend to become inaccurate because mechanical friction losses reduce pump efficiency.

Rules-of-Thumb. Approximate HP requirement can be estimated by our "rule of 1500" which states that 1 HP is required for each multiple of 1500 when multiplying PSI x GPM. For example, a 5 GPM pump at 1500 PSI would require 5 HP, or at 3000 PSI would require 10 HP. A 10 GPM pump at 1000 PSI would require 6-2/3 HP or the same pump at 1500 PSI would require 10 HP, etc.

Another rule-of-thumb states that about 5% of the pump maximum rated HP is required to idle the pump when it is "unloaded" and the full flow is circulating at near 0 PSI. This amount of power is consumed in flow losses plus mechanical friction losses in bearings and pumping elements.

Figures in Body of Table are HP's Required to Drive a Hydraulic Pump (A Pump Efficiency of 85% is Assumed)

GPM	500 PSI	750 PSI	1000 PSI	1250 PSI	1500 PSI	1750 PSI	2000 PSI	2500 PSI	3000 PSI	3500 PSI	4000 PSI	5000 PSI
3	1.03	1.54	2.06	2.57	3.09	3.60	4.12	5.15	6.18	7.21	8.24	10.3
5	1.72	2.57	3.43	4.29	5.15	6.00	6.86	8.58	10.3	12.0	13.7	17.2
7½	2.57	3.86	5.15	6.43	7.72	9.01	10.3	12.9	15.4	18.0	20.6	25.7
10	3.43	5.15	6.86	8.58	10.3	12.0	13.7	17.2	20.6	24.0	27.5	34.3
12½	4.29	6.43	8.58	10.7	12.9	15.0	17.2	21.4	25.7	30.0	34.3	42.9
15	5.15	7.72	10.3	12.9	15.4	18.0	20.6	25.7	30.9	36.0	41.2	51.5
17½	6.01	9.01	12.0	15.0	18.0	21.0	24.0	30.0	36.0	42.0	48.0	60.1
20	6.86	10.3	13.7	17.2	20.6	24.0	27.5	34.3	41.2	48.0	54.9	68.6
22½	7.72	11.6	15.4	19.3	23.2	27.0	30.9	38.6	46.3	54.1	61.8	77.2
25	8.58	12.9	17.2	21.4	25.7	30.0	34.3	42.9	51.5	60.1	68.6	85.8
30	10.3	15.4	20.6	25.7	30.9	36.0	41.2	51.5	61.8	72.1	82.4	103
35	12.0	18.0	24.0	30.0	36.0	42.0	48.0	60.1	72.1	84.1	96.1	120
40	13.7	20.6	27.5	34.3	41.2	48.0	54.9	68.6	82.4	96.1	110	137
45	15.4	23.2	30.9	38.6	46.3	54.1	61.8	77.2	92.7	108	124	154
50	17.2	25.7	34.3	42.9	51.5	60.1	68.6	85.8	103	120	137	172
55	18.9	28.3	37.8	47.2	56.6	66.1	75.5	94.4	113	132	151	189
60	20.6	30.9	41.2	51.5	61.8	72.1	82.4	103	124	144	165	206
65	22.3	33.5	44.6	55.8	66.9	78.1	89.2	112	134	156	178	223
70	24.0	36.0	48.0	60.1	72.1	84.1	96.1	120	144	168	192	240
75	25.7	38.6	51.5	64.3	77.2	90.1	103	129	154	180	206	257
80	27.5	41.2	54.9	68.7	82.4	96.1	110	137	165	192	220	275
85	29.2	43.8	58.3	72.9	87.5	102	117	146	175	204	233	292
90	30.9	46.3	61.8	77.2	92.7	108	124	154	185	216	247	309
100	34.3	51.5	68.6	85.8	103	120	137	172	206	240	275	343

Oversizing or Undersizing of an Induction-Type Electric Motor

Optimum results are obtained if HP rating of an electric motor is neither too far oversize or undersize for the job. Some effects of motor power mismatch are:

Oversize Motor. Using a 20 HP motor to do a 10 HP job, for example, will give good results as far as running the fluid power system is concerned, but it will consume a little more current for the same 10 HP output. It will also cause the power factor of the plant electric system to be poorer, with higher power cost. Idling current will be higher so more power will be wasted during periods in the cycle when the motor is running under idle conditions.

Undersize Motor. A 3-phase induction motor can usually be safely overloaded during peak parts of the cycle as explained on Page 88, but during these peak periods, motor current will be all out of proportion to the extra power being produced with a considerable amount of overheating. If the motor is too far undersize it will, of course, burn out in a short time.

MATCHING THE SIZE OF A BROKEN PUMP OR MOTOR

To select a replacement for a broken or worn out hydraulic pump or motor which has no nameplate or has no rating marked on the case, the formulae on this page may be used after making physical measurements of the internal elements.

When replacing a pump, catalog ratings will usually be shown in GPM at a specified shaft speed. On motors, catalog ratings will usually be shown as C.I.R. (cubic inches displacement per shaft revolution). Formulae are shown below for calculating either GPM at 1800 RPM or calculating C.I.R. Use the formula which is appropriate.

Make all measurements in inches, and as accurately as possible. Convert fractional dimensions into decimal equivalents for use in the formulae.

When selecting a new pump or motor, be sure the catalog pressure rating is adequate for your application, and in the case of a pump, be sure the direction of shaft rotation is the same as for the old pump.

GEAR PUMPS AND MOTORS

1. Measure gear width. This is W in the formulae.
2. Measure bore diameter of one of the gear chambers. This is D in the formulae.
3. Measure distance across both gear chambers. This is L in the formulae.

$$\text{GPM @ 1800 RPM} = 47 \times W \times (2D - L) \times \frac{(L - D)}{2}$$

A speed of 1800 RPM is used in the formula. At other speeds, GPM is proportional to RPM.

$$\text{C.I.R Displ.} = 6 \times W \times (2D - L) \times \frac{(L - D)}{2}$$

VANE PUMPS AND MOTORS

(Balanced type, not variable displacement)

1. Measure width of rotor. This is W in the formulae.
2. Measure shortest distance across bore. This is D.
3. Measure longest distance across bore. This is L.

$$\text{GPM @ 1800 RPM} = 94 \times W \times \frac{(L + D)}{4} \times \frac{(L - D)}{2}$$

A speed of 1800 RPM is used in the formula. At other speeds, GPM is proportional to speed.

$$\text{C.I.R. Displ.} = 12 \times W \times \frac{(L + D)}{4} \times \frac{(L - D)}{2}$$

PISTON PUMPS OR MOTORS

1. Find piston area from piston diameter. This is A in formulae.
2. Measure length of stroke. This is L in formulae.
3. Count the number of pistons. This is N in formulae.

$$\text{GPM @ 1800 RPM} = A \times L \times N \times 1800 \div 231$$

A speed of 1800 RPM is used in this formula. At other speeds, GPM is proportional to speed.

$$\text{C.I.R. displ.} = A \times L \times N$$

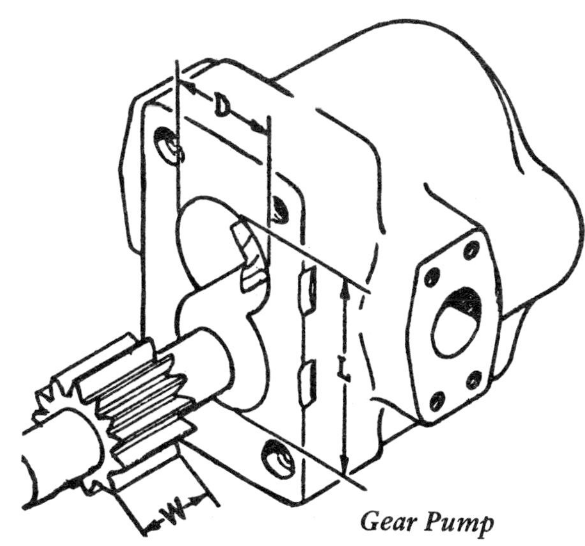

Gear Pump

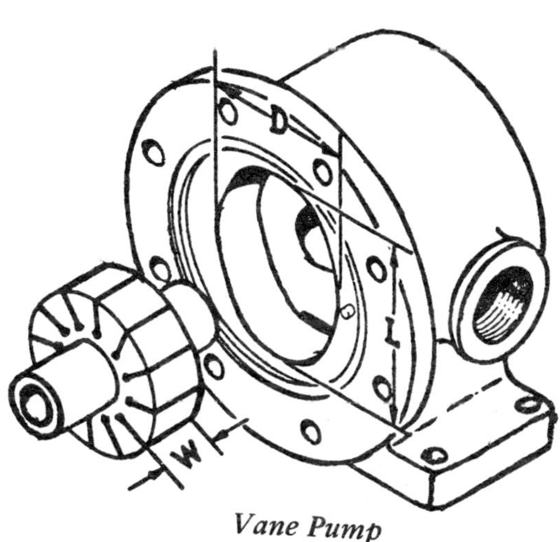

Vane Pump

After finding GPM rating of a pump or C.I.R displacement of a motor, choose a replacement with as near as possible the same catalog rating at zero pressure.

If a pump with higher GPM has to be used, it will require more HP at the same pressure and cylinders in the system will move faster. If one with lower GPM has to be used, the system will have plenty of power but cylinders will move slower than originally.

If a motor with greater displacement has to be used, it will deliver greater torque at the same pressure. It will not require additional HP from the pump but will rotate at a lower RPM than the original motor. If it has less displacement it will have less torque than the original motor but will rotate at a higher speed with the same oil flow from the pump. It will deliver the same HP as the original motor but at a ratio of higher speed and lower torque.

NFPA STANDARD DIMENSIONS ON MOUNTING FLANGES AND SHAFTS

For Positive Displacement Hydraulic Pumps and Motors

For many years the only standardization on pump and motor shafts and mounting flanges has been the SAE (Society of Automotive Engineers) Standard J744a. It was widely accepted by other industries as well as the automotive and is still a valid standard. The NFPA (National Fluid Power Association) started work in 1962 on a set of standards which would more directly apply to industrial fluid power pumps and motors, and would be more complete. This set of standards was unanimously approved by the NFPA Board of Directors and issued in 1965 as NFPA Recommended Standard STD T3.9.65.1. In 1966 it was also adopted by the ANSI (American National Standards Institute) and issued as ANSI Standard B93.6-1966. It has since been revised and now carries the numbers ANSI B93.6-1972, and NFPA T3.9.2 R1. Copies may be purchased from the NFPA.

The purpose of the standard is to encourage manufacturers to use interchangeable dimensions on shafts and mounting flanges as far as possible, to simplify dimensional interchangeability for the user. No standards exist at this time on foot mounting dimensions. No performance specifications exist although recommended methods of testing pumps and motors, and the manner of presenting test data are given in NFPA Recommended Standard T3.9.17-1971.

The new NFPA and ANSI standards are similar to the SAE standards but differ in these respects: Additional shaft diameters and an alternate long length shaft have been added to the straight-shaft-without-thread listings. Additional mounting flange sizes have been added to provide a wider selection for the designer. NFPA has not assigned any horsepower ratings, making it the responsibility of the pump designer. However, the SAE horsepower ratings are shown for mounting flanges.

Drawings and charts are only for identification in the field. Complete decimal dimensions and tolerances are contained in the NFPA standards, and copies of the standards mentioned may be purchased from the NFPA.

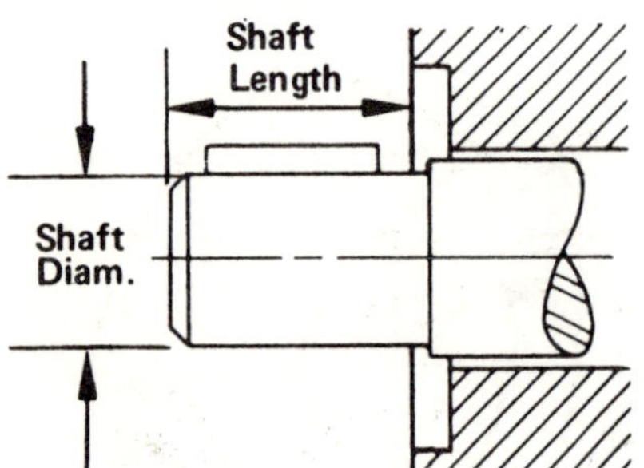

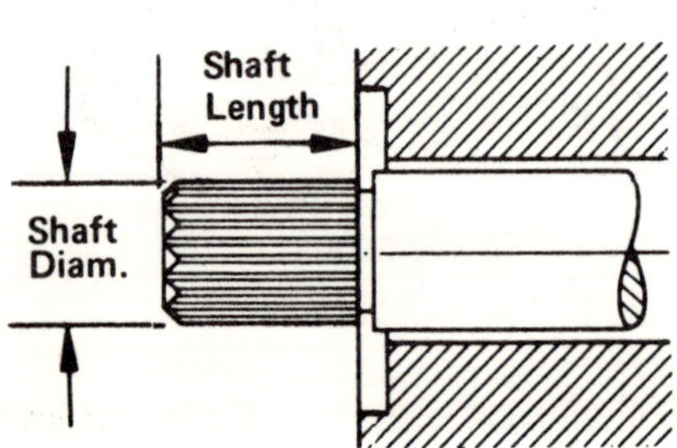

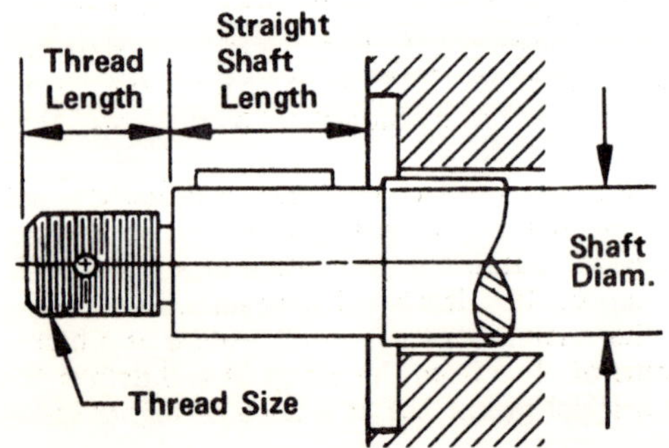

NFPA STANDARD
STRAIGHT SHAFTS WITHOUT THREAD

Shaft Ident. Code	Shaft Diam.	Short Shaft Lgth.	Long Shaft Lgth.	Key Width Inches	SAE Reference*
13-1	1/2"	3/4"	- - - -	1/8"	- - - -
16-1	5/8	15/16	2"	5/32	A
22-1	7/8	1-5/16	2-1/2	1/4	B
25-1	1	1-1/2	2-3/4	1/4	- - - -
32-1	1-1/4	1-7/8	3	5/16	C
38-1	1-1/2	2-1/8	3-1/4	3/8	- - - -
44-1	1-3/4	2-5/8	3-5/8	7/16	D,E

NFPA STANDARD
30-DEGREE INVOLUTE SPLINE SHAFTS

Shaft Code	Shaft Diam.	Shaft Lgth.	Spline Specifications	SAE Ref.*
13-4	1/2"	3/4"	9T, 20/40 DP	- - - -
16-4	5/8	15/16	9T, 16/32 DP	A
22-4	7/8	1-5/16	13T, 16/32 DP	B
25-4	1	1-1/2	15T, 16/32 DP	- - - -
32-4	1-1/4	1-7/8	14T, 12/24 DP	C
38-4	1-1/2	2-1/8	17T, 12/24 DP	- - - -
44-4	1-3/4	2-5/8	13T, 8/16 DP	D,E
50-4	2	3-1/8	15T, 8/16 DP	- - - -

*Indicates matching SAE front flange for each shaft diameter

NFPA STANDARD
STRAIGHT SHAFTS WITH THREAD

Shaft Code	Shaft Diam.	Str. Shaft	Th'd Size	Th'd Lgth.	Key Width
13-2	1/2"	3/4"	3/8 – 24	9/16"	1/8"
16-2	5/8	15/16	1/2 – 20	23/32	5/32
22-2	7/8	1-5/16	5/8 – 18	29/32	1/4
25-2	1	1-1/2	3/4 – 16	1-1/16	1/4
32-2	1-1/4	1-7/8	1 – 12	1-7/32	5/16
38-2	1-1/2	2-1/8	1-1/8-12	1-3/8	3/8
44-2	1-3/4	2-5/8	1-1/4-12	1-9/16	7/16

NFPA STANDARD
TAPERED SHAFTS WITH THREAD

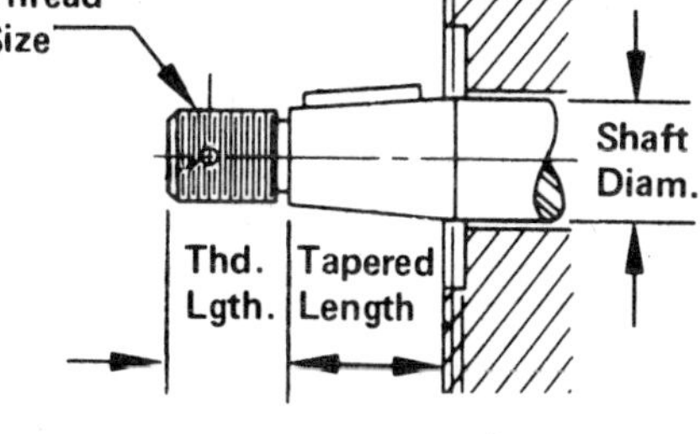

Shaft Ident. Code	Shaft Diam.	Tprd. Shaft Lgth.	Th'd Shaft Lgth.	Th'd Size	Key Width
13-3	1/2''	11/16''	1/2''	5/16-32	1/8''
16-3	5/8	11/16	23/32	1/2-20	5/32
22-3	7/8	1-1/8	29/32	5/8-18	1/4
25-3	1	1-3/8	1-1/16	3/4-16	1/4
32-3	1-1/4	1-3/8	1-7/32	1-12	5/16
38-3	1-1/2	1-7/8	1-3/8	1-1/8-12	3/8
44-3	1-3/4	2-1/8	1-9/16	1-1/4-12	7/16
50-3	2	2-7/8	1-9/16	1-1/4-12	1/2

NFPA STANDARD
TWO-BOLT MOUNTING FLANGES

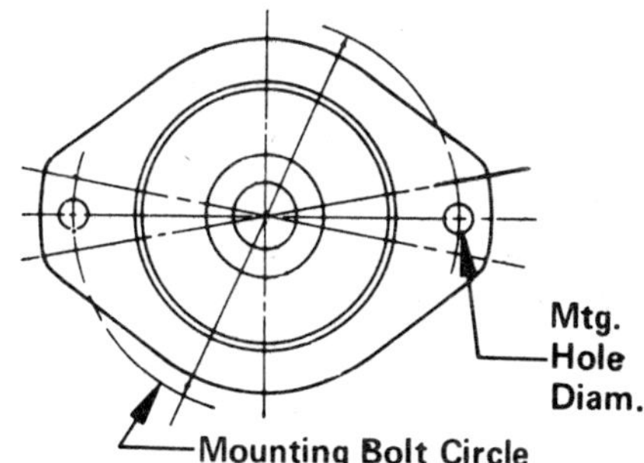

Flange Code	SAE No.	SAE HP Rating	M'tg. Bolt Circle	M'tg. Hole Diam.	Pilot Diam.	Pilot Hgt.
50-2	- - - -	- - - -	3-1/4''	13/32''	2''	1/4''
82-2	A	10	4-3/16	7/16	3-1/4	1/4
101-2	B	25	5-3/4	9/16	4	3/8
127-2	C	50	7-1/8	11/16	5	1/2
152-2	D	100	9	13/16	6	1/2
165-2	E	200	12-1/2	1-1/16	6-1/2	5/8
177-2	F	300	13-25/32	1-1/16	7	5/8

NFPA STANDARD
FOUR-BOLT MOUNTING FLANGES

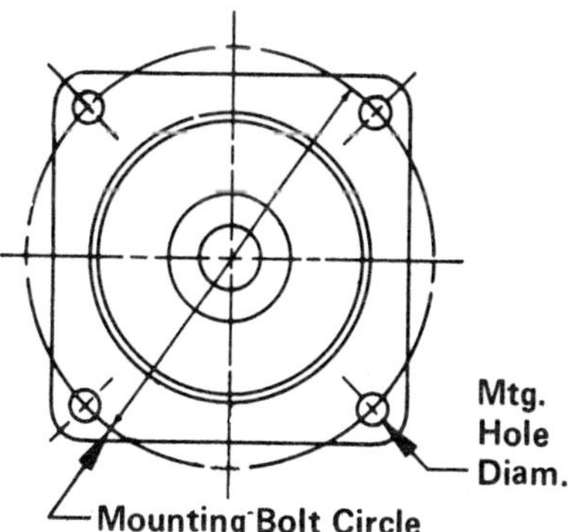

Flange Code	SAE No.	SAE HP Rating	M'tg. Bolt Circle	M'tg. Hole Diam.	Pilot Diam.	Pilot Hgt.
101-4	B	25	5''	9/16''	4	3/8''
127-4	C	50	6-3/8	9/16	5	1/2
152-4	D	100	9	13/16	6	1/2
165-4	E	200	12-1/2	13/16	6-1/2	5/8
177-4	F	300	13-25/32	1-1/16	7	5/8

VISCOSITY CONVERSION — SSU AND CENTISTOKES

In the U.S.A. oil viscosity is usually expressed in units of SSU (Saybolt Seconds Universal). This is called the kinematic viscosity and includes fluid friction plus effect of mass or weight of the fluid. It is a measure of the "thickness" of the fluid, or its resistance to flow. Since oil viscosity changes with changes in temperature, SSU viscosity is usually expressed at a reference temperature of 100°F.

The SSU viscosity of a fluid is based on the number of seconds required for 60 cc of the fluid at 100°F to flow through a standard orifice of 0.0695'' diameter. For example if it takes 150 seconds, the oil is rated at 150 SSU, etc.

For international usage in S.I. units, the centistoke is the unit of viscosity. It is derived from metric units (1 stoke = 100 centistokes). For use in certain formulae in this book the viscosity must be expressed in centistokes, not in SSU. The chart gives equivalent centistoke values for SSU viscosities.

There are a number of other viscosity systems in use in various parts of the world, including SSF, Redwood No. 1, Ford No. 3, and Engler. These units will also be replaced by the centistoke for international standard usage.

SSU Saybolt Universal Seconds	Cs Centi-stokes	SSU Saybolt Seconds Universal	Cs Centi-stokes
9,000	1950	500	110
8,000	1700	400	87
7,000	1500	300	65
6,000	1300	200	43
5,000	1050	100	20.8
4,000	850	90	18.3
3,000	630	80	15.8
2,000	420	70	13.3
1500	315	60	10.5
1000	220	55	8.9
900	195	50	7.5
800	170	45	5.9
700	150	40	4.3
600	130	35	2.7

DIMENSIONS OF SAE PORT PADS FOR HYDRAULIC PUMPS AND MOTORS

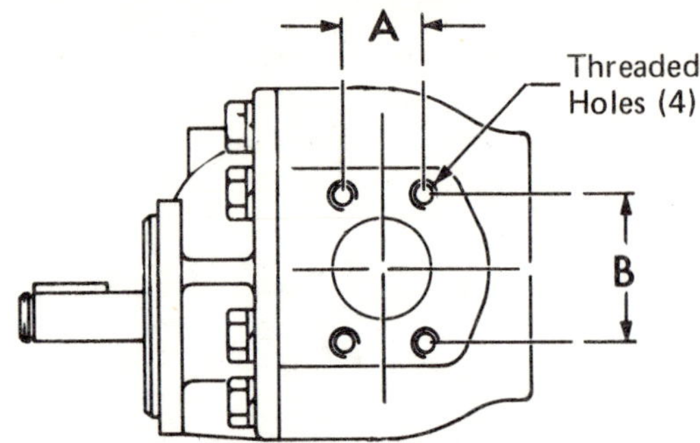

Figure 1. Many hydraulic components have 4-bolt port pads with standard SAE rectangular bolt spacing.

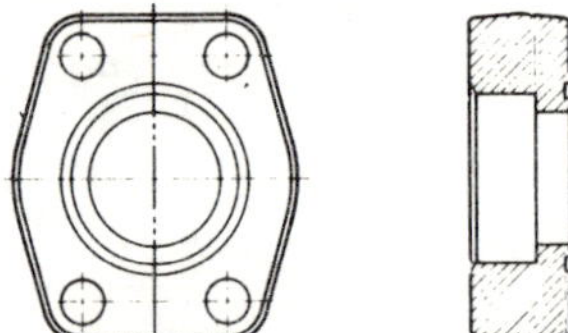

Figure 2. Typical solid flange with weld socket for connecting pipe to an SAE port pad.

Information on SAE port pads was obtained primarily from SAE J518 specification. For additional dimensions, that source may be consulted.

Figure 1. Many hydraulic components, particularly those of large size and those operating at high pressure, are manufactured with SAE 4-bolt port pads with rectangular bolt hole mounting pattern. Advantages of this type of fluid connection are that it can operate at high pressure with zero leakage; it can be connected and disconnected many times without developing a leak; the O-ring seal is replaceable; it forms a union joint for disconnection without disturbing other plumbing; the mating flange can be attached with tools much smaller than required for pipe thread, straight thread, or swivel fittings; and the sealing element, usually a standard O-ring, can be easily replaced if damaged, or changed to another seal compound. Originally, the 4-bolt pad was designed for connections of 1-inch size and larger, although it is now available in smaller sizes.

There are two series of bolt patterns. One (usually referred to as the 3000 PSI Series) is for lower pressure, in the range of 500 to 5000 PSI depending on the size. The other size (referred to as the high pressure or 6000 PSI Series) is larger. It will accept larger and longer bolts, more widely spaced. It is rated for 6000 PSI in all available sizes. Mounting bolt pattern is different for the two series. Connecting flanges designed for one series will not fit the bolt pattern of the other series. There are several types of connecting flanges which may be used to connect into these port pads.

Pipe Connections to an SAE Port Pad

Figure 2. Solid flanges may be used for connecting pipe or tubing into an SAE port pad. These flanges are available either with dryseal (NPTF) threads for pipe, or with a plain round bore for welding to pipe or tubing. In all cases, the hole diameter through the flange and pad is the same as the rated size of the port.

The O-ring seal is contained in a circular groove on the inside surface of the solid flange and is pressed flat against the flat surface of the pad as the bolts are tightened.

A mounting kit for a solid flange consists of four high tensile cap screws, four lockwashers, and one O-ring.

Hose Connections to an SAE Port Pad

Figure 3. These port pads are especially desirable for connecting hose to avoid any twist in the hose during assembly.

Hoses should have an SAE flange end either straight-thru or angle type. The O-ring is contained in a circular groove on the end surface of the flange. The flange is held against the port pad with two half-flanges (called a split flange). Hose flange ends are available in straight-thru type or with angles of 22½, 30, 45, 60, 67½, or 90 degrees. Since high pressure hose will not bend in a sharp radius, it is important to carefully plan lengths and angles before ordering.

The mounting kit for each hose end consists of four high tensile cap screws, four lockwashers, two split flange halves, and one O-ring. Split flange mounting kits can usually be purchased from the hose supplier when ordering hose.

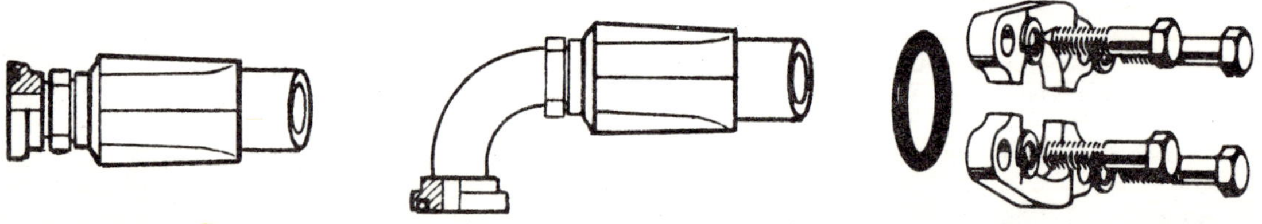

Hose Flange - Straight-Thru   Hose Flange - 90° Angle   Split Flange Mounting Accessories

Figure 3. Hoses may be ordered with an SAE flange fitting. They can be bolted to the port pad with a split flange.

Notes on SAE Flanges

a. Solid flanges in the high pressure series are not usually offered with pipe thread connections — only with weld sockets.

b. One disadvantage of SAE pads, especially with split flanges is the possibility of human error in attachment. When mounting a flange to a port pad, the bolts should be finger-tightened to be sure the flange is going on straight. Then all four screws should be tightened to the same torque, going in rotation and tightening each one a little at a time.

Also, a mechanic may forget to install the O-ring seal before bolting the flange on to the pad.

c. SAE flange unions for joining two pieces of pipe or tubing are made to the same bolt spacing as shown in the tables. They use one standard solid flange with clearance holes for mounting bolts, and with contained O-ring seal. This is mated to a "companion flange" of similar dimensions which has threaded bolt holes but without the O-ring groove and seal.

d. Some hydraulic components have bolt-on port pads with square mounting hole pattern. This is not an SAE standard; rather, it apparently originated with Vickers many years ago and was adopted by a few other manufacturers. Connection flanges for these ports may be purchased from the component manufacturer.

Low Pressure SAE 4-Bolt Mounting Pads

All dimensions are in inches. They refer to Figure 1 on the preceding page.

Nominal Size	PSI Pres. Rating	Bars Pres. Rating	Dim. A Inches	Dim. B Inches	Mounting Bolt Size, Inches*	Standard O-ring Size†
1/2	5000	345	0.688	1.500	5/16-18 x 1¼	1 x 3/4
3/4	5000	345	0.875	1.875	3/8-16 x 1¼	1-1/4 x 1
1	5000	345	1.031	2.062	3/8-16 x 1¼	1-9/16 x 1-5/16
1¼	4000	276	1.188	2.312	7/16-14 x 1½	1-3/4 x 1-1/2
1½	3000	207	1.406	2.750	1/2-13 x 1½	2-1/8 x 1-7/8
2	3000	207	1.688	3.062	1/2-13 x 1½	2-1/2 x 2-1/4
2½	2500	172	2.000	3.500	1/2-13 x 1¾	3 x 2-3/4
3	2000	138	2.438	4.187	5/8-11 x 1¾	3-5/8 x 3-3/8
3½	500	34	2.750	4.750	5/8-11 x 2	4-1/8 x 3-7/8
4	500	34	3.062	5.125	5/8-11 x 2	4-5/8 x 4-3/8
5	500	34	3.625	6.000	5/8-11 x 2¼	5-5/8 x 5-3/8

*Length shown is minimum.   †All O-rings have 1/8" section.

High Pressure SAE 4-Bolt Mounting Pads

Pressure rating all sizes: 6000 PSI, 414 bar. O-ring sizes are the same as for the low pressure series. Dimensions refer to Figure 1 on the preceding page.

Rated Size	Dim. A Inches	Dim. B Inches	Mounting Bolt Size, Inches	Weld Socket Diam.
1/2	0.718	1.594	5/16-18 x 1½	0.855
3/4	0.937	2.000	3/8-16 x 1½	1.063
1	1.093	2.250	7/16-14 x 1¾	1.328
1¼	1.250	2.625	1/2-13 x 2¼	1.672
1½	1.437	3.125	5/8-11 x 2¾	1.923
2	1.750	3.812	3/4-10 x 3	2.406
2½	2.312	4.875	7/8-9 x 3½	2.906
3	2.812	6.000	1-7/8-7 x 4½	3.547

O-RING COMPATIBILITY

Several types of fittings use an O-ring seal. These types include the SAE 4-bolt rectangular flange, the SAE and AND straight thread fittings, and certain proprietary types such as the 4-bolt square flange. The following chart, published by Aeroquip, gives the compound material considered best for the listed fluid over the temperature range shown.

Material	Temperature	Fluids
Buna-N	–65 to +275°F	Air, gasoline, mineral oil, water, JP-4, JP-5, MIL-L-7808C, MIL L 5606
Butyl	–65 to +250°F	Steam, Pydraul, Skydrol
Rubber	–40 to +200°F	Ammonia, hydrochloric acid, methyl alcohol, sodium hydroxide, steam, water
Neoprene	65 to +300°F	Salt water, Freon 12, Freon 22, MIL-L-7808C
Silicones	–65 to +550°F	Gases, hot air, high aniline oils
Teflon	–100 to +500°F	Acids, alkalies, gases, steam Pydraul, Skydrol
Viton	–40 to +450°F	Air, exh. gases, MIL-O-8200, MIL-O-8515, MIL-H-7808C, OS-45, DC200, JP-4, JP-5

TOOL CLEARANCE

Recommended minimum spacing between holes for tool clearance, pipe, thread, or straight thread portholes.

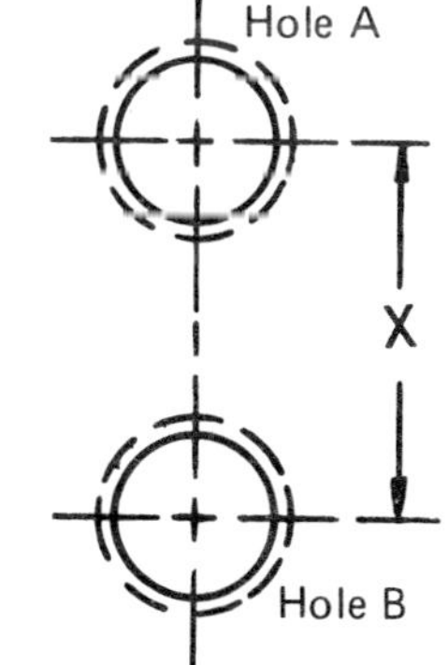

Hole A Size	Hole B Size	Min. Distance X
1" NPT	1" NPT	2-3/4"
1" NPT	3/4 NPT	2-1/2
3/4 NPT	3/4 NPT	2-1/4
3/4 NPT	1/2 NPT	2-1/8
1/2 NPT	1/2 NPT	2
1/2 NPT	3/8 NPT	1-3/4
1/2 NPT	1/4 NPT	1-5/8
3/8 NPT	3/8 NPT	1-3/4
3/8 NPT	1/4 NPT	1-5/8
1/4 NPT	1/4 NPT	1-3/8

NEMA FRAME ASSIGNMENTS — 3-PHASE INDUCTION-TYPE ELECTRIC MOTORS

Open Dripproof Frame Electric Motors

HP	Speed, RPM	NEMA Frame	Shaft Diam.	Shaft Length	Shaft Height
1	1200	145T	7/8	2-1/4	3-1/2
1	1800	143T	7/8	2-1/4	3-1/2
1½	1200	182T	1-1/8	2-3/4	4-1/2
1½	1800	145T	7/8	2-1/4	3-1/2
1½	3600	143T	7/8	2-1/4	3-1/2
2	1200	184T	1-1/8	2-3/4	4-1/2
2	1800	145T	7/8	2-1/4	3-1/2
2	3600	145T	7/8	2-1/4	3-1/2
3	1200	213T	1-3/8	3-3/8	5-1/4
3	1800	182T	1-1/8	2-3/4	4-1/2
3	3600	145T	7/8	2-1/4	3-1/2
5	1200	215T	1-3/8	3-3/8	5-1/4
5	1800	184T	1-1/8	2-3/4	4-1/2
5	3600	182T	1-1/8	2-3/4	4-1/2
7½	1200	254T	1-5/8	4	6-1/4
7½	1800	213T	1-3/8	3-3/8	5-1/4
7½	3600	184T	1-1/8	2-3/4	4-1/2
10	1200	256T	1-5/8	4	6-1/4
10	1800	215T	1-3/8	3-3/8	5-1/4
10	3600	213T	1-3/8	3-3/8	5-1/4
15	1200	284T	1-7/8	4-5/8	7
15	1800	254T	1-5/8	4	6-1/4
15	3600	215T	1-3/8	3-3/8	5-1/4
20	1200	286T	1-7/8	4-5/8	7
20	1800	256T	1-5/8	4	6-1/4
20	3600	254T	1-5/8	4	6-1/4
25	1200	324T	2-1/8	5-1/4	8
25	1800	284T	1-7/8	4-5/8	7
25	3600	256T	1-5/8	4	6-1/4
30	1200	326T	2-1/8	5-1/4	8
30	1800	286T	1-7/8	4-5/8	7
30	3600	284TS	1-5/8	3-1/4	7
40	1200	364T	2-3/8	5-7/8	9
40	1800	324T	2-1/8	5-1/4	8
40	3600	286TS	1-5/8	3-1/4	7
50	1200	365T	2-3/8	5-7/8	9
50	1800	326T	2-1/8	5-1/4	8
50	3600	324TS	1-7/8	3-3/4	8
60	1200	404T	2-7/8	7-1/4	10
60	1800	364TS	1-7/8	3-3/4	9
60	3600	326TS	1-7/8	3-3/4	8
75	1200	405T	2-7/8	7-1/4	10
75	1800	365TS	1-7/8	3-3/4	9
75	3600	364TS	1-7/8	3-3/4	9
100	1200	444T	3-3/8	8-1/2	11
100	1800	404TS	2-1/8	4-1/4	10
100	3600	365TS	1-7/8	3-3/4	9
125	1200	445T	3-3/8	8-1/2	11
125	1800	405TS	2-1/8	4-1/4	10
125	3600	404TS	2-1/8	4-1/4	10
150	1800	444TS	2-3/8	4-3/4	11
150	3600	405TS	2-1/8	4-1/4	10
200	1800	445TS	2-3/8	4-3/4	11
200	3600	444TS	2-3/8	4-3/4	11
250	3600	445TS	2-3/8	4-3/4	11

Totally Enclosed Motors

HP	Speed, RPM	NEMA Frame
1	1200	145T
1	1800	143T
1½	1200	182T
1½	1800	145T
1½	3600	143T
2	1200	184T
2	1800	145T
2	3600	145T
3	1200	213T
3	1800	182T
3	3600	182T
5	1200	215T
5	1800	184T
5	3600	184T
7½	1200	254T
7½	1800	213T
7½	3600	213T
10	1200	256T
10	1800	215T
10	3600	215T
15	1200	284T
15	1800	254T
15	3600	254T
20	1200	286T
20	1800	256T
20	3600	256T
25	1200	324T
25	1800	284T
25	3600	284TS
30	1200	326T
30	1800	286T
30	3600	286TS
40	1200	364T
40	1800	324T
40	3600	324TS
50	1200	365T
50	1800	326T
50	3600	326TS
60	1200	404T
60	1800	364TS
60	3600	364TS
75	1200	405T
75	1800	365TS
75	3600	365TS
100	1200	444T
100	1800	405TS
100	3600	405TS
125	1200	445T
125	1800	444TS
125	3600	444TS
150	1800	445TS
150	3600	445TS

Note: TS indicates standard short shaft. For belt or gear drive, the standard long shaft, Symbol T, may be preferred.

WIRE SIZE AND FUSE RATINGS FOR 3-PHASE INDUCTION MOTORS

Motors are usually protected by both fuses (or circuit breakers) and by heater coils in a magnetic starter. Fuses open the circuit quickly in case of a massive overload or short circuit. Heater coils provide a delay and open the circuit if the average current, over a period of time, is greater than the circuit is designed for.

In some cases it may be necessary to use delayed action fuses. These provide a very short delay to prevent blowing a fuse during the short interval while the motor is accelerating up to its normal speed after being started.

Current values in this chart are approximate, and are compiled from data published by several motor manufacturers. They may be a little high or low for a specific motor. For selection of magnetic starter heater coils, it is better to follow the nameplate current rating of the actual motor to be used rather than depending on this or any other table.

Wire and fuse sizes are listed for reference only, and may vary with type of insulation, number of conductors in a cable and other factors. For a new design, requirements of the NEC (National Electrical Code) should be followed. Copies of the code can be ordered through most book stores. Other local ordinances may also apply.

HP	Speed RPM	230-Volt Service			460-Volt Service		
		Full Load Amps	Wire Size	Fuse Amps	Full Load Amps	Wire Size	Fuse Amps
1	1200	3.76	14	10	1.88	14	6
1	1800	3.56	14	10	1.78	14	6
1	3600	2.80	14	10	1.40	14	6
1½	1200	5.28	14	15	2.64	14	10
1½	1800	4.86	14	15	2.43	14	10
1½	3600	4.36	14	15	2.18	14	10
2	1200	6.84	15	20	3.42	14	10
2	1800	6.40	14	20	3.20	14	10
2	3600	5.60	14	20	3.00	14	10
3	1200	10.2	14	25	5.12	14	15
3	1800	9.40	14	25	4.70	14	15
3	3600	8.34	14	25	4.17	14	15
5	1200	15.8	12	30	7.91	14	20
5	1800	14.4	12	30	7.21	14	20
5	3600	13.5	12	30	6.76	14	20
7½	1200	21.8	10	40	10.9	14	20
7½	1800	21.5	10	40	10.7	14	20
7½	3600	19.5	10	40	9.79	14	20
10	1200	28.0	8	60	14.0	12	30
10	1800	26.8	8	60	13.4	12	30
10	3600	25.4	8	60	12.7	12	30
15	1200	41.4	6	80	20.7	10	40
15	1800	39.2	6	80	19.6	10	40
15	3600	36.4	6	80	18.2	10	40
20	1200	52.8	4	110	26.4	8	60
20	1800	51.2	4	110	25.6	8	60
20	3600	50.4	4	110	25.2	8	60

HP	Speed RPM	230-Volt Service			460-Volt Service		
		Full Load Amps	Wire Size	Fuse Amps	Full Load Amps	Wire Size	Fuse Amps
25	1200	65.6	3	120	32.8	6	180
25	1800	64.8	3	120	32.4	6	80
25	3600	60.8	3	120	30.4	6	80
30	1200	78.8	1	150	39.4	6	80
30	1800	75.6	1	150	37.8	6	80
30	3600	73.7	1	150	36.8	6	80
40	1200	102	0	200	50.6	4	110
40	1800	101	0	200	50.4	4	110
40	3600	96.4	0	200	48.2	4	110
50	1200	126	000	250	63.0	3	120
50	1800	124	000	250	62.2	3	120
50	3600	120	000	250	60.1	3	120
60	1200	150	000	300	75.0	2	150
60	1800	149	000	300	74.5	2	150
60	3600	143	000	300	71.7	2	150
75	1200	184	300	350	92.0	0	200
75	1800	183	300	350	91.6	0	200
75	3600	179	300	350	89.6	0	200
100	1200	239	500	500	120	000	250
100	1800	236	500	500	118	000	250
100	3600	231	500	500	115	000	250
125	1200	298	- - -	- - -	149	0000	300
125	1800	293	- - -	- - -	147	0000	300
125	3600	292	- - -	- - -	146	0000	300
150	1200	350	- - -	- - -	174	300	350
150	1800	348	- - -	- - -	174	300	350
150	3600	343	- - -	- - -	174	300	350

For selecting ampere rating of heater coils for magnetic motor starters, select the standard coil closest to the rating on the motor nameplate. If the motor operates in a cold environment, the coil with the next lower rating may be preferred. If in a hot environment, the coil with next larger current rating may work better. More detailed information on magnetic motor starters including wiring diagrams will be found in the Womack book "Electrical Control of Fluid Power". See inside back cover of this book for listings of Womack books.

The National Electrical Code (NEC) has been adopted and is published by several agencies. Copies may be purchased by writing to the American National Standards Institute, Inc. (ANSI), 1430 Broadway, New York, N.Y. 10018, or to National Fire Protection Association (NFPA), 60 Batterymarch St., Boston, MA 02110. An amplified version with explanations accompanying the text is published by McGraw-Hill and may be ordered through any book store.

Index

Other Womack Books

on Industrial Fluid Power

VOLUME 1 — INDUSTRIAL FLUID POWER. The first of a series of three textbooks on fluid power as used in the industrial plant and on mobile equipment. This is the basic textbook, covering the full range of hydraulic, compressed air, and vacuum usage.

This Third Edition has been completely re-written, with more photos of components, more circuit diagrams, and about 40 more pages of design calculations, troubleshooting information, and design tables and charts. The new metric S. I. system of units is explained and charts are included for sizing metric bore cylinders.

Subjects covered include laws and terms relating to fluid power; simple layouts for plant hydraulic and air systems; fluid flow through pipes; selection of pipe size; air and hydraulic cylinders — direction control, speed control, and sizing for power; air and hydraulic direction, pressure, and speed control valves; hydraulic pumps — how they operate and how to use them; trio units for compressed air; air dryers; hydraulic filters; hydraulic power units; hydraulic accumulators and heat exchangers.

This book is excellent for home study, with review questions at the end of each chapter. It is now being used by several hundred vocational/technical schools and is the official course textbook in many of these schools. Also used for in-plant employee training by hundreds of nationally known companies in many industries. Recommended as a training manual by the NFPA (National Fluid Power Association), and the FPS (Fluid Power Society). It has been selected as the most suitable training manual by the FPDA (Fluid Power Distributors Association) and is used as a part of their audio/visual program which was prepared for training employees of their member companies.

VOLUME 2 — INDUSTRIAL FLUID POWER. This is one of the more advanced textbooks. It uses the foundation laid in Volume 1 and goes on into certain subjects to a greater depth. Covers both hydraulic and air fluid power.

Some of the subjects include: how to calculate many types of cylinder loads, vertical, horizontal, and at an angle. Direction, force, and speed control of cylinders using some of the more sophisticated types of 4-way directional, pressure compensated flow control, and pressure control valves such as sequence, by-pass, unloading, reducing, and counterbalance valves. Pressure intensification with piston type and rotary type intensifiers. Sample circuit diagrams, design charts, and formulae are shown for each application. This book is limited to applications using air or hydraulic cylinders to deliver power output. Rotary mechanical output is covered in Volume 3.

Other material includes a chapter on air-over-oil circuits in which compressed shop air is used for power to produce a hydraulic oil flow. The last chapter shows many ideas, some novel, for the use of cylinders.

VOLUME 3 — INDUSTRIAL FLUID POWER. Although covering many advanced applications which require rotary mechanical output, the presentation is simple as it is in all the Womack books, and can be readily understood by anyone with mechanical or electrical aptitude. Mathematics are limited to very simple formulae, and in most cases the charts and tables can be used instead of the formulae.

The format includes first an explanation of each component and how it works, using photos and sectional views. Then circuit diagrams plus design charts show how to design the components into circuits.

Among the components covered in this particular book are common types of hydraulic motors — gear, vane, and piston types. Shows how to match a hydraulic motor to the load; circuits for direction and speed control; closed loop hydrostatic transmissions; motor starting and running torque, efficiency, HP, life expectancy, and installation; air motor operation and circuit design; power steering design; rotary and spool-type flow dividers and how to use them; bootstrapping principles and circuits for saving power; and many other subjects.

ELECTRICAL CONTROL OF FLUID POWER. This is a companion book for students of the textbooks who may not be experts in the design of electrical circuits to control fluid power. It is not a course on electricity or electrical theory, but it does show how to draw electrical ladder diagrams to control many kinds of fluid power circuits, both hydraulic and air. In each application the electrical diagram is shown alongside the fluid diagram.

First, the common electrical components, switches, relays, timers, etc., are described, along with the method of drawing and reading ladder diagrams. These and many more applications are described: directional control and sequencing of cylinders, peck drilling, cylinder dwell, deceleration at the end of the stroke, index table operation, automatic clamp and work, feed and drill, one-cycle control, multiple indexing, counting and stacking, turnover of parts, jogging, safety circuits, etc.

Other chapters cover operation of electric motors and motor starters, remote motor control, etc. Any technician, using this book, should be able to draw and wire an electrical circuit for most machines even though he may not understand electrical theory.

HOW TO ORDER . . . Send personal or company check to publisher. Prices include postage. Texas residents add 6% sales tax. Foreign shipments $1.00 per book extra. Prices are in U.S.A. funds. Company purchase orders for more than $35.00 accepted for 30-day invoicing, and may be placed by phone. Call (214) 357-3871 for information or to place book orders. Prices may change without notice. Request latest price list.